Nuclear Reprogramming

METHODS IN MOLECULAR BIOLOGY™

John M. Walker, SERIES EDITOR

METHODS IN MOLECULAR BIOLOGY™

Nuclear Reprogramming

Methods and Protocols

Edited by

Steve Pells

Department of Gene Function and Development
Roslin Institute
Roslin, Midlothian, UK

HUMANA PRESS ❋ TOTOWA, NEW JERSEY

This publication is printed on acid-free paper. ∞
ANSI Z39.48-1984 (American Standards Institute)

Permanence of Paper for Printed Library Materials.

Cover design by Patricia F. Cleary

Cover illustrations: *(Foreground)* Reprogramming of DNA methylation during mouse preimplantation development: from the two-cell stage onwards, there is passive loss of DNA methylation attributable to exclusion of the maintenance DNA methyltransferase from the nucleus; de novo methylation is evident in the inner cell mass at the blastocyst stage. The upper row shows 5-Methylcytosine staining, and the lower row shows DNA staining in the same embryo (Fig. 1, Chap. 11; *see* complete figure and caption on p. 130 and discussion on pp. 130–131). *(Background)* Piezo-assisted injection of a cumulus cell nucleus into an MII oocyte (Fig. 3, Chap. 10; *see* complete caption and discussion on pp. 122–123).

For additional copies, pricing for bulk purchases, and/or information about other Humana titles, contact Humana at the above address or at any of the following numbers: Tel.: 973-256-1699; Fax: 973-256-8341; E-mail: orders@humanapr.com; or visit our Website: www.humanapress.com

Printed in the United States of America. 10 9 8 7 6 5 4 3 2 1

eISBN 1-59745-005-7
ISSN 1064-3745

Library of Congress Cataloging-in-Publication Data

Nuclear reprogramming : methods and protocols / edited by Steve Pells.
   p. ; cm. -- (Methods in molecular biology ; 325)
   Includes bibliographical references and index.ISBN 1-58829-379-3 (alk. paper)
    1. Cell nuclei--Transplantation--Laboratory methods.[DNLM: 1. Cell Nucleus--physiology. 2. Cell Fusion--methods. 3. Transplantation--methods. QU 350 N965 2006] I. Pells, Steve. II. Series: Methods in molecular biology (Clifton, N.J.) ; v. 325.
QH442.2.N84 2006
660.6'5--dc22

                                                              2005020464

Preface

Starting in the 1950s, Briggs and King, and later John Gurdon and colleagues, showed that the transfer of a nucleus into an enucleated amphibian oocyte reconstitutes a cell that is capable of development at least part way to mature adulthood. This body of work showed that as an animal develops, despite the apparent loss of phenotypic potential that accompanies differentiation, the cell nucleus still contains sufficient genetic information to allow the cell to behave like a zygote and undergo embryonic development. We now know that nearly every animal cell, other than such odd exceptions as mature erythrocytes or lymphocytes with rearranged antigen receptors, contains a complete complement of the genetic information required to build another copy of the animal in question. However, the vast majority of cells in the adult are terminally differentiated with limited patterns of gene expression. These patterns of gene expression are conferred by both limitation of the repertoire of transcription factors expressed by the cell, and also by the chromatin configuration of the nucleus in that cell. This "nuclear program" defines the range of genes expressed by the cell, and hence its phenotype and function. Many of the molecular modifications conferring this program, DNA methylation, for example, are inherited epigenetically; that is, their inheritance is independent of the base pair sequence of the nuclear DNA. The sum total of all epigenetic marks in a cell comprises its epigenome.

Despite the early amphibian successes, it appeared that the more differentiated the cell that donated the nucleus, the poorer the efficiency of tadpole recovery, so that adult, terminally differentiated cells were much less efficient nuclear donors than early blastula cells. Early work in mammals, using enucleated mouse zygotes, reinforced this picture and for a long time it was thought to be impossible to transform the epigenome configuration of an adult terminally differentiated cell to that of a fertilized oocyte and thus reverse the linear, one-way process of differentiation to create a pluripotent, embryonic cell. More recently, the cloning from adult somatic cells using enucleated oocytes of several mammalian species, most famously the creation of the sheep "Dolly" at the Roslin Institute in 1997, showed that reversal of this process of increasing specification is possible. Because cloning of mammalian adults is achieved by nuclear transfer, we can conclude that the nucleus of the somatic donor provides all of the genetic information required, and that the mammalian oocyte contains an activity that acts on the donor nucleus to reprogram an embryonic state. This reassignment of a cell's nucleus from a somatic to an embryonic program is an example of "nuclear reprogramming." Nuclear reprogramming is a complex process involving the restructuring of the chromatin in the nucleus, remodeling of such DNA modifications as methylation, and a major change in gene expression patterns. Nuclear reprogramming is obviously a major function of the oocyte, as it combines two haploid genomes from different sources into one diploid genome and then reprograms this new genome to form a pluripotent cell that then begins development. The oocyte is thus the canonical reprogramming cell. Nuclear reprogramming can also occur after fusion of cells of different types, where (usually) the less-differentiated cell reprograms the more-differentiated cell. Since this usually results in the loss of specific differentiated functions, this phenomenon was originally called extinction. In addition, events called transdifferentiation have been described, in which cells of one, possibly well-differentiated, type appear to differentiate into cells of another lineage, sometimes one derived from a different embryonic germ layer. Trans-differentiation events also almost certainly involve some level of nuclear reprogramming to change the phenotype and potential of the cell. At present, nuclear reprogramming is largely a "black box" and many of the events involved in a reprogramming event are either completely unknown or poorly characterized. How nuclear reprogramming is controlled is also unknown, but clues are beginning to emerge and it is a very exciting time to be working in this field.

It is therefore particularly well-timed of Humana Press to release *Nuclear Reprogramming: Methods and Protocols* devoted to research techniques in nuclear reprogramming. I hope that this book will be of interest to a variety of people, be they cloners interested in the generation of live animals from cells, perhaps endangered species, or medics interested in the generation of stem cells or other cell types for therapeutic purposes, or biochemists or molecular biologists interested in the mechanism of the reprogramming process itself. This volume includes chapters describing various methods of nuclear reprogramming, including nuclear transfer in several different species, both amphibian and mammalian; fusion achieved both by chemical treatment and by electrically shocking cells; quantitative fusion and reprogramming by in vitro treatment of cells with cell extracts. Isolation of an adult stem-cell type is included. Several different methods of monitoring nuclear reprogramming are described, including the use of transgenic markers to follow reprogramming after nuclear transfer, activation of telomerase as an ES-specific marker, observation of structural changes in the nucleus by both light and electron microscopy, and verification of stem cells' surface marker expression and differentiation potential. With respect to the biochemistry of nuclear repro-gramming, methods for the examination of chromatin protein modifications, nucleosomal footprinting, and transcription factor binding are included here, as are methods for the study of DNA methylation changes both at the specific locus level and, by microscopy, at the level of the whole nucleus.

I have enjoyed working on this project and I wish to take the opportunity here to express my gratitude to John Walker for the invitation to compile and edit this book, to Tom Lanigan and his team at Humana Press, and to the many scientists who have generously contributed chapters to this volume.

Steve Pells

Contents

Contributors

VISAR BELEGU • *Kennedy Krieger Institute, Johns Hopkins University, Baltimore, MD*

DEAN H. BETTS • *Department of Biomedical Sciences, Ontario Veterinary College, University of Guelph, Guelph, Ontario, Canada*

PRZEMYSLAW BLYSZCZUK • *In Vitro Differentiation Group, Institute of Plant Genetics and Crop Plant Research (IPK), Gatersleben, Germany*

CONSTANZE BONIFER • *Molecular Medicine Unit, University of Leeds; St. James's University Hospital, Leeds, United Kingdom*

ANDREW C. BOQUEST • *Institute of Medical Biochemistry, University of Oslo, Oslo, Norway*

JAN E. BRINCHMANN • *Institute of Immunology, Rikshospitalet University Hospital and University of Oslo, Oslo, Norway*

ROBERT H. BROYLES • *Oklahoma Medical Research Foundation, University of Oklahoma Health Sciences Center, Oklahoma City, OK*

PETER CHEUNG • *Department of Medical Biophysics, University of Toronto; Ontario Cancer Institute, Toronto, Ontario, Canada.*

WANG L. CHEUNG • *Department of Pathology, Johns Hopkins School of Medicine, Baltimore, MD*

PETER N. COCKERILL • *Molecular Medicine Unit, University of Leeds, St. James's University Hospital, Clinical Sciences Building, Leeds, United Kingdom*

PHILIPPE COLLAS • *Institute of Medical Biochemistry, University of Oslo, Oslo, Norway*

WENDY DEAN • *Laboratory of Developmental Genetics and Imprinting, The Babraham Institute, Cambridge, United Kingdom*

MARIA P. DE MIGUEL • *Cell Therapy Laboratory, La Paz Hospital, Madrid, Spain; Institute for Cell Engineering, The Johns Hopkins University, Baltimore, MD*

PETER J. DONOVAN • *Institute for Cell Engineering, The Johns Hopkins University, Baltimore, MD*

BRUCE A. FENDERSON • *Department of Pathology, Anatomy, and Cell Biology, Thomas Jefferson University; Jefferson Medical College, Philadelphia, PA*

JACQUES-E. FLÉCHON • *Biologie du Developpement et Reproduction, INRA, Jouy en Josas, France*

ED J. GALLAGHER • *MRC Technology, Western General Hospital, Edinburgh, United Kingdom*

SHAORONG GAO • *Fels Institute for Cancer Research and Molecular Biology, Temple University School of Medicine, Philadelphia PA*

xi

KRISTINE G. GAUSTAD • *Institute of Medical Biochemistry, University of Oslo, Oslo, Norway*

DAVID M. GILBERT • *Department of Biochemistry and Molecular Biology, SUNY Upstate Medical University, Syracuse, NY*

JOHN B. GURDON • *Wellcome Trust/Cancer Research United Kingdom Gurdon Institute, Cambridge, United Kingdom*

ANNE-MARI HÅKELIEN • *Institute of Medical Biochemistry, University of Oslo, Oslo, Norway*

YONG-MAHN HAN • *Laboratory of Development and Differentiation, Korea Research Institute of Bioscience and Biotechnology (KRIBB), Daejeon, Korea*

LEA HARRINGTON • *The Department of Medical Biophysics, The University of Toronto; Princess Margaret Hospital, Toronto, Ontario, Canada*

GABRIELA KANIA • *In Vitro Differentiation Group, Institute of Plant Genetics and Crop Plant Research (IPK), Gatersleben, Germany*

YONG-KOOK KANG • *Laboratory of Development and Differentiation, Korea Research Institute of Bioscience and Biotechnology (KRIBB), Daejeon Korea*

SEOK-HO KIM • *Laboratory of Development and Differentiation, Korea Research Institute of Bioscience and Biotechnology (KRIBB), Daejeon, Korea*

TAE KOOK KIM • *Department of Biological Sciences, Korea Advanced Institute of Science and Technology; Biomedical Research Center, Daejeon, South Korea; Harvard Cancer Center, Harvard Medical School, Boston, MA*

W. ALLAN KING • *Department of Biomedical Sciences, Ontario Veterinary College, University of Guelph, Guelph Ontario, Canada*

PASCAL LEFEVRE • *University of Leeds, St. James's University Hospital, Clinical Sciences Building, Molecular Medicine Unit, Leeds United Kingdom*

JIM MCWHIR • *Department of Gene Function and Development, Roslin Institute, Roslin, Midlothian, United Kingdom*

RICHARD R. MEEHAN • *Department of Biomedical and Clinical Laboratory Sciences, University of Edinburgh, Edinburgh; Human Genetics Unit, MRC, Western General Hospital, Edinburgh, United Kingdom*

HANNAH R. MOORE • *Division of Gene Function and Development, Roslin Institute, Roslin United Kingdom; and Department of Biomedical and Clinical Laboratory Sciences, University of Edinburgh, Hugh Robson Building, George Square, Edinburgh, United Kingdom; and Wellcome Centre for Cell Biology, University of Edinburgh, Swann Building, Kings Buildings, Edinburgh, United Kingdom*

PETER MOUNTFORD • *Stem Cell Sciences Ltd, Collingwood, Victoria, Australia*
MEGAN MUNSIE • *Stem Cell Sciences Ltd, Collingwood, Victoria, Australia*
JENNIFER NICHOLS • *Centre Development in Stem Cell Biology, Institute for Stem Cell Research, University of Edinburgh, Edinburgh, United Kingdom*
STEVE PELLS • *Department of Gene Function and Development, Roslin Institute, Roslin, Midlothian, United Kingdom*
STEVEN PERRAULT • *Department of Biomedical Sciences, Ontario Veterinary College, University of Guelph, Guelph, Ontario, Canada*
APRIL D. PYLE • *The Johns Hopkins University, Institute for Cell Engineering, Baltimore, MD*
WILLIAM A. RITCHIE • *Department of Gene Function and Development, Roslin Institute, Roslin, Midlothian, United Kingdom*
ALEXANDRA ROLLETSCHEK • *In Vitro Differentiation Group, Institute of Plant Genetics and Crop Plant Research (IPK), Gatersleben, Germany*
AUSTIN C. ROTH • *Free Radical Biology and Aging Research Program, Oklahoma Medical Research Foundation, Oklahoma City, OK*
FÁTIMA SANTOS • *Laboratory of Developmental Genetics and Imprinting, The Babraham Institute, Cambridge, United Kingdom*
ABOULGHASSEM SHAHDADFAR • *Institute of Immunology Rikshospitalet University Hospital and University of Oslo, Rikshospitalet, Oslo, Norway*
MING HONG SHEN • *MRC Human Genetics Unit, Western General Hospital, Edinburgh, United Kingdom*
STEPHEN SULLIVAN • *Wellcome Trust/Cancer Research United Kingdom Gurdon Institute of Cancer and Developmental Biology, University of Cambridge, Cambridge, United Kingdom*
MASAKO TADA • *ReproCELL Inc., Tokyo; Stem Cell Engineering, Institute for Frontier Medical Sciences, Kyoto University, Kyoto, Japan*
TAKASHI TADA • *Stem Cell Engineering, Institute for Frontier Medical Sciences, Kyoto University, Kyoto, Japan*
HIROMI TAGOH • *Molecular Medicine Unit, St. James's University Hospital, University of Leeds, Leeds United Kingdom*
ANNA V. TERRY • *Department of Biochemistry and Molecular Biology, SUNY Upstate Medical University, Syracuse, NY*
MAIREAD TODD • *Department of Biochemistry and Molecular Biology, University of Oklahoma Health Sciences Center, Oklahoma City, OK*
FIONA B. TURNER • *Department of Microbiology and Immunology, Virginia Commonwealth University, Richmond, VA*
MARTIN WATERFALL • *Scottish Stem Cell Network Sorting Facility, Roslin Institute, Roslin, Midlothian, United Kingdom*

CORNELIA WIESE • *In Vitro Differentiation Group, Institute of Plant Genetics and Crop Plant Research (IPK), Gatersleben, Germany*

ANNA M. WOBUS • *In Vitro Differentiation Group, Institute of Plant Genetics and Crop Plant Research (IPK), Gatersleben, Germany*

JAEJOON WON • *Biomedical Research Center, Department of Biological Sciences, Korea Advanced Institute of Science and Technology, Daejeon, South Korea*

RONG WU • *Department of Biochemistry and Molecular Biology, SUNY Upstate Medical University, Syracuse, NY*

JIAN YANG • *Institute for Stem Cell Research, University of Edinburgh, Edinburgh, United Kingdom*

LORRAINE E. YOUNG • *Division of Obstetrics and Gynaecology and Institute of Genetics, University of Nottingham, Queens Medical Centre, Nottingham, United Kingdom*

1

Nuclear Transplantation in *Xenopus*

John B. Gurdon

Summary

Nuclear transplantation in amphibia started in 1952. By this is meant, sexually mature cloned frogs can be obtained from the nuclei of embryo cells, differentiating cells, and larval-differentiated cells. Transplanted nuclei are reprogrammed to entirely new patterns of gene expression. In this chapter, the methods used to transplant living nuclei into enucleated eggs of *Xenopus* are described. A method also is described for transplanting multiple somatic cell nuclei into nonenucleated oocytes, a procedure that achieves reprogramming of gene expression in the absence of cell division.

Key Words: Nuclear transplantation; *Xenopus*; ultraviolet enucleation; ovarian oocytes; oocyte germinal vesicle.

1. Introduction

During the last half-century, nuclear transfer experiments in *Xenopus* have tended to fall into two categories. One involves the transplantation of single somatic nuclei into enucleated unfertilized eggs, from which combination embryos and adults may proceed. The other involves the transfer of multiple nuclei of various kinds of cells into nonenucleated oocytes, the growing ovarian egg cells. These oocytes are in the prophase of meiosis. They are tetraploid and cannot be fertilized. When nuclei are injected into them, they undergo no morphological changes but can cause dramatic switches in the pattern of gene expression. The primary purpose of nuclear transplantation experiments into unfertilized eggs is to determine the genetic content and developmental capacity of the nuclei of differentiating or differentiated somatic cells. Such experiments also have led to an analysis of the mechanism and control of deoxyribonucleic acid (DNA) replication. However, the aim of nuclear transfer into oocytes primarily is to determine the mechanism of transcriptional gene reprogramming. Because oocytes do not induce DNA synthesis or chromo-

From: *Methods in Molecular Biology, vol. 325: Nuclear Reprogramming: Methods and Protocols*
Edited by: S. Pells © Humana Press Inc., Totowa, NJ

some replication but change or enhance transcription, they are particularly well suited for attempts to identify reprogramming molecules and mechanisms.

Methods of nuclear transfer in *Xenopus* have been reviewed in some detail in previous publications. For nuclear transfer into unfertilized enucleated eggs, *see* Gurdon 1991 *(1)*; for details of nuclear transfer to oocytes, *see* Gurdon 1977 *(2)*. The present account summarizes the main features of these two types of nuclear transfer experiments in *Xenopus* and adds some relevant technical information brought into use since the aforementioned works were published.

2. Materials

2.1. Composition of Salt Solutions Used in Nuclear Transplantation Experiments[a] (Based on [1])

	Modified Barth's saline (MBS) (mM)	High-salt MBS (mM)	Normal amphibian Ringer's (NAM) (mM)	OR2 (mM)
NaCl	88	110	110	83
KCl	1	2	2	2.5
CaCl$_2$	0.41	–	1	1
Ca(NO$_3$)$_2$	0.33	–	–	–
MgSO$_4$	0.82	1	1	1
NaHCO$_3$	2.4	2	0.5	–
Na$_2$HPO$_4$	–	0.5	–	1
HEPES (NaOH), pH 7.4	10	–	–	5
Tris base, pH 7.6, acetic acid	–	15	–	–
EDTA	–	–	0.1	–
Sodium phosphate, pH 7.4	–	0.5	1	1.25
Penicillin, benzyl	10 mg/L	–	100 units/mL	–
Streptomycin sulfate	10 mg/L	–	60 units/mL	–
Nystatin	–	–	2 mg/L	–

[a]For MBS, *see* **ref. 2**. For NAM, *see* **ref. 3**. For OR2, *see* **ref. 4**. For CaMg-free MBS, omit CaCl$_2$, Ca(NO$_3$)$_2$, and MgSO$_4$.

2.2. Composition of SuNaSp and SuNaSp Bovine Serum Albumin (BSA; Based on [5])

SuNaSp

Sucrose	0.25 M
NaCl	75 mM
Spermicline trihydrochloride	0.5 mM
Spermine tetrahydrochloride	0.15 mM

For SuNaSp BSA, make more than 3% with respect to bovine serum albumin fraction V.

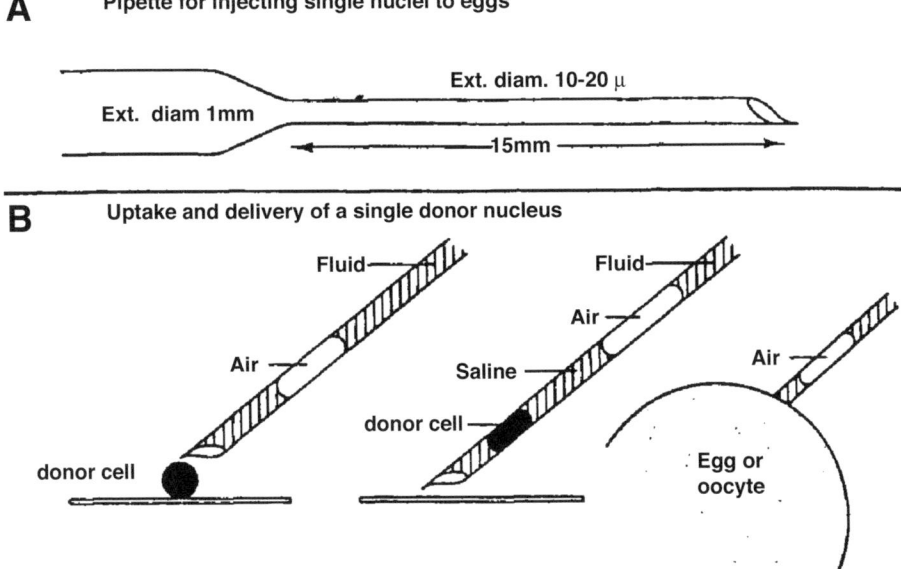

A Pipette for injecting single nuclei to eggs

Ext. diam. 10-20 μ

Ext. diam 1mm

15mm

B Uptake and delivery of a single donor nucleus

Fluid

Air

donor cell

Fluid

Air

Saline

donor cell

Air

Egg or oocyte

Fig. 1. Pipet for injecting single nuclei to eggs. (Reproduced from **ref. 1**.)

2.3. Diagram of Nuclear Transplantation Pipets

Figure 1A is a diagram of a nuclear transplantation pipet suitable for *Xenopus* embryo nuclei. **Figure 1B** is a diagram of a procedure for drawing a donor cell into the pipet, in such a way as to rupture the cell wall but not disturb the cytoplasm from around the nucleus *(1)*.

2.4. Diagram of Ultraviolet Light Sources

Figure 2 shows examples of ultraviolet (UV) light apparatus suitable for enucleating unfertilized Xenopus eggs (**Fig. 2A**, Hanovia UVS-100; **B**, Mineralite UVSL-15 *[1]*).

3. Methods
3.1. Nuclear Transfer to Unfertilized Enucleated Eggs

Three steps are involved in the nuclear transfer to unfertilized enucleated eggs, namely the preparation of donor cells, the preparation of enucleated recipient eggs, and the nuclear transfer procedure itself.

3.1 1. Donor Cells (see **Note 1**)

Prepare a single-cell suspension by any method that provides single isolated viable cells (*see* **Note 2**).

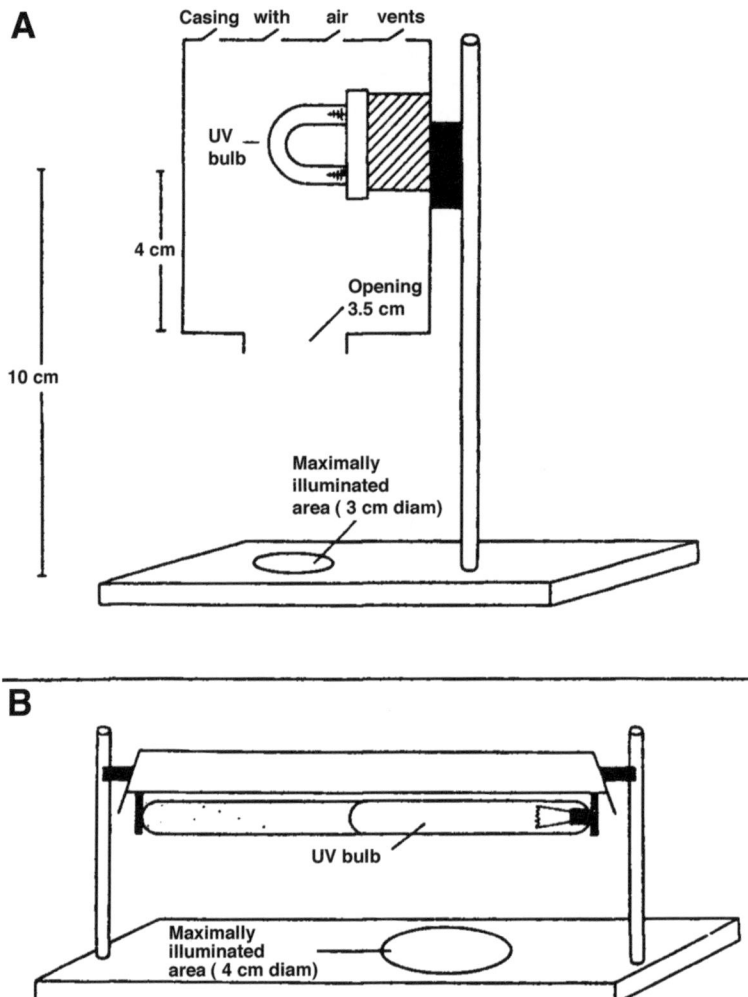

Fig. 2. Examples of UV light apparatus that are suitable for enucleating unfertilized *Xenopus* eggs **(A)** Hanovia UVS-100; **(B)** Mineralite UVSL-15. (Reproduced from **ref. *1*.**)

1. Prepare microinjection pipets (*see* **Subheading 2.3.**), 3.5-cm plastic dishes with a lining of 1% agarose in CaMg-free MBS (*see* **Subheading 2.1.**), and 3.5-cm plastic culture dishes containing MBS.
2. Dissect a donor embryo (such as a late blastula) to isolate the pigmented layer of cells called an animal cap (*see* **Fig. 3**). Place the animal cap in the agarose-lined culture dish containing CaMg-free MBS. 1 m*M* ethylene diamine tetraacetic acid (EDTA) can be added to accelerate cell dissociation.

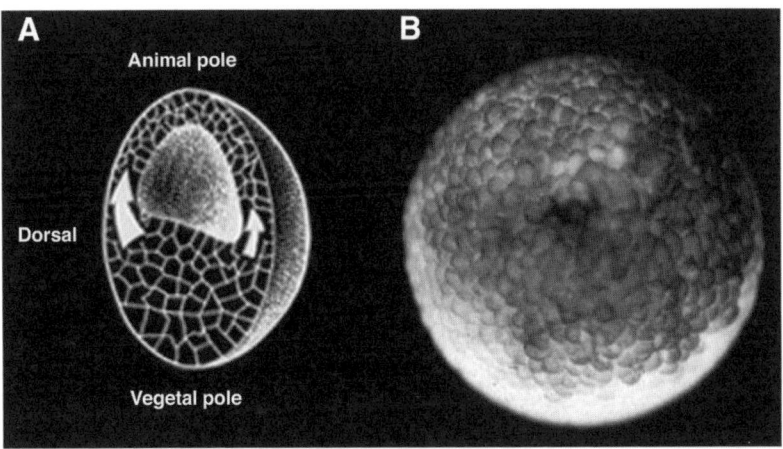

Fig. 3. (**A**) Schematic of the early embryo showing animal and vegetal poles. (**B**) Late-stage blastula showing the pigmented (darker) animal pole at the top of the figure.

3. After approx 10 to 15 min, the cells of the donor embryo fragment should be largely dissociated from each other.
4. Gently blow the dissociated cells with a pipet until they are spread out in the agarose-lined dish.
5. Check that the microinjection needle is suitable for sucking a donor cell into it in such a way as to break the cell membrane but not to disperse the contents of the cell (*see* **Subheading 2.3.**).

3.1.2. Recipient Eggs (see **Note 3**)

Unfertilized eggs must be enucleated. The egg chromosomes are in second meiotic metaphase and are located on a spindle near the centre of the white spot in the pigmented animal hemisphere of eggs. The best method of enucleating *Xenopus* eggs is by UV light (*see* **Note 4**; for an alternative, *see* **Note 5**).

1. Obtain unfertilized eggs laid within the last half-hour from an ovulating Xenopus female.
2. Place approximately three of these eggs on damp blotting paper on a microscope slide.
3. Remove excess medium, so that the eggs are surrounded by jelly but not by any medium. The jelly of the eggs will ensure that they stick in position on the blotting paper.
4. Use forceps to ensure that the unfertilized recipient eggs have their small white spot at the animal pole pointing vertically upward.
5. Place the microscope slide carrying the unfertilized eggs under the beam of a Mineralite UV source. Approximately 30 s is appropriate if set up as shown above (*see* **Fig. 2**).

6. Transfer the eggs on the slide to the Hanovia light beam; 6 to 8 s is appropriate if set up as detailed previously (*see* **Note 6**).
7. Immediately transfer the UV-irradiated recipient eggs on their slide to the microscope.

3.1.3. Nuclear Transfer Procedure

The original method that has continued to be used for many years and that still works very well for embryo donor cells depends upon sucking a donor cell into a micropipet that is too small so that the donor cell membrane is broken and the whole broken cell injected into the recipient egg (*see* **Fig. 1** and **Notes 7–9**).

1. Carry out nuclear transfer with the microinjection pipet as described previously. Note: the irradiated eggs will begin to dry after approx 3 min and, therefore, only the number of eggs should be placed on a slide for nuclear transfer that can be injected in that time.
2. As soon as the eggs have been injected, transfer the blotting paper with attached eggs to a culture dish containing MBS, immersing the paper and eggs in the medium. Make sure the eggs are under the medium so that they don't dry out.
3. Transfer the culture dish containing eggs stuck to the blotting paper to an incubator at approx 15°C. At this point, the nuclear transplanted eggs are in MBS medium.
4. After approx 3 or 4 h, remove any dead or abortively cleaved eggs from the culture dish so as not to contaminate those that are dividing correctly.
5. During the late blastula stage (approx 8 h), replace the MBS medium with 0.1X MBS medium. 1X MBS medium inhibits gastrulation.
6. At intervals of at least once every 12 h, remove dead or dying embryos so as to enable the survivors to develop without contamination.

3.2. Nuclear Transfer to Oocytes

3.2.1. Donor Nuclei

1. Start with a suspension of separated cells (either naturally so or by trypsinization; *see* **Subheading 3.1.1.**), and transfer them into a non-BSA-containing medium such as SuNaSp (*see* **Subheading 2.2.**) by washing and centrifugation. Finish with a final concentration of 10^7 cells per milliliter.
2. Add 40 U of streptolysin 0 to 50 µL of donor cells at 10^7 cells per milliliter in a volume just more than 50 µL. At room temperature, we obtain approx 95% permeabilization (Trypan blue) after 4 min.
3. Arrest the permeabilization by addition of an excess (1 mL) of SuNaSp BSA. Centrifuge at 2000*g* for 2 min to concentrate the cells to 10^7 per milliliter.
4. Keep the cells in this condition on a slide at a low temperature (15°C).

3.2.2. Nuclear Transfer Procedure

Individual oocytes are kept in a bath containing MBS. A suitable micropipet is loaded with the thick suspension of permeabilized cells (~15,000 per microliter). For most purposes, it is desirable to inject nuclei into the germinal vesicle of oocytes. In this case, it is best to inject no more than 10 nL to avoid destruction of the germinal vesicle. We prefer to penetrate an oocyte (held by forceps) at a point in the middle of the pigmented hemisphere and then to aim the pipet toward the center of the oocyte, thereby having a good chance of penetrating the germinal vesicle (*see* **Note 11**).

1. Place permeabilized donor cells in a drop of approx 10 μL of SuNaSp BSA on a siliconized slide so that the nuclear suspension sticks up from the suspension. Note that the nuclear suspension must be very concentrated (10^7 cells/mL). The recipient oocytes for injection must remain immersed in SuNaSp BSA medium throughout the procedure.
2. Suck up the donor cell suspension with a suitable injection pipet so marked that approx 10 nL of nuclear suspension can be delivered.
3. Inject the nuclear suspension into each oocyte with the microinjection pipet. The oocyte must be held with forceps so that the injection pipet enters the oocyte tangentially and aims to deposit the nuclei along the animal–vegetal axis (brown pole to white pole) approx 30% of the distance from the animal pole to vegetal pole.
4. When completed, the injected oocytes are placed into a 3.5-cm Petri dish containing MBS medium in an incubator at a desired temperature, from 15 to 20°C.

4. Notes

1. In nuclear transfer to egg experiments, it is desirable to use a genetic marker for donor cells to prove that the resulting embryos, which should be diploid and should carry the genetic marker, are derived solely from the transplanted nucleus and not with any contribution from the egg pronucleus. Currently, it is common to use donor cells carrying the GFP marker so that this can be seen in each cell of the resulting embryos. The presence of a GFP nuclear marker does not exclude a contribution from the egg pronucleus to make triploids. However, a GFP marker in a diploid set of chromosomes is good evidence for the successful elimination of the egg pronucleus.
2. Preparation of a viable single cell suspension may require trypsinization or further treatment of specialized tissues or, for amphibian embryos, incubation in a standard saline medium lacking calcium and magnesium ions but containing 1 mM EDTA at pH 8.0 (*see* **Subheading 2.1.**). It is preferable to keep isolated cells in a saline solution, such as Ca-free MBS, to which 0.1% bovine serum albumin has been added; this can be present during EDTA dissociation. The preparation of dissociated cells may best be kept on a bed of 1% agarose in saline solution. For most purposes, the donor cell preparation is kept cold, for instance, at 15°C for amphibian cells, but the temperature can be lower for many other cells, including tissue culture cells.
3. It is best to use unfertilized eggs as recipients because they carry only one set of (maternal) chromosomes. We prefer to use eggs naturally deposited by laying females in high salt MBS (*see* **Subheading 2.1.**), rather than the enforced squeezing of ovulating females. Unfertilized eggs laid in high-salt MBS do not appear to lose activity as recipients for at least an hour, if kept cool (15°C).

4. We use a Mineralite UV source beaming downwards 5 cm above the platform on which eggs are placed (*see* **Subheading 2.4.**). Using UV light for 30 s is sufficient to enucleate eggs so long as these have been placed with the white egg spot facing upwards and therefore directly under the UV light. We prefer to conduct this stage of UV irradiation using nondejellied eggs that are stuck to filter paper by their jelly (egg spot facing upwards). The slide carrying the filter paper and eggs is not under water because this can absorb UV light. This source of UV does not dissolve the jelly around the eggs and this must be removed or weakened for easy penetration of the microinjection pipet. We have used, for many decades, a second UV light whose characteristics dissolve the jelly on top of the egg. This Hanovia UV source (described by Gurdon in 1997 *[2]* and 1991 *[1]*) dissolves the jelly and weakens the vitelline membrane so that pipet penetration is particularly harmless. We normally use this source, which is located 10 cm above the eggs that are still stuck to the filter paper and not covered with water. The exact dosage of light needs to be tested for each batch of eggs, but is commonly in the range of 6 to 8 s. After this, the enucleated and jelly-depleted eggs will soon dry. However, there is time to transplant nuclei into them before drying happens. As soon as the nuclear transfer has been completed, these eggs need to be immersed, still stuck to the filter paper, in a suitable medium such as MBS.

5. It is also possible to penetrate unfertilized eggs that have not been treated with the Hanovia light by dejellying the eggs by the usual cysteine hydrochloride method and injecting them under water. This is the procedure used by Amaya and Kroll *(6)* for transgenesis in *Xenopus*. However, penetration by this method may involve some damage to the eggs, although this is tolerated when eggs are subsequently fertilized by sperm.

6. If a Hanovia light source is not available, dejellied eggs (2% cysteine hydrochloride in MBS, 4 min) can be used, and only the Mineralite, but not Hanovia, UV source is required.

7. This requires a certain level of technical skill because it is important that the donor cell should be sufficiently broken that the donor nucleus is in contact with egg cytoplasm; if an unbroken donor cell is injected into an egg it completely fails to respond to the egg cytoplasm and the nuclear transfer fails. However, too much disruption of the donor cell exposes its nucleus to the saline medium, which is not designed as a nuclear medium and the nucleus, as well as embryo survival, suffers. This procedure is conducted with the donor cells resting on an agarose-covered slide in calcium- and magnesium-free medium, and the recipient eggs are uncovered by medium, still stuck to their filter paper.

8. An alternative procedure developed by Chan and Gurdon *(7)* is to permeabilize the donor cell membrane chemically. This can be performed with streptolysin O, lysolecithin, or other agents. The aim is to cause the least amount of permeabilization by these agents whose activity is suppressed by the addition of 0.1% BSA. The percent of permeabilization can be checked by trypan blue entry to cells. A preparation of permeabilized cells may be kept on a cooled agarose-covered slide for at least an hour. This permeabilization procedure avoids the need to physi-

cally distort donor cells, which can be injected painlessly in a wide pipet, and is especially suitable for very small donor cells.

9. The position in a recipient egg at which the donor nucleus is deposited is not critical because eggs will tend to move implanted nuclei to their desired position slightly above the center, just as they do with egg and sperm pronuclei. As described previously, it is convenient to include an air bubble in the microinjection pipet so as to follow the ejection of a nucleus into the egg, which is opaque.

10. In the United Kingdom, Home Office permission is needed for this to be performed under anesthesia but not if the oocyte donor is terminally anesthetized. Consult your animal procedures office for local regulations.

11. Practice at the successful germinal vesicle penetration of oocytes can be achieved by injecting trypan blue dye and immediately opening the oocyte. Once injected, the oocytes are cultured in MBS or OR2 for as long as required. Such injected oocytes can survive for several days. At the end of the desired incubation time, groups of oocytes can be frozen and analyzed as required.

References

1. Gurdon, J. B. (1991) Nuclear transplantation in *Xenopus*, in: *Methods of Cell Bioliogy*. Academic Press, New York, pp. 299–309.

2. Gurdon, J. B. (1977) Methods for nuclear transplantation in Amphibia, in *Methods of Cell Bioiogy* (Stein, G., Stein, J., and Kleinsmith, L. J., eds). Academic Press, New York, pp. 125–139.

3. Slack, J. M. and Forman, D. (1980) An interaction between dorsal and ventral regions of the marginal zone in early amphibian embryos. *J. Embryol. Exp. Morphol.* **56,** 283–299.

4. Wallace, R. A., Jared, D. W., Dumont, J. N., and Sega, M. W. (1973) Protein incorporation by isolated amphibian oocytes. 3. Optimum incubation conditions. *J. Exp. Zool.* **184,** 321–333.

5. Gurdon, J. B. (1976) Injected nuclei in frog oocytes: fate, enlargement, and chromatin dispersal. *J. Embryol. Exp. Morphol.* **36,** 523–540.

6. Amaya, E. and Kroll, K. L. (1999) A method for generating transgenic frog embyros. *Methods Moll. Biol.* **97,** 393–414.

7. Chan, A. P. and Gurdon, J. B. (1996) Nuclear transplantation from stably transfected cultured cells of *Xenopus*. *Int. J. Dev. Biol.* **40,** 441–451.

2

Nuclear Transfer in Sheep

William A. Ritchie

Summary

Somatic cell nuclear transfer is a complex and intricate procedure with a low success rate. Despite advances in embryo culture and the production of specialized tools and equipment success rate has remained poor. The procedure remains basically the same despite technical innovations but is now easier to learn and requires less technical expertise. Oocytes now come from an in vitro system, which may make the procedure more economical and flexible than before.

Oocytes are produced from ovaries collected at an abattoir. The cumulus oocyte complexes (COCs) are recovered from the ovary with hypodermic needle and syringe. Selected COCs are matured for 18 to 20 h in medium that is supplemented with hormones. Cumulus cells are stripped from the oocytes using hyaluronidase, and mature oocytes with polar bodies, selected for enucleation. These oocytes are treated with a cytoskelatal inhibitor to prevent lysis of the oocyte during enucleation and with a DNA-specific dye to visualize the chromosomes with the aid of ultraviolet light. These permit removal of the metaphase II chromosomes and polar body prior to reconstruction of the embryo. A diploid cell is injected under the zona pellucida of the enucleated oocyte, which can then be fused to the cytoplast using an electrical pulse. The fused embryos are activated and allowed to develop in culture to the morula or blastocyst stage, when they are surgically implanted into previously prepared synchronized, recipient animals.

Key Words: Micromanipulation; in vitro maturation; recipient animals; enucleation; electrofusion; embryo culture; embryo transfer.

1. Introduction

Nuclear transfer (NT) is an inefficient method of replicating animals. Originally, cleavage-stage embryos were used, usually before compaction of the blastomeres, and this meant that it was only possible to produce small numbers of identical animals (*1*). This changed in 1995, when animals were produced from cells that had been grown in culture (*2*), greatly simplifying the production of identical animals. These experiments were conducted on sheep oocytes

From: *Methods in Molecular Biology, vol. 325: Nuclear Reprogramming: Methods and Protocols*
Edited by: S. Pells © Humana Press Inc., Totowa, NJ

because they were relatively cheap to use and because efficient methods of handling the animals existed at the Roslin Institute. The cells used as nuclear donors were cultured from the inner cell mass (ICM) of sheep embryos. The following year, similar experiments were performed using three different cell types: ICM cells, fetal cells, and cells from the udder of a 6-yr-old adult animal. The three cell lines produced four lambs, two lambs, and one lamb, respectively; the animal produced from the udder cells was "Dolly" (*3*).

Current methods of NT have not changed to any great extent, but the process has become easier as a result of the development of specific sheep culture medium (*4*) and the production of specialized tools and equipment. NT is still inefficient with somatic cell nuclear transfer (SCNT) efficiency running at 1–5%, but it is much easier because of the development of new technology (*5,6*).

NT consists of the removal of the maternal chromosomes and the replacement of these chromosomes with those from a diploid cell. The oocytes are usually used at the metaphase II stage; they have their maternal deoxyribonucleic acid (DNA) removed using cytoskeletal inhibitors and a DNA-specific dye. A cell is placed in the perivitelline space and the cell is then fused to the cytoplast using an electric current. The reconstructed embryo is then activated and allowed to develop before the transfer into a surrogate mother (*7*).

2. Materials and Equipment

2.1. Media

1. Synthetic oviduct fluid + amino acids + bovine serum albumin (SOFaaBSA; *see* **Table 1** *[4]*).
2. Calcium-free synthetic oviduct fluid + amino acids + bovine serum albumin (SOFaaBSA-Ca), which is the same as SOFaaBSA except for the omission of $CaCl_2$.
3. Hepes-buffered synthetic oviduct fluid (HSOF; *see* **Table 2**).
4. Hepes-buffered calcium-free synthetic oviduct fluid. (HSOF–Ca), which is the same as HSOF except for the omission of $CaCl_2$.
5. Hepes-buffered synthetic oviduct fluid + 1% fetal calf serum (HSOF/FCS).
5. Phosphate-buffered saline + 1% fetal calf serum (PBS + 1% FCS).
6. Fusion medium: 0.3 *M* manitol, 0.1 m*M* $MgSO_4$, and 0.05 m*M* $CaCl_2$ in double-distilled water (dd water). Osmolarity adjusted to 280 mosmoles with dd water.
7. Cell culture medium: Dulbecco's modified essential medium (DMEM) + 10% FCS.
8. Serum deprivation medium: DMEM + 0.5% FCS.
9. Oocyte wash medium (*see* **Table 3**).
10. Maturation medium (*see* **Table 4**).

2.2. Additional Reagents

1. Hyaluronidase stock. Aliquots of 300 U in 20 µL of dd water.

Table 1
SOFaaBSA Composition

Component	Sigma cat. no.	g (or mL)/1000 mL
NaCl	S-5886	6.29
KCl	P-5405	0.534
KH_2PO_4	P-5655	0.162
$MgSO_4 \cdot 7H_2O$	M-1880	0.182
Sodium lactate	L-7900	0.6 mL
Penicillin	P-4687	0.06
$NaHCO_3$	S-5761	2.1
Phenol red	P-5530	0.01
Na pyruvate	P-4562	0.0357
$CaCl_2 \cdot 2H_2O$	C-7902	0.262
L-Glutamine	G-5763	0.3
β-Mercaptoethanol	B-6766	20 mL
MEM Nonessential amino acid solution	M-7145	10 mL
Bovine serum albumin	A-6003	4

Table 2
HSOF Composition

Component	Sigma cat. no.	g (or mL)/1000 mL
NaCl	S-5886	6.29
KCl	P-5405	0.534
KH_2PO_4	P-5655	0.162
$MgSO_4 \cdot 7H_2O$	M-1880	0.182
Sodium lactate	L-7900	0.6 mL
Penicillin	P-4687	0.06
$NaHCO_3$	S-5761	0.42
Phenol Red	P-5530	0.01
Na pyruvate	P-4562	0.0357
$CaCl_2 \cdot 2H_2O$	C-7902	0.262
HEPES	H-3784	5.208
BSA	A-6003	4

2. Cytochalasin B stock. Aliquots of 7.5 μg in 10 μL of dimethyl sulfoxide.
3. Bisbenzimide (Hoechst 33342) stock; 5 μg in 1 mL of dd water.
4. Petroleum jelly.
5. Sigmacote.
6. Mannitol.

Table 3
Oocyte Wash Medium

Component	Sigma cat. no.	g (mL or U)/1000 mL
TCM199[a]	M-0650	100 mL
NaHCO$_3$	S-5761	0.400
HEPES	H-6147	2.98
Heparin	H-3149	2500 U
Sheep serum	Prepared from our animals	20 mL

[a]Tissue Culture Medium 199.

Table 4
Maturation Medium

Component	Sigma cat. no.	g (or mL)/1000 mL
TCM199[a]	M-0650	100 mL
NaH CO$_3$	S-5761	2.100
FSH (follicle-stimulating hormone)	Ovagen[b]	0.005
LH (ovine leutinizing hormone)	L-5269	0.005
Estradiol 17-β	E-1024	0.001
Sheep serum	Collected from our animals	200 mL

[a]Tissue Culture Medium 199.
[b]Ovagen; Ovine Pituitary Extract, UK Distributor: David Maharg, Wiltshire, England.

7. Dimethyl sulfoxide.
8. Veramix sponges from Upjohn Ltd., for the synchronization of ewes.

2.3. Equipment

2.3.1. Microscope and Manipulators (see **Fig. 1**)

A Nikon TE300 inverted microscope fitted with differential interference contrast (DIC) optics and epiflourescence capability is used. This is used with 4× phase contrast, 10× phase contrast, 20× DIC, and 40× DIC lenses (*see* **Note 1**). The microscope should be fitted with two Nikon Narishige MO-188 "Joystick Hydraulic Micromanipulators" and two IM-188 Microinjectors. The volume per turn of the injector can be altered by replacing the 3-mL syringe fitted with a 250-µL pipet for the enucleation pipet and a 500-µL pipet for the holding pipet.

A three-way tap (Vigon VG1) in the hydraulic line between the microinjector and the tubing allows the system to be filled using a suitable syringe thus over-

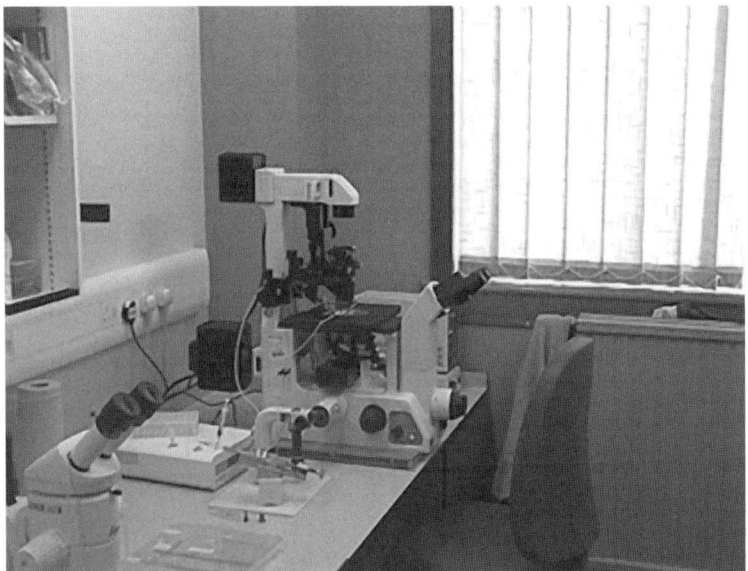

Fig. 1. Equipment used for NT: inverted microscope, two micromanipulators, and two microinjectors.

coming any need to take the equipment apart for filling (*see* **Notes 2** and **3**). Embryos are handled using Leica M7.5 and M12.5 microscopes fitted with transmitted light stands and Linkam CO102 temperature controllers and warm stages.

2.3.2. Miscellaneous Equipment

1. Siliconized glass slides.
2. Glass strips approx 20 × 3 × 2 mm.
3. Glass holding pipet.
4. Glass enucleation pipet (18 μm, Humagen).
5. Gilson or similar micropipettors with 10-μL, 20-μL, 100-μL, 200-μL, and 1000-μL capacities.
6. A BLS CF 150/B cell fusion machine is used for electrofusion of the cell couplets. The fusion machine is connected to a BLS fusion chamber that can be placed in a sterile 100-mm disposable Petri dish for fusion.

2.4. Isolation and In Vitro Maturation of Oocytes

1. Collect ovaries from the slaughterhouse in warm PBS in a vacuum flask to maintain their temperature.
2. Wash the ovaries in clean PBS and return to the flask to maintain their temperature.
3. Take a small quantity of ovaries from the flask and replace the lid to maintain the temperature.

4. Using an 18-gage hypodermic needle and a 10-mL syringe, aspirate the follicles on the ovaries to recover the cumulus oocyte complexes (COCs).

5. Insert the needle into the ovary and push the point into the follicles, at the same time maintaining gentle suction on the plunger of the syringe (*see* **Note 4**).

6. Once an appropriate volume of follicular fluid has been extracted and collected in the syringe remove the needle and gently expel this into tubes containing warm oocyte wash medium (*see* **Table 3** and **Note 5**).

7. Pour the fluid into a 90-mm Petri dish that has been previously marked with a grid pattern to assist in searching for the COCs.

8. If necessary, add more oocyte wash, and select the COCs that have at least three layers of cumulus cells (*see* **Note 6**).

9. Wash the COCs three times in oocyte wash medium and once in maturation medium (*see* **Table 4**) before placing 40 COCs in 800 µL of prewarmed and gassed maturation medium in Nunc 4-well plates.

10. Cover the medium with oil and mature for 18 to 20 h at 38.5°C in a humidified atmosphere of 5% CO_2 in air (*8*).

3. Method

3.1. Prepare the Recipient Animals

Recipient animals: prepare in advance because the procedure takes several days to complete.

1. Select ewes with reasonable body condition (condition score >3) and are sound in mouth and udder.

2. At day 0, insert a 60-mg medroxyprogesterone acetate sponge (Veramix /Upjohn) into the vagina of each ewe.

3. Remove the sponge on day 13.

4. Check for estrus between days 14 and 16 using a vasectomized ram.

5. NT (*see* **Subheading 3.2.** onwards) will take place on day 15, which is heat day of the recipient.

6. Starve the animal on the afternoon of day 21.

7. Conduct embryo transfer on the morning of day 22.

3.2. Media Preparation

On the day of the procedure, make up the following culture media volumes/ dishes, i.e., SOFaaBSA culture medium dishes for enucleation, fusion, and culture, and place in the incubator (37°C, 5% CO_2 in air) to gas.

1. Enucleation medium dish: prepare 25-µL drops of SOFaaBSA-Ca under oil in a 60-mm culture dish.

2. Fusion medium dish: to 1 mL of SOFaaBSA, add 4.5 µL of Cytochalasin B stock. Prepare 25-µL drops of SOFaaBSA under oil in a 60-mm culture dish.

3. Culture medium dish: place 800 µL of SOFaaBSA in a 4-well Nunc culture plate.

4. HSOF culture medium for the removal of residual cumulus cells (Hyaluronidase treatment: *see* **Subheading 3.4., steps 1–3**), washing, micromanipulation, and Hoechst staining (These medium dishes should be warmed in a temperature-controlled "Hot box" set at 37°C).
5. Add 5 µL of hyaluronidase stock to 500 µL of HSOF–Ca for hyaluronidase treatment.
6. Washes: add 500 µL of fetal calf serum (FCS) to 4.5 mL of HSOF-Ca.
7. Micromanipulation medium: to 1 mL of HSOF-Ca/FCS (*see* **Subheading 2.1.** and **Table 2**), add 4.5 µL of Cytochalasin B stock.
8. Hoechst/CB medium dish: take 500 µL of HSOF-Ca/FCS + CB from **step 8** and add 5 µL of Hoechst stock.

3.3. Manipulation Chamber Preparation (see Fig. 2)

1. Clean a siliconized glass slide with a paper towel and 70% alcohol.
2. Apply a small quantity of petroleum jelly along both sides of the slide for a distance corresponding to the length of the glass strips.
3. Using sterile technique place a glass strip on each of the areas with the petroleum jelly.
4. Apply another small quantity of petroleum jelly along the top of the glass strips.
5. Using a pipettor, place 300 µL of the micromanipulation medium in the middle of the glass on the slide.
6. Place a clean sterile cover slip on the top of the chamber and push down to make sure that a liquid-tight seal has been made.
7. Fill each end of the chamber with mineral oil or Dow Corning Silicone Fluid.
8. Mount the slide on the microscope.
9. Mount a holding pipet on the left manipulator arm of the micromanipulator workstation. Fluorinert is used as a hydraulic fluid in this system (*see* **Note 7**).
10. Flush all of the air from the system and push the end of the pipet into the chamber (*see* **Note 8**).
11. Suck back slightly to put a little medium into the pipet.
12. Mount an enucleation pipet in the right manipulator arm of the micromanipulator workstation. Fluronert is used as a hydraulic fluid in this system.
13. Remove the pipet from its packaging using sterile technique and mount in the tool holder.
14. Expel all of the air from the microinjection system.
15. Insert the pipet into the manipulation chamber.
16. Rotate the pipet until the bevel of the pipet is seen, and tighten the tool holder in this position (*see* **Notes 9** and **10**).

3.4. Preparation of the Oocytes

1. Remove any residual cumulus cells from the oocytes by culturing in the hyaluronidase treatment medium (*see* **Subheading 3.2., step 6**) for up to 10 min in the hot box until the cumulus cells begin to loosen.
2. Pipet repeatedly with an automatic pipettor to remove the cumulus cells (*see* **Note 11**).

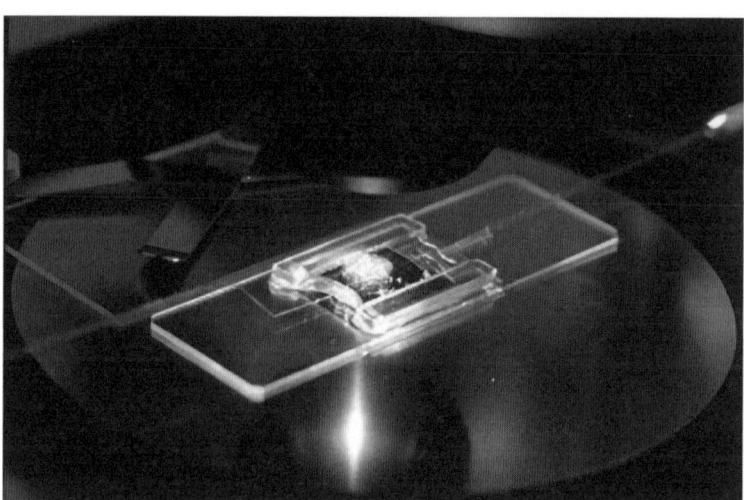

Fig. 2. Micromanipulation chamber with the holding pipet entering the chamber from the left side and the enucleation pipet entering from the right side.

3. Wash several times in wash medium to leave the cumulus cells behind.
4. Examine the oocytes under a dissecting microscope while turning the oocytes over to select mature oocytes with polar bodies.
5. Lift these oocytes out of the dish and store in the enucleation medium dish until required.
6. Take a manageable group of oocytes and place in the Hoechst/CB dish and culture for 15 min in the hot box.

3.5. Enucleation (see Fig. 3)

1. Start the mercury vapor lamp on the microscope.
2. Remove the group of oocytes from the Hoechst + CB dish and place in the manipulation chamber.
3. Lift an oocyte from the group and find an area in the chamber in which there are no other oocytes visible.
4. Using the enucleation pipet, turn the oocyte until the polar body is at 1-o'clock or 5-o'clock position, depending on whether the bevel of the pipet is facing up or down.
5. Insert the pipet through the zona pellucida (ZP), taking care to avoid damaging the cytoplasm of the oocyte.
6. Aspirate the polar body and the adjoining cytoplasm into the pipet.
7. Remove the holding pipet with the oocyte out of the field of view.
8. Expose the enucleation pipet to the ultraviolet light. If low light levels are used on the microscope, it is not necessary to turn the light off. The polar body will glow brightly, and the metaphase chromosomes less brightly.
9. Turn off the ultraviolet light.

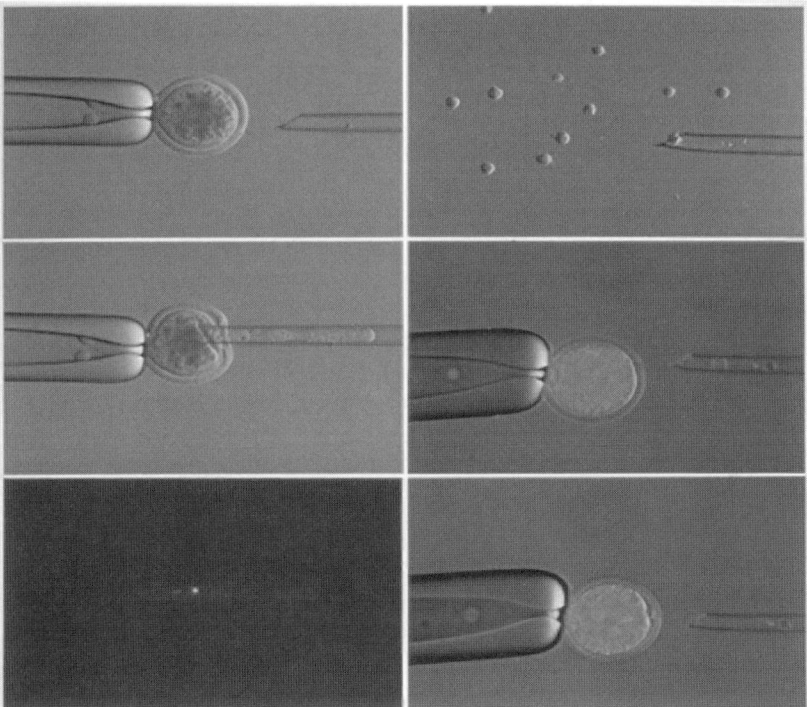

Fig. 3. Left column: Enucleation of oocyte; removal of the polar body and surrounding cytoplasm; checking to see that the polar body and metaphase chromosomes
are fluorescing in the pipet. Right column: cell transfer; picking up cells; transfer of a
cell to the PV space of an enucleated oocyte.

10. If the metaphase chromosomes and polar body cannot be seen, the process of
    checking enucleation can be repeated with the light turned off.
11. Repeat the enucleation process if the metaphase chromosomes are not present in
    the pipet.
12. Deposit the successfully enucleated oocytes in a group at the right side of the
    manipulation chamber.
13. After enucleation of each group, return them to culture in the "Enucleation" dish.
14. Repeat with suitably sized groups until complete.

3.6. Preparation of Nuclear Donor Cells

1. Use suitable cells for reconstruction of the couplet. These cells may be of many
   different lines, and these may require different strategies to make them suitable
   for supplying the donor nucleus.
2. Thaw the cells and culture in DMEM with serum.
3. Serum-deprive the cells (*see* **Note 12 [9]**).

4. Create a single-cell suspension by Trypsin treatment and resuspension in DMEM.
5. Store the trypsinized cells in the appropriate conditions for the cell type being used.

3.7. Cell Transfer

1. Prepare the manipulation chamber as before (*see* **Subheading 3.3.**) but with HSOF only.
2. Deposit a small number of cells in the top right corner of the chamber.
3. Deposit a small number of enucleated oocytes in the centre of the chamber.
4. Use an appropriately sized pipet on the right tool holder to pick up cells without damage. An 18-μm pipet is suitable for most cell types (*see* **Note 13**).
5. Pick up an enucleated oocyte with the holding pipet.
6. Push the pipet through the ZP at a position opposite the holding pipet.
7. Deposit a single cell in the perivitelline (PV) space (*see* **Note 14**).
8. Make sure that the cell is in good contact with the cytoplasm.
9. Repeat the process for all of the enucleated oocytes.

3.8. Electrofusion

1. Immediately after cell transfer, conduct the electrofusion (*see* **Note 15**).
2. Wash the couplets in fusion buffer to remove excess ions, making sure that the minimum amount of medium is transferred with the couplets (*see* **Note 16**).
3. Move the couplets individually between the electrodes and align the couplets either manually or using an AC pulse of electricity.
4. Give a fusion pulse of 0.25 kV/cm AC for a few seconds followed by three pulses of 1.25 kV/cm DC for 80 μs.
5. Immediately lift the pulsed couplet out of the fusion buffer and place in HSOF to prevent lysis.
6. Repeat the procedure, taking care not to introduce ions into the fusion chamber, until all of the couplets have been pulsed.
7. Transfer the fusing couplets into SOFaaBSA and culture for 1 h in a 5% CO_2 incubator.
8. Check for fusion of the couplets after 1 h.
9. Repeat the fusion procedure on unfused couplets if appropriate.

3.9. Culture

1. Culture the fused oocyte-cell couplets in SOFaaBSA in Nunc 4-well plates overlaid with mineral oil in a humidified 5% CO_2: 5% O_2: 90% N_2 gas atmosphere.
2. On day 7 of culture, select those embryos that have developed normally for transfer to surrogate recipient animals.

3.10. Embryo Transfer

1. Anesthetize the animal using a short-acting barbiturate.
2. Intubate the animal and maintain anesthesia using a mixture of Halothane, nitrous oxide, and oxygen.
3. Conduct a mid-ventral laparotomy to expose the uterus.

4. Using a blunt needle (16 gage × 1 in.) puncture the uterus near the utero-tubal junction.
5. Using a positive displacement pipet (Drummond 20-µL pipet) transfer the NT blastocysts through the hole made by the blunt needle (*see* **Note 17**).

4. Notes

1. It is easier to arrange the lenses in the nosepiece so that those which are used most often are adjacent to each other. This prevents having to move the nose-piece past lenses which are not in use. I keep the 4× and 20× lenses adjacent to each other as these are the most frequently used lenses for this technique.
2. I control the holding pipet mounted on the left of the chamber with my right hand and the enucleation pipet mounted on the right side of the chamber with my left hand. This allows both the joystick and injector on one side to be used simultaneously.
3. Make sure that there are no air bubbles in the tubing because this can prevent the pipet from working smoothly. Any trapped air can be compressed like a spring, making it difficult to control the system. The metal pipet holder can also obscure air bubbles so flush a little of the hydraulic fluid through the system to prevent this from happening.
4. Take care when collecting oocytes from the ovaries, because the follicles are small. It is advisable to insert the needle a few millimeters from the follicle so that the bevel of the needle is completely covered, thus preventing air from entering the syringe.
5. Always remove the needle from the syringe before emptying the syringe of follicular fluid as squeezing the COCs through the needle at high pressures can remove some of the cumulus cells.
6. When selecting COCs, discard those that have too few cumulus cells, which appear to have very patchy cytoplasm, or which show other abnormalities.
7. Prepare the holding pipet by pulling a GC10 glass capillary over a small gas flame to make a long parallel length of glass of approx 150 mm. Bend the glass in a micro forge four times to make the pipet fit under the cover slip of the manipulation chamber. Fire polish the end of the pipet until it is approx 20-µm wide.
8. When setting up the microscope for manipulation use the 4× lens to set up the glass microtools, then use the 20× lens for micromanipulation.
9. Successful micromanipulation relies on control of the glass micropipet. Keep the hydraulic fluid near the end of the pipet, so that the viscosity of the fluid and the narrow diameter of the pipet give friction which helps to control the suction.
10. If there is not sufficient control over suction in the enucleation pipet, pick up some drops of oil. The oil increases the friction in the pipet and so increases control of suction.
11. When stripping cumulus cells from the oocytes using a pipettor, keep the pipettor at an angle so that the oocytes are not squeezed against the bottom of the dish.
12. Deprive the cells of serum until they enter G0. The time for this to occur is variable, depending on the cell line used. It must therefore be determined empirically for each new cell line. The status of the cells can be determined by using proliferating cell nuclear antigen (i.e., PCNA) staining.

13. Select bright cells without vacuoles and even cytoplasm. Several cells can be aspirated into the pipet, leaving a small gap between each of the cells.
14. When transferring cells into the PV space, make sure that there is not excess pressure that can force the cell out of the hole. This can be prevented by withdrawing the pipet slowly, allowing any pressure to dissipate.
15. Change the fusion buffer frequently or flood the fusion chamber with fusion buffer so that the fusion buffer does not change osmolarity.
16. Wash reconstructed couplets prior to electrofusion to prevent ions being carried into the fusion buffer and causing lysis. Make sure that no medium is transferred into the fusion chamber as this can cause lysis of the couplets.
17. When transferring NT embryos using a positive displacement pipet, suck up a small quantity of air followed by the embryos in a small amount of medium and then more air in the end of the pipet. The first air bubble prevents the embryos from sticking to the piston, and the air bubble at the end prevents the embryos from coming out when the end of the pipet touches the uterus of the animal.

Acknowledgments

With thanks to all of my colleagues past and present at the Roslin Institute, also for all of the financial support from The Roslin Institute, MAFF, DEFRA, EC, and others without whom none of this work could have taken place. My thanks especially go to Professor Ian Wilmut for the facilities to carry out the work, to Dr. Jane Taylor for expert help, and Marjorie Ritchie for critical reading and checking of the manuscript.

References

1. Willadsen, S. M. (1986) Nuclear transplantation in sheep embryos. *Nature* **320,** 63–65.
2. Campbell, K. H. S., McWhir, J., Ritchie, W. A., and Wilmut, I. (1996) Sheep cloned by nuclear transfer from a cultured cell line. *Nature* **380,** 64–66.
3. Wilmut, I, Schnieke, A. E., McWhir, J., Kind, A. J., and Campbell, K. H. S. (1997) Viable offspring derived from foetal and adult mammalian cells. *Nature* **385,** 810–813.
4. Walker, S. K., Hill, J. L., Kleemann, D. O., and Nancarrow, C. D. (1996) Development of ovine embryos in synthetic oviductal fluid containing amino acids at oviductal fluid concentrations. *Biol. Reprod.* **55,** 703–708.
5. Denning, C., Burl, S., Ainslie, A., Bracken, J., Dinnyes, A., Fletcher, J., et al. (2001) Deletion of the alpha (1,3) galactosyl transferase (G. G.T. A.1) gene and the prion protein (PrP) gene in sheep. *Nat. Biotechnol.* **19,** 559–562.
6. Schnieke, A., Kind, A. J., Ritchie, W. A., Mycock, K., Scott, A. R., Ritchie, M., et al. (1997) Human factor I. X. transgenic sheep produced by transfer of nuclei from transfected foetal fibroblasts. *Science* **278,** 2130–2133.
7. Young, L. E., Sinclair, K. D., and Wilmut, I. (1998) Large offspring syndrome in cattle and sheep. *Rev. Reprod.* **3,** 155–163.

8. Wells D. N., Misica P. M., Day A. M., and Tervit, H. M. (1997) Production of cloned lambs from an established embryonic cell line: a comparison between in vitro and in vivo matured cytoplasts. *Biol. Reprod.* **57,** 385–393.
9. Campbell K. H. S., Loi P., Otaegui P. J., and Wilmut, I. (1996) Cell cycle co-ordination in embryo cloning by nuclear transfer. *Rev. Reprod.* **1,** 40–46.

3

Protocols for Nuclear Transfer in Mice

Shaorong Gao

Summary

Cloning by nuclear transfer in mammals has revealed the remarkable ability of an oocyte to reprogram transferred cell nuclei and induce them to recapitulate the developmental program. This chapter summarizes the method used since 1998 for mouse cloning, which differs from that for large animal cloning. A Piezo-drill micromanipulator allows direct injection of nuclei into enucleated mice oocytes instead of the electrofusion method used for introducing a nucleus into an enucleated oocyte of a large animal. After activation, reconstructed embryos are allowed to develop to the morulae or blastocyst stage before transfer into surrogate mothers.

Key Words: Oocyte; cloning; nuclear transfer; mouse; reprogramming.

1. Introduction

In 1996, a landmark breakthrough was made by Ian Wilmut and his collegues. A sheep named Dolly was successfully cloned from an adult somatic cell by electrofusion of a nuclear donor cell from the mammary gland to an enucleated metaphase II oocyte *(1)*. This advance demonstrated that cellular commitment in the adult was not irreversible and has led to the further development of this important technology. Compared with large animals, mice have shorter gestation and generation periods and more is known about the genetic background of mice. It is possible to use the mouse as a model to study the basic mechanism of reprogramming that follows nuclear transfer and to observe features such as gene expression, regulation, and chromatin structure alteration. Cloning of mice also provides an alternative means of producing transgenic or knockout mice, even though very few laboratories have succeeded since successful mouse cloning was first reported *(2)*. We have successfully cloned mice by using cumulus cells and embryonic stem cells as donor cells *(3,4)*, and here the detailed procedures for cloning mice will be discussed.

From: *Methods in Molecular Biology, vol. 325: Nuclear Reprogramming: Methods and Protocols*
Edited by: S. Pells © Humana Press Inc., Totowa, NJ

These include oocyte enucleation, donor nucleus injection, reconstructed oocyte activation, cloned embryo culture, and embryo transfer.

2. Materials

2.1. Equipment

1. Olympus or Nikon stereomicroscopes with a gimbal-mounted mirror are suitable for oocyte collection.
2. A Nikon inverted microscope equipped with differential interference contrast (DIC) optics and objectives can be used for nuclear transfer. Alternatively, inverted Olympus microscopes equipped with Hoffman Modulation Optics also are suitable for visualizing the metaphase II spindle.
3. The micromanipulator used for nuclear transfer is purchased from either the Narishige or Eppendorf companies.
4. The injector used for injection of the cell nucleus into the oocyte is an IM-9B, purchased from Narishige or a Cell-Tram Vario from Eppendorf.
5. The Piezo-drill micromanipulation controller PMM-150 can be purchased from Prime Tech Ltd. (Ibaraki, Japan).
6. A Flaming–Brown-type pipet puller (Sutter Instruments, Novato, CA) is used to pull pipets.
7. A microforge (De Fonbrune style) is used for cutting a blunt pipet to perform either enucleation or injection.
8. Mercury (Fisher) is placed in the pipet with a 1- to 2-mm long bead for the piezo operation.
9. The mouth control aspiration system for collecting oocytes is purchased from Sigma. The pipet used for collecting oocytes is pulled from a Pasteur pipet (Fisher).
10. Culture dishes (Falcon 1008, Falcon 3500) are used for culturing the cloned embryos and purchased from Fisher.
11. Mineral oil used for the culture of oocytes and cloned embryos is purchased from Fisher.

2.2. Medium Preparation

1. CZB–glucose (CZBG), HCZBG, and calcium-free CZBG media are used for culture of oocytes, performing nuclear transfer, and activation of reconstructed oocytes. The protocol for preparation of these media has been described previously (3). The compositions of the media for oocyte manipulation, culture and activation are listed in the tables (see Tables 1–4).
2. PVP (Sigma or ICN; molecular weight = 360 kDa) solution used for suspending cells is prepared by dissolving PVP in either HCZBG or CZBG. To prepare PVP solutions, place 10 mL of HCZBG into a Falcon dish and carefully pour 1 g or 0.3 g of PVP into the medium-containing dish. Seal the dish with parafilm and keep in a refrigerator for at least 2 d before use. PVP solution prepared this way has given us consistent results (see Note 1).

Table 1
Composition of the CZB Stock

Component	Weight (mg)/1000 mL
MilliQ water	990 mL
NaCl (S-5886)	4760
KCl (P-5405)	360
$MgSO_4 \cdot 7H_2O$ (M-1880)	290
EDTA $\cdot$ 2Na (E-6635)	40
Na–Lactate (L-7900)	5.3 mL
D-Glucose (G-6152)	1000
KH_2PO_4 (P-5655)	160

The stock will be 1000 mL after dissolving all components in 990 mL of MilliQ water.

Table 2
Composition of the CZBG

Component	Weight (mg)/100 mL
CZB stock	99 mL
$NaHCO_3$ (S-5761)	211
$CaCl_2 \cdot 2H_2O$ 100× stock (C-7902)	1 mL
Pyruvate (P-4562)	3
Glutamine (Gibco-BRL, 21051-024)	15
BSA (A-3311)	500

Table 3
Composition of the HCZBG

Component	Weight (mg)/100 mL
CZB stock	99 mL
Hepes (H-4034)	476
$NaHCO_3$ (S-5761)	42
$CaCl2 \cdot 2H_2O$ 100× stock (C-7902)	1 mL
Pyruvate (P-4562)	3
Glutamine (Gibco-BRL, 21051-024)	15
PVA (P-8136)	10

The pH of HCZBG medium should be adjusted to pH 7.4 after mixing these chemicals together.

Table 4
Composition of the Calcium-Free CZBG

Component	Weight (mg)/100 mL
CZB stock	100 mL
NaHCO$_3$ (S-5761)	211
Pyruvate (P-4562)	3
Glutamine (Gibco-BRL, 21051-024)	15
BSA (A-3311)	500

3. The medium for culturing cloned embryos after activation depends on the donor cell type. Minimum essential medium (MEM-α; Sigma), supplemented with BSA (same quantity as in CZBG; *see* **Table 2**) and sodium pyruvate (same quantity as in CZBG; *see* **Table 2**) has been successfully used for the culturing of cumulus cell-derived cloned embryos, and more than 60% blastocyst-stage development and 2% term development can be achieved consistently *(3)*. Either M16 (Sigma) or CZBG can be used for culturing embryonic stem (ES) cell-derived cloned embryos *(4,5)*. F10/DMEM (Gibco-BRL) has been proved efficient for the culture of myoblast cell-derived cloned embryos in our study *(6)*.

4. Prepare a stock solution of cytochalasin B (Sigma) by dissolving cytochalasin B in 100% ethanol to a concentration of 5 mg/mL and store 10-μL aliquots at –70°C before use.

2.3. Oocyte Collection

B6D2F1 (C57BL/6 × DBA/2) female mice 8- to 10-wk-old are used to collect recipient oocytes for nuclear transfer. (Either somatic cells or ES cells can be used as donor cells.) Because cloning is critically dependent on the oocyte for success, care must be taken when collecting the oocytes. The method used for collecting oocytes was as described previously *(3)*.

1. Superovulate the female mice by injecting 5 U of pregnant mare serum gonadotrophin (PMSG) (Calbiochem) and 5 U of hCG (Sigma) ip about 48 h apart.
2. Prepare CZBG culture drops in a Petri dish and cover with oil. Keep the culture dishes in the incubator for at least 30 min before collecting oocytes.
3. Collect the oviducts from sacrificed mice (14–15 h after hCG injection) into an M2 (Sigma) drop. Collect cumulus oocyte complexes (COCs) into M2 medium containing 50 to 100 U/mL of hyaluronidase (ICN) at room temperature (without oil covering).
4. Pipet the COCs using a mouth-control pipet (inner diameter of 200–300 μm) after 5 min of enzyme treatment and remove the cumulus cells from the oocytes by extensive washing with three to four drops of M2 medium without oil covering.

5. Transfer the oocytes to CZBG medium and keep in the incubator before enucleation (removal of spindle–chromosome complex [SCC]).

2.4. Donor Cell Isolation

Successful full-term development has been observed in our studies using both cumulus cells and ES cells (*see* **Note 2**) as nuclei donors. The following procedure is used for collecting cumulus cells:

1. Collect cumulus cells and place in M2 hyaluronidase medium in a 1.5-mL Eppendorf tube as described in **Subheading 2.3., step 2**. Add 1 mL of HCZBG and wash once by centrifugation at 250g.
2. Resuspend the cell pellet in a small volume of 3% PVP solution (30 µL) and keep the cells on ice before nuclear transfer.

3. Methods
3.1. Removal of the SCC

To prepare the enucleation dish, place one 50-µL drop of 10% PVP in HCZBG and six to eight 20-µL drops of HCZBG containing 3 µg/mL CB in a Falcon 1029 tissue culture dish and cover the drops with mineral oil. Remove the SCC as follows:

1. Before enucleation (removal of SCC), preculture the metaphase II oocytes for at least 10 min in CZBG medium at 37°C with 5% CO_2 in air.
2. Transfer a group of 20 to 30 oocytes to one drop of HCZBG medium containing CB for a couple of minutes before enucleation.
3. The inner diameter of the pipet used for the removal of the SCC should be approx 10 µm. Draw 1 to 2 mm of mercury into the pipet and then connect the pipet to the microinjector. Fill the microinjector with MilliQ water instead of oil (this allows much smoother control than oil).
4. Push the mercury forward to the tip of the pipet in the 10% PVP drop and suck a tiny volume of PVP into the pipet. Move the pipet to the enucleation drop and aspirate a tiny volume of HCZB into the pipet.
5. Keep the oocytes in HCZBG/CB for at least 2 min before enucleation.
6. Finding the spindle is most important for correct enucleation. The spindle area of the MII oocyte can be visualized as a transparent area under both DIC and Hoffman lenses. Rotation of the oocyte using the enucleation pipet will assist in locating the spindle.
7. Hold the oocyte by the holding pipet once the spindle is visualized.
8. Apply several piezo pulses to the tip of the enucleation pipet against the zona pellucida. Gentle suction to the zona will help the pipet to make a hole. Stop the piezo pulse immediately once a hole is made in the zona.
9. Push the pipet forward to touch the oolemma (oocyte membrane). Suck up the SCC using the pipet and pull back the pipet to pinch off the SCC.

10. Enucleate one group of oocytes within 10 to 15 min and transfer the enucleated oocytes back to the CZBG culture drops before injection of donor nuclei.

3.2. Nucleus Injection

To make an injection dish for cumulus cell injection, three kinds of drops should be prepared. These drops comprise one drop of 10% PVP for pipet washing, 6 to 10 drops of 3% PVP for cell suspension, and 6 to 10 drops of HCZBG for oocyte injection. The volume of the drops is approx 30 µL. To make an injection dish for ES cell injection, the 3% PVP drops can be omitted and 6 to 10 drops of 10% PVP can be prepared.

A well-made injection pipet is critical for successful injection of the cell nucleus into the cytoplasm without lysing the oocyte. The inner diameter of the pipet should be approx 5 µm and the pipet should be cut very flat. The method of filling the mercury into the injection pipet is the same as the enucleation pipet. The injection procedure is the same when using ES cells as nuclear donors, except that ES cells can be suspended in 10% PVP. Injection of the cell nuclei into the enucleated oocytes is performed as follows:

1. Wash the pipet thoroughly with 10% PVP before collecting the cell nuclei.
2. For cumulus cells, mix one drop of concentrated cumulus cells with one drop of 3% PVP drop by drop using the mouth control pipet.
3. Transfer a group of 10 to 20 enucleated oocytes to one drop of HCZBG.
4. Move the washed pipet to the 3% PVP drop to load it with cumulus cell nuclei. Select round cumulus cells with a diameter of 10 µm and aspirate into the pipet. It is usually necessary to apply a piezo pulse to the tip of the pipet to assist the pipet in breaking the cell membrane.
5. After 6 to 10 cell nuclei are lined up inside the pipet, move the injection pipet to the injection drop containing enucleated oocytes. Inject the nuclei one by one into the enucleated oocytes. To improve survival rate after injection, the pipet should push the membrane inwards while applying gentle suction. Then, use the piezo pulse to break the membrane. Once the membrane is broken, inject the cell nucleus into the cytoplasm with the smallest possible quantity of PVP. Withdraw the injection pipet quickly (*see* **Note 3**).

3.3. Activation of Reconstructed Oocytes

After injecting cell nuclei into enucleated oocytes, allow the injected oocytes to recover for 5 min on the stage of the microscope before transferring them back to CZBG medium and culturing them for 1 to 3 h before activation treatment. During this time, chromosome condensation ocurrs *(1,7)*.

1. Equilibrate calcium-free CZBG medium in the incubator for 10 min.
2. Prepare Activation Medium by adding CB and 100X or 1000X strontium stock to 1 mL of prewarmed calcium-free CZB medium to a final concentration of 5 µg/mL CB and 10 mM of strontium (*see* **Note 4**).

3. Prepare 10 drops (100 μL/drop) of activation medium in the culture dish and cover with mineral oil.
4. Keep the culture dish in the incubator for 30 min before transferring the injected oocytes to the drops.
5. Wash the oocytes through the drops and keep them in the final drop for 5 to 6 h.

3.4. Cloned Embryo Culture and Embryo Transfer

1. Collect the activated oocytes in which pseudo-pronuclei formation has occurred and culture them in appropriate culture medium. For cumulus cell-derived cloned embryos, use MEM-α supplemented with sodium pyruvate and BSA for culture. Most embryo culture media can be used for the culturing of ES cell-derived cloned embryos. Use 5% CO_2 in air and 37°C for the culture of cloned embryos for 3.5 d (*see* **Note 5**).
2. Collect cloned embryos that have developed to morulae and early blastocyst stages for embryo transfer.
3. Transfer 5 to 10 embryos to each uterine horn of 2.5 d post coitus (dpc) pseudo-pregnant mice.
4. Kill the recipient mothers at 19.5 dpc and quickly remove the pups from the uteri. After cleaning fluid away from their air passages, place the pups in a warm box supplied with oxygen for up to 20 min,, after which the pups should be able to breathe normally. We have a tube connected to the box from an oxygen tank to raise the oxygen concentration, but because the door of the box is not closed, the pups are in a raised oxygen concentration environment rather than 100% oxygen.
5. Foster the surviving pups onto lactating mothers.

4. Notes

1. No cleavage of cloned embryos: in some cases, the cloned embryos showed no cleavage after transfer from activation medium into culture medium for 20 h and the embryos remain at the one-cell stage. This was observed with cumulus cell-derived cloned embryos made by most novices. Incomplete dissolution of the PVP and a high PVP concentration causes this problem. Care must be taken when preparing PVP solutions because undissolved PVP is toxic to cloned embryos. Properly dissolving the PVP and reducing the PVP concentration will improve cleavage rate. Both 10% and 3% PVP have been used successfully in our cloning studies.
2. If using ES cells as nuclear donors, culture ES cells to high density and then harvest the ES cells by normal methods (trypsinization, neutralization, and centrifugation). Wash the cells once with the culture medium and resuspend the cells in a small volume of culture medium. Keep the cell suspension on ice before the cloning experiment.
3. Oocytes lysed during injection: more than 80% of oocytes manipulated by an experienced researcher should survive after injection of cell nuclei into the enucleated oocytes. However, in some cases, especially for those learning the technique, very few oocytes survived injection and most oocytes lysed after

injection. It is important to make correctly sized and flat pipets to reduce the chance of oocytes lysing. Thorough washing of the injection pipet with 10% PVP will also help. The setting of frequency and intensity of the piezo pulse to break the oolemma should be as low as possible, to minimize damage to the oocyte or cell nucleus. The quantity of mercury backfilled in the pipet and the size of the pipet both affect the piezo setting. Less mercury will produce much stronger power and make it easier for the pipet to break the membrane. A pipet of smaller size will break the membrane more easily. A pipet of smaller size filled with a small amount of mercury is therefore ideal for microinjection of cell nuclei. A piezo setting of *frequency 1* and *intensity 1* is sufficient to allow the pipet to break the oocyte membrane and inject the cell nucleus into the oocyte, and yet permit the survival of the injected oocytes.

4. Strontium precipitation during activation of reconstructed oocytes: this problem occurs when activation medium is prepared imcorrectly. The strontium should be prepared as a 100X or 1000X stock. Before adding the strontium to the calcium-free CZB medium, the medium should be warmed in the incubator for at least 10 min to inhibit precipitation.

5. Culture conditions for cloned embryos: oxygen concentration during culture is a potential factor affecting development of cloned embryos. Low oxygen concentration (5%) has been shown to be inappropriate for cloned mouse embryos *(5)*.

Acknowledgments

The author would like to thank Dr. Ian Wilmut, Michelle McGarry, Tricia Ferrier, and Judy Fletcher from Roslin Institute for their kind help. I also acknowledge Drs. Keith E Latham, Young Gie Chung, and Zhiming Han from Fels Institute for their assistance. The studies were supported in part by NIH grant to K.E.L. (HD38381) and by Geron Biomed.

References

1. Wilmut, I., Schnieke, A. E., McWhir, J., Kind, A. J., and Campbell, K. H. S. (1997) Viable offspring derived from fetal and adult mammalian cell. *Nature* **285,** 810–813.
2. Wakayama, T., Perry, A. C. F., Zuccotti, M., Johnson, K. R., and Yanagimachi, R. (1998) Full-term development of mice from enucleated oocytes injected with cumulus cell nuclei. *Nature* **394,** 369–374.
3. Gao, S., McGarry, M., Latham, K. E., and Wilmut, I. (2003) Cloning of mice by nuclear transfer. *Cloning Stem Cells* **5,** 287–294.
4. Gao, S., McGarry, M., Ferrier, T., Pallante, B., Gasparrini, B., Fletcher, J., et al. (2003) Effect of cell confluence on production of cloned mice using an inbred embryonic stem cell line. *Biol. Reprod.* **68,** 595–603.
5. Gao, S., McGarry, M., Priddle, H., et al. (2003) Effect of donor oocytes and culture conditions on development of cloned mice embryos. *Mol. Reprod. Dev.* **66,** 126–133.

6. Gao, S., Chung, Y. G., Williams, J. W., Riley, J., Moley, K., and Latham, K. E. (2003) Somatic cell-like features of cloned mouse embryos prepared with cultured myoblast nuclei. *Biol. Reprod.* **69,** 48–56.
7. Gao, S., Gasparrini, B., McGarry, M., Ferrier, T., Fletcher, J., Harkness, L., et al. (2002) Germinal vesicle material is essential for nucleus remodeling after nuclear transfer. *Biol. Reprod.* **67,** 928–934.

4

Isolation of Stromal Stem Cells From Human Adipose Tissue

Andrew C. Boquest, Aboulghassem Shahdadfar,
Jan E. Brinchmann, and Philippe Collas

Summary

The stromal compartment of mesenchymal tissues is thought to harbor stem cells that display extensive proliferative capacity and multilineage potential. Stromal stem cells offer a potentially large therapeutic potential in the field of regenerative medicine. Adipose tissue contains a large number of stromal stem cells, is relatively easy to obtain in large quantities, and thus constitutes a very convenient source of stromal stem cells. Importantly, the number of stem cells obtained is compatible with extensive analyses of the cells in an uncultured, freshly isolated, form. This chapter describes procedures for isolating millions of highly purified stromal stem cells from human adipose tissue and methods of establishing polyclonal and monoclonal cultures of adipose tissue-derived stem cells.

Key Words: Adipose tissue; mesenchyme; stem cell; stroma.

1. Introduction

The stromal compartment of mesenchymal tissues is thought to harbor stem cells that display extensive proliferative capacity and multilineage potential. Often called mesenchymal stem cells or stromal stem cells, these cells have been isolated from several mesodermal tissues, including bone marrow *(1)*, muscle *(2)*, perichondrium *(3)*, and adipose tissue *(4–6)*. Stromal stem cells isolated from various mesodermal tissues share key characteristics, including the ability to adhere to plastic to form fibroblastic-like colonies (called CFU-F), extensive proliferative capacity; the ability to differentiate into several mesodermal lineages, including bone, muscle, cartilage, and fat; and the ability to express several common cell surface antigens. Recent evidence suggests that these cells can also form nonmesodermal tissues, including neuron-like cells *(5,7,8)*. Because of these characteristics, stromal stem cells may potentially be useful in the field of regenerative medicine.

From: *Methods in Molecular Biology, vol. 325: Nuclear Reprogramming: Methods and Protocols*
Edited by: S. Pells © Humana Press Inc., Totowa, NJ

To date, extensive characterization of stromal stem cells has been limited for the most part to cultured cells because stromal stem cells are rare and, therefore, are difficult to isolate in numbers compatible with extensive analyses in an uncultured form. For example, in human bone marrow, only 0.01 to 0.001% of nucleated cells form colony-forming units–fibroblasts (CFU-F *[1]*). Until now, their ability to adhere to plastic and proliferate has been used as the conventional isolation method. Therefore, methods to isolate uncultured stromal stem cells in significant numbers are needed to extensively study the biology of these cells in their native form. It is likely that such knowledge will enhance research efforts toward clinical use of these cells.

Adipose tissue contains a large number of stromal stem cells *(4)*. Because it is easy to obtain in large quantities, adipose tissue constitutes an ideal source of uncultured stromal stem cells. We have recently isolated and extensively characterized uncultured human stem cells from the stroma of adipose tissue *(9)*. Cells with a $CD45^-CD31^-CD34^+CD105^+$ surface phenotype freshly isolated from adipose tissue form CFU-F, proliferate extensively, and can be differentiated toward several lineages, including osteogenic, chrondrogenic, adipogenic, and neurogenic lineages, under appropriate conducive conditions. Here, we describe methods for isolating stromal stem cells from human adipose tissue as we perform them in our laboratory.

2. Materials

2.1. Isolation of Adipose Tissue Stromal Stem Cells

1. Lipoaspirate, commonly obtained from a clinic. The lipoaspirate can be kept at room temperature overnight before stem cell isolation.
2. Hanks balanced salt solution (HBSS, without phenol red; cat. no. 14025-050, Gibco-BRL; Paisley, UK).
3. Fetal bovine serum (FBS; cat. no. 10099-141, Gibco-BRL). Heat-inactivate at 55°C for 30 min. Aliquot and store frozen. Thaw at 4°C.
4. Collagenase A type I (cat. no. C-0130, Sigma-Aldrich Co; St. Louis, MO).
5. Antibiotics: penicillin–streptomycin mix (100X solution; cat. no. 15140-122, Gibco-BRL).
6. Fungizone (100X solution; Gibco-BRL).
7. Histopaque-1077 (cat. no. 1077-1, Sigma-Aldrich).
8. Red blood cell lysis buffer: 2.06 g/L Tris base, pH 7.2, 7.49 g/L NH_4Cl. Sterile filter after preparation. Can be kept at room temperature for 4 wk.
9. Bench media: HBSS with 2% FBS and antibiotics.
10. Falcon 100-μm cell strainers (cat. no. 352360, Becton Dickinson,; Franklin Lakes, NJ).
11. Falcon 40-μm cell strainers (cat. no. 352360, Becton Dickinson).
12. MACS® anti-CD45 FITC-conjugated antihuman monoclonal antibody (cat. no. 130-080-202, Miltenyi Biotec, Bergisch Gladbach; Germany).

13. Anti-CD31 FITC-conjugated mouse antihuman monoclonal antibody (cat. no. MCA 1738F, Serotec; Oxford, UK).
14. MACS anti-FITC Microbeads (cat. no. 130-048-701, Miltenyi Biotec).
15. Column buffer: phosphate-buffered saline (PBS; pH 7.2), 0.5% FBS, 2 mM ethylene diamine tetraacetic acid (EDTA).
16. MACS LD columns (cat. no. 130-042-901, Miltenyi Biotec).
17. MidiMACS separation unit (cat. no. 130-048-701, Miltenyi Biotec)
18. 50-mL and 15-mL Plastic conical tubes (Corning; Corning, NY).
19. 162-cm^2 Cell culture flasks (Corning).
20. Empty 500-mL Sterile medium bottles (Gibco-BRL).
21. 50-mL, 25-mL, and 10-mL Disposable plastic pipets (Corning).
22. A swing-out centrifuge with buckets for 50-mL and 15-mL tubes.
23. Fluorescence microscope equipped with a blue light filter for FITC viewing.

2.2. Culture of Adipose Tissue Stromal Stem Cells

1. Cell incubator set at 100% humidity, 37°C and 5% CO_2 in air.
2. DMEM:F12 (1:1; cat. no. 31331-028, Gibco-BRL).
3. 25-cm^2 cell culture flasks (Corning).

3. Methods
3.1. Collection and Storage of Lipoaspirate

We have found that stromal stem cells can be isolated from lipoaspirate that has been collected from several regions of the body, including the hip, thigh, and abdominal regions. At least 300 mL of lipoaspirate should be collected into a sterile container to isolate uncultured stem cells in significant numbers (million-range). Using the technique described below, we routinely isolate as many as 10^7 adipose stromal stem cells with greater than 95% purity from 300 mL of lipoaspirate. However, yields can vary widely between patients. The actual adipose tissue volume used for digestion after the washing steps usually is two-thirds of the collected lipoaspirate volume. If overnight storage of lipoaspirate is required, we recommend storage at room temperature. We have found overnight storage at 4°C reduces the enzyme digestibility of adipose tissue from the lipoaspirate.

3.2. Separation of the Stromal Vascular Fraction
3.2.1. Lipoaspirate Washing

It is necessary to wash the lipoaspirate extensively to remove the majority of the erythrocytes and leukocytes. The following procedures should be performed under aseptic conditions, ideally in a laminar air flow cabinet.

1. Place a maximum of 300 mL of lipoaspirate into a used sterile medium bottle.
2. Allow the adipose tissue to settle above the blood fraction.

3. Remove the blood using a sterile 25-mL pipet.
4. Add an equivalent volume of HBSS with antibiotics and fungizone and firmly tighten the lid.
5. Shake vigorously for 5 to 10 s.
6. Place the bottle on the bench and allow the adipose tissue to float above the Hanks' balanced salt solution (HBSS). This will take 1 to 5 min, depending on the sample.
7. Carefully remove the HBSS using a 50-mL pipet.
8. Repeat the aforementioned washing procedure (**steps 4–7**) three times.
9. Medium from the final wash should be clear. If it is still red, wash again by repeating **steps 4** to **7**.

3.2.2. Collagenase Digestion

Dispersion of adipose tissue is achieved by collagenase digestion. Collagenase has the advantage over other tissue digestive enzymes that it can efficiently disperse adipose tissue while maintaining high cell viability.

1. Make up collagenase solution just prior to digestion. The final volume required is half that of the washed adipose tissue volume. Add powdered collagenase to HBSS at a final concentration of 0.2%. We dissolve the required amount of collagenase into 40 mL of HBSS, then filter sterilize into the remaining working volume. Add antibiotics and fungizone.
2. Add the washed adipose tissue to large cell culture flasks (100 mL per 162-cm^2 flask).
3. Add collagenase solution.
4. Resuspend the adipose tissue by shaking the flasks vigorously for 5 to 10 s.
5. Incubate at 37°C on a shaker for 1 to 2 h, manually shaking the flasks vigorously for 5 to 10 s every 15 min.
6. During the digestion, prepare Histopaque gradients by dispensing 15 mL of Histopaque-1077 into 50-mL tubes. Two gradients are required for each 100 mL of washed adipose tissue. The gradients must be equilibrated at room temperature before use. Prepare 200 mL of washing medium consisting of HBSS containing 2% FBS, antibiotics, and fungizone.
7. On completion of the digestion period, the digested adipose tissue should have a "soup-like" consistency.
8. Add FBS to a final concentration of 10% to stop collagenase activity.

3.2.3. Separation of the Stromal Vascular Fraction

After digestion, the ability of lipid-filled adipocytes to float is used to separate them from the stromal vascular fraction (SVF).

1. Dispense the collagenase-digested tissue into 50-mL tubes. Avoid dispensing undigested tissue. Centrifuge at room temperature at 400g for 10 min.
2. After centrifugation, use a 50-mL pipet to aspirate the floating adipocytes, lipids, and the digestion medium. Leave the SVF pellet in the tube.

3.3. Separation of Stromal Stem Cells From the SVF

The SVF predominantly contains erythrocytes, leukocytes, endothelial cells, and stromal stem cells. Erythrocytes are removed first, using the red blood cell lysis buffer.

3.3.1. Removal of Erythrocytes

1. Resuspend thoroughly each SVF pellet in 20 mL of cell lysis buffer at room temperature.
2. Incubate at room temperature for 10 min.
3. Centrifuge at 300*g* for 10 min and aspirate the cell lysis buffer.

3.3.2. Removal of Cell Clumps and Remaining Undigested Tissue

It is essential to obtain a cell suspension free from undigested tissue and cell clumps to effectively separate stromal stem cells from other cell types using antibody-conjugated magnetic beads. The strategies used to achieve this are separation of gross undigested tissue using gravity, straining of cells, and gradient separation.

1. Resuspend SVF pellets thoroughly in 2 mL of washing medium using a 1-mL pipet.
2. Pipet the cells up and down several times to reduce clumping.
3. Pool the pellets into two 50-mL tubes.
4. Allow undigested tissue clumps to settle by gravity for approx 1 min.
5. Aspirate and pass the suspended cells through 100-μm cell strainers.
6. Pass the filtered cells through 40-μm cell strainers.
7. Add extra washing buffer so that the final volume is equivalent to that of the gradients (i.e., for four gradients, the volume of cells in washing buffer should be 60 mL).
8. Hold each tube containing Histopaque at a 45° angle and carefully add the cells by running the suspension along the inside wall of the tube at a flow rate of approx 1 mL/s. Careful layering of cells onto the gradients is essential for successful cell separation.
9. Centrifuge gradients at exactly 400*g* for 30 min.
10. Carefully remove the medium (~10 mL) above the white band of cells found at the gradient interface and discard.
11. Carefully remove the white band of cells (~5 mL) by careful aspiration and place into a new 50-mL tube.
12. Add an equivalent volume of washing medium and centrifuge at 300*g* for 10 min using a low brake setting.
13. Aspirate and resuspend each pellet in 25 mL of washing medium.
14. Centrifuge at 300*g* for 10 min using a low brake setting.

3.3.3. Separation of Stromal Stem Cells From Endothelial Cells and Leukocytes by Magnetic Cell Sorting

Stromal stem cells are separated from remaining cells using magnetic cell sorting. Unwanted endothelial cells (CD31$^+$) and leukocytes (CD45$^+$) are magnetically labeled and eliminated from the cell suspension when applied to a column under a magnetic field. Magnetically labeled cells are retained in the column, while unlabeled stem cells with a CD45$^-$CD31$^-$ phenotype pass through the column and are collected. To this end, CD31$^+$ and CD45$^+$ cells are labeled with FITC-conjugated anti-CD31 and anti-CD45 antibodies. The stained cells are magnetically labeled by the addition of anti-FITC-conjugated magnetic microbeads. This approach presents the advantage that cell purity after separation can be assessed by flow cytometry or fluroescence microscopy. For the following steps, use cold buffer and work on ice to reduce cell clumping.

1. Resuspend and pool the sedimented pellets in 10 mL of column buffer (PBS containing 2 m*M* EDTA and 0.5% BSA).
2. Remove all remaining cell clumps by passing the suspension through a 40-μm cell strainer.
3. Perform a cell count.
4. Transfer cells to a 15-mL tube and centrifuge at 300*g* for 10 min at 4°C using a low brake setting.
5. Resuspend the cell pellet in column buffer and label with anti-CD31 FITC-conjugated and anti-CD45 FITC-conjugated antibodies according to the manufacturer's recommendations. We resuspend cells in 100 μL of column buffer and add 10 μL of each antibody per 10^7 cells.
6. Mix well and incubate for 15 min in the dark at 4°C (resuspend the cells after 7 min of incubation).
7. Wash the cells to remove unbound antibody by adding 2 mL of column buffer per 10^7 cells. Centrifuge at 300*g* for 10 min at 4°C using a low brake setting.
8. Aspirate the supernatant completely and resuspend the cell pellet in 90 μL of column buffer per 10^7 cells. Add 10 μL of MACS anti-FITC magnetic microbeads per 10^7 cells.
9. Mix well and incubate for 15 min at 4°C (resuspend the cells after 7 min of incubation).
10. Wash the cells to remove unbound beads by adding 2 mL of column buffer per 10^7 cells. Centrifuge at 300*g* for 10 min at 4°C using a low brake setting.
11. Aspirate the supernatant completely and resuspend the cell pellet in 500 μL of column buffer.
12. For magnetic cell separation, we use the MACS LD column specifically designed for the depletion of unwanted cells. Place a MACS LD column onto the MidiMACS separation unit or onto a compatible unit.
13. Prepare the column by washing with 2 mL of column buffer.
14. Apply the cell suspension to the column and collect the flow-through unlabeled cells in a 15-mL tube.

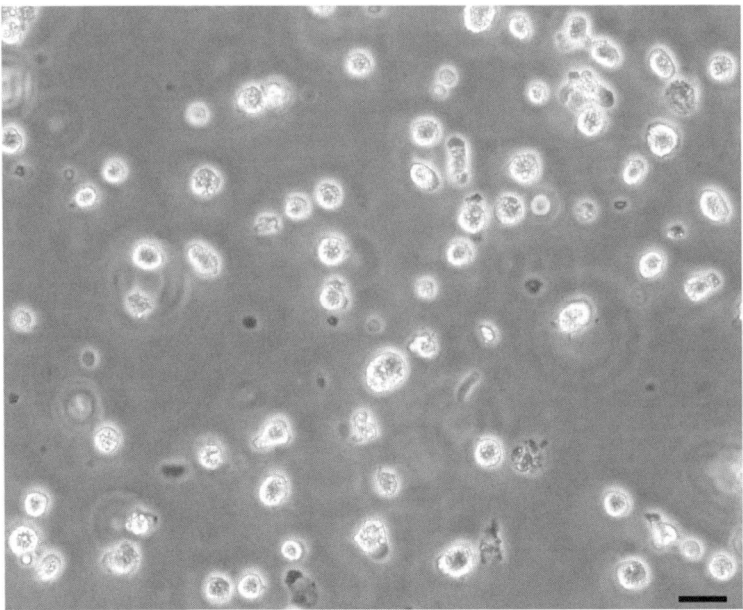

Fig. 1. Stromal stem cells purified from collagenase-digested and strained liposuction material. The cells have an even, round phenotype. Bar, 20 μm.

15. Wash unlabeled cells through the column by twice adding 1 mL of column buffer. Collect the total effluent.
16. Check for stem cell purity as described in **Subheading 3.3.4.**
17. If higher purity is required, centrifuge the collected cells at 300*g* for 10 min at 4°C using a low brake setting and repeat **steps 11–16**.
18. Perform a cell count.
19. Centrifuge at 300*g* for 10 min at 4°C using a low brake setting.
20. Use the cells as required or freeze the cells according to standard protocols.

3.3.4. Assessment of Stem Cell Purity

We have found that the success of obtaining pure stromal stem cell samples of high purity varies between donors for unclear reasons. It is therefore important to assess the purity of the sample using a fluorescence-based assay. We find that fluorescence microscopy is sufficient to evaluate purity. A more accurate assessment can be made by flow cytometry; however, this assay requires many more cells.

1. Place 5 μL of the collected cell fraction onto a glass slide.
2. View under white light. Stromal stem cells have an evenly round phenotype (*see* **Fig. 1A**), whereas endothelial cells have an irregular shape (*see* **Fig. 1B**).
3. Observe the samples under epifluorescence.

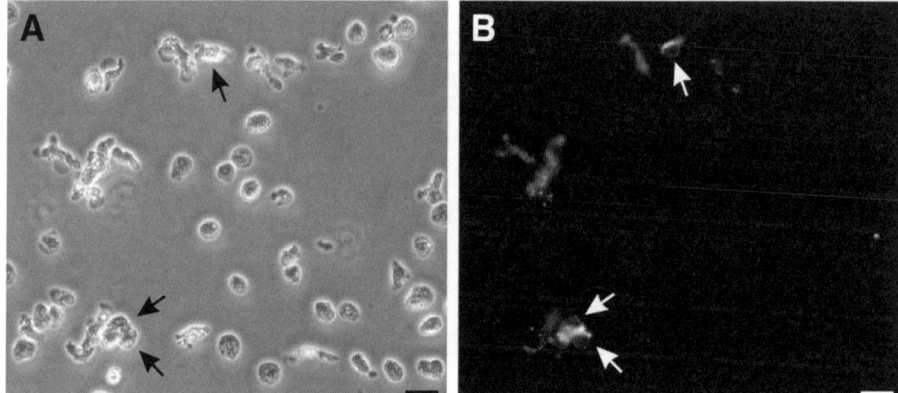

Fig. 2. Identification of contaminating CD45⁺ and CD31⁺ cells among stem cells isolated from adipose tissue. Cells recovered from liposuction material were labeled using anti-CD31 FITC- and anti-CD45 FITC-conjugated antibodies prior to purification of CD45⁻ and CD31⁻ cells by negative selection. Unwanted endothelial cells (CD31⁺) and leukocytes (CD45⁺) were eliminated from the cell suspension. (**A**) Cells were examined by phase contrast microscopy. (**B**) As a quality control assessment, fluorescence microscopy examination of the flow-through cells enables the identification of a low proportion of contaminating CD45⁺CD31⁺ cells (arrows). Bars, 20 μm.

4. Determine the percentage of fluorescent cells under five fields of view (*see* **Fig. 2A,B**). The average represents the percentage of contamination of nonstem cells in the sample.

3.4. Culture of Stromal Stem Cells

Culture also can be used to further validate the successful isolation of stromal stem cells from adipose tissue using the above procedure. Stromal stem cells, when cultured, adhere to plastic and acquire a fibroblastic-like morphology. It may take several days before all adherent cells change their morphology. We find that approx 50% of cells isolated as above will plate under the correct culture conditions. However, plating efficiency can vary substantially between donors. To encourage adherence, we plate isolated stem cells in medium containing 50% FBS in a volume sufficient to smear the medium across the surface of a cell culture flask.

3.4.1. Validation of Stem Cell Isolation by Cell Culture

1. Resuspend 10⁵ freshly isolated stromal stem cells in 2 mL of DMEM:F12 medium containing 50% FBS, antibiotics, and fungizone.
2. Add the cell suspension to a 25-cm² cell culture flask.

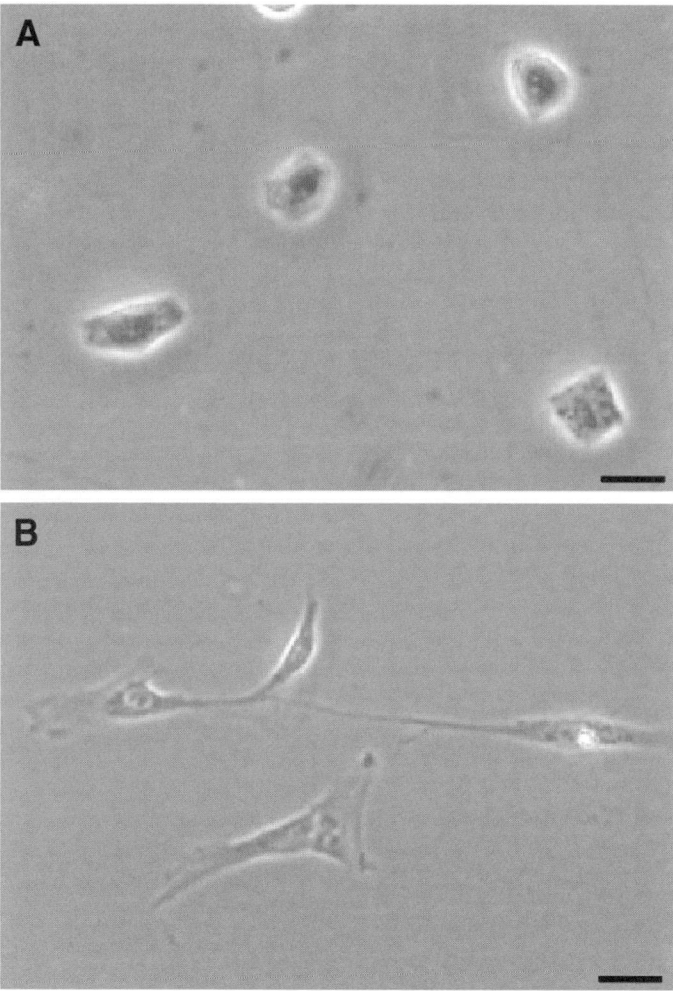

Fig. 3. Morphology of isolated and seeded stem cells derived from adipose tissue. **(A)** Cells adhere to the plastic surface and, **(B)** after 7 d of culture, acquire a fibroblastic-like morphology. Bars, 20 µm.

3. Smear the cell suspension across the entire surface by rocking the flask.
4. Incubate in a humidified incubator at 37°C, 5% CO_2.
5. Observe the cells daily using an inverted phase contrast microscope.
6. It usually takes several days before those cells which form a fibroblastic morphology start dividing.
7. Estimate the percentage of cells which adhere to the plastic surface (*see* **Fig. 3A**) and form a fibroblastic-like morphology after 7 d of culture (*see* **Fig. 3B**).

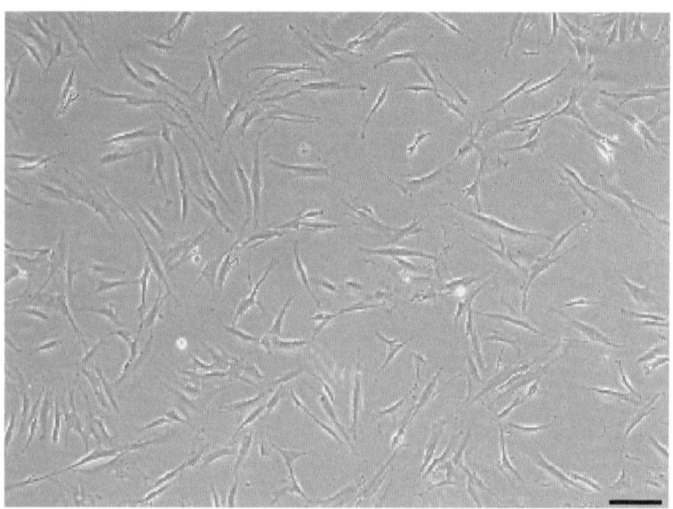

Fig. 4. Monoclonal culture derived from a single adipose tissue-derived stem cell. Bar, 50 μm.

3.4.2. Generation of a Stable Stromal Stem Cell Line

The generation of stable adipose stem cell lines is required to evaluate their differentiation capacity and proliferative ability. We have kept lines of stem cells generated by the above isolation method for longer than 6 mo without loss attributable to senescence (*see* **Note 1**).

1. After at least 7 d of culture, replace the medium with DMEM:F12 containing antibiotics, 20% FBS, and no fungizone.
2. Subculture cells using standard methods of trypsinization after a further week of culture to form a stable cell line (*see* **Fig. 4**).
3. Split the cells weekly at a ratio of 1:3.

4. Note

1. To evaluate the "stemness" of adipose stem cell lines established, we recommend differentiating the cells towards various mesodermal lineages according to Zuk et al. *(4)*. For adipogenic differentiation, incubate cultures in DMEM:F12 medium containing 10% FBS, 0.5 μM 1-methyl-3 isobutylxanthine, 1 μM dexamethasone, 10 μg/mL insulin, and 100 μM indomethacin for 3 wk. Change the medium every 4 d. To visualize lipid droplets, fix the cells with 4% formalin and stain with Oil-Red O (*see* **Fig. 5A**). For osteogenic differentiation, incubate the cells in DMEM:F12 medium containing 10% FBS, 100 nM dexamethasone, 10 mM β-glycerophosphate, and 0.05 mM ʟ-ascorbic acid-2-phosphate for 3 wk. Change the medium every 4 d. Mineralization of the extracellular matrix is visualized by staining with Alzarin Red (*see* **Fig. 5B**).

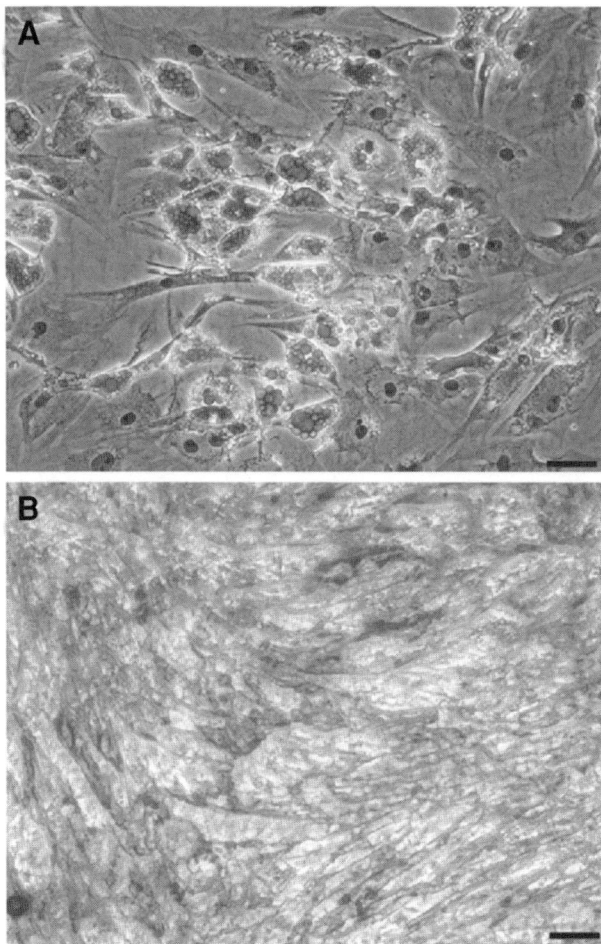

Fig. 5. In vitro differentiation of cultured human adipose tissue-derived stem cells towards **(A)** the adipogenic pathway and **(B)** the osteogenic pathway. (A) To visualize intracellular lipid droplets, we fixed the cells and stained them with Oil-Red O. (B) Mineralization of the extracellular matrix was visualized by staining with Alzarin Red. Bars, 50 μm.

References

1. Pittenger, M. F., Mackay, A. M., Beck, S. C., Jaiswal, R. K., Douglas, R., Mosca, J. D., et al. (1999) Multilineage potential of adult human mesenchymal stem cells. *Science* **284,** 143–147.
2. Howell, J. C., Lee, W. H., Morrison, P., Zhong, J., Yoder, M. C., and Srour, E. F. (2003) Pluripotent stem cells identified in multiple murine tissues. *Ann. N. Y. Acad. Sci.* **996,** 158–173.

3. Arai, F., Ohneda, O., Miyamoto, T., Zhang, X. Q., and Suda, T. (2002) Mesenchymal stem cells in perichondrium express activated leukocyte cell adhesion molecule and participate in bone marrow formation. *J. Exp. Med.* **195,** 1549–1563.

4. Zuk, P. A., Zhu, M., Ashjian, P., De Ugarte, D. A., Huang, J. I., Mizuno, H., et al. (2002) Human adipose tissue is a source of multipotent stem cells. *Mol. Biol. Cell* **13,** 4279–4295.

5. Zuk, P. A., Zhu, M., Mizuno, H., Huang, J., Futrell, J. W., Katz, A. J., et al. (2001) Multilineage cells from human adipose tissue: implications for cell-based therapies. *Tissue Eng.* **7,** 211–228.

6. Gronthos, S., Zannettino, A. C., Hay, S. J., Shi, S., Graves, S. E., Kortesidis, A., et al. (2003) Molecular and cellular characterisation of highly purified stromal stem cells derived from human bone marrow. *J. Cell Sci.* **116,** 1827–1835.

7. Safford, K. M., Hicok, K. C., Safford, S. D., Halvorsen, Y. D., Wilkison, W. O., Gimble, J. M., et al. (2002) Neurogenic differentiation of murine and human adipose-derived stromal cells. *Biochem. Biophys. Res. Commun.* **294,** 371–379.

8. Woodbury, D., Reynolds, K., and Black, I. B. (2002) Adult bone marrow stromal stem cells express germline, ectodermal, endodermal, and mesodermal genes prior to neurogenesis. *J. Neurosci. Res.* **69,** 908–917.

9. Boquest A. C., Shahdadfar, A., Frønsdal, K., Sigurjonsson, O., Tunheim, S. H., Collas, P., et al. (2005) Isolation and transcription profiling of purified uncultured human stromal stem cells: alteration of gene expression following in vitro cell culture. *Mol. Biol. Cell.* **16,** 1131–1141.

5

Nuclear Reprogramming by Cell Fusion

Robert H. Broyles, Austin C. Roth, Mairead Todd, and Visar Belegu

Summary

The use of cell fusion to study exchange of information at the molecular level between the nucleus and the cytoplasm of cells during regulation of gene expression was pioneered by Harris and Ringertz more than three decades ago. The ability to make heterokaryons with cells from different species or genetic strains is especially useful because genetic differences in gene products allow the origin of *trans*-acting regulatory factors to be determined. Heterokaryons between adult nucleated erythroid cells of one species and embryonic/larval nucleated erythroid cells of another species, for example, show cross-induction between the two types of nuclei, resulting in reprogramming of the adult nucleus to embryonic/larval globin gene expression and/or reprogramming of the embryonic/larval cell nucleus to adult globin expression. These experiments provided definitive evidence that developmental program switching is mediated by *trans*-acting factors. Other possible uses of this cell fusion protocol in stem cell biology and transplantation of genetically engineered cells for tissue regeneration are briefly discussed.

Key Words: Cell fusion; heterokaryons; cell hybrids; nuclear reprogramming; membrane fusion; fusisomes; globin genes; developmental hemoglobin switching; gene regulation; *trans*-acting factors; polyethylene glycol.

1. Introduction

Heterokaryons—cells containing the nuclei of two or more cell types—are created by cell fusion, either in vitro by techniques such as that described in this chapter or in vivo by natural means in tissues such as liver and developing muscle *(1–3)*. In vivo cell fusion, rather than stem cell transdifferentiation, has been suggested to account for donor-marked liver regeneration or muscle expansion in some of the experiments in which animals were transfused with hematopoietic stem cells (HSC *[4–8]*). In the case of liver, myeloid-restricted cells derived from a single HSC have been reported to fuse with liver cells and to create a hybrid stem cell with a reprogrammed nucleus that is able to differentiate into fully functional hepatocytes *(9)*.

From: *Methods in Molecular Biology, vol. 325: Nuclear Reprogramming: Methods and Protocols*
Edited by: S. Pells © Humana Press Inc., Totowa, NJ

In the laboratory, cell fusion can be performed much more efficiently than the cell fusion that has been observed to occur in vivo, and the extent of nuclear programming can be controlled by altering the donor cell type and/or the ratio of donor to recipient cells *(10–14)*. In addition, a small number of cell hybrids will ultimately develop and survive from the initial, transient heterkaryons, through stable nuclear fusion; these hybrid cells may be grown as clones and examined/selected for expression of desired genes *(15,16)*.

By fusing erythroid cells of different developmental programs from closely related or from divergent vertebrate species (**Fig. 1**), we and others have shown that erythroid nuclei can be reprogrammed with respect to developmental hemoglobin switching *(17–24)*. For example, this gene switch can be reversed in adult erythroid nuclei so that larval/fetal globins are expressed; or, a switch can be prematurely induced in fetal/embryonic erythroid nuclei such that adult globins are expressed. Formation of developmental erythroid heterokaryons between species by cell fusion provided direct evidence that this developmental programming is caused by *trans*-acting factors *(19)*. Heterokaryons between erythroid cells of different vertebrate classes have shown that the gene switching factors are highly conserved and can be exchanged between species as well as between developmental stages *(21,23)*.

Here, we describe in detail a very efficient method of creating transient heterokaryons in high yields. References cited show that nuclei in these heterokaryons are rapidly reprogrammed within 24 h, before nuclear fusion occurs *(17,19–23)*. This method may prove useful in creating stem cells as therapeutic agents, tailored to specific needs for regeneration of a variety of tissues in degenerative diseases, such as cirrhosis of the liver, muscular dystrophy, or Parkinson's disease *(25–28)*. Cell fusion technology combined with the ability to control cell differentiation may be used to create tissue-specific, genetically engineered stem cells to be used in treatments or cures for other diseases such as diabetes, arteriosclerosis and cancer *(26–32)*.

The method described below typically produces high cell fusion efficiencies, ranging from 20 to 80% of cells in a mixture undergoing fusion *(17)*. The

---

Fig. 1. *(opposite page)* Photographs of cell fusion mixtures at 30 min after initiating fusion, stained with Giemsa as described under **Subheading 3.2.** (**A**) Developmental erythroid heterokaryons from two species of amphibians (*Rana catesbeiana* and *Xenopus laevis*). *Rana* larval erythroblasts (oval cells with large, dark, expanded nuclei) were obtained from tadpoles recovering from a phenylhydrazine-induced anemia *(23)*, and *Xenopus* adult erythrocytes (elliptical cells with a lightly staining, small, centrally located nucleus) were obtained from a nonanemic adult frog. Both homo- and heteropolykaryons can be seen in this photograph. Approximately one-half of the cells in this field have more than one nucleus. (**B**) Developmental erythroid heterokaryons of frog (*Rana*) and mouse (MEL cells) origin. The small, round, dark

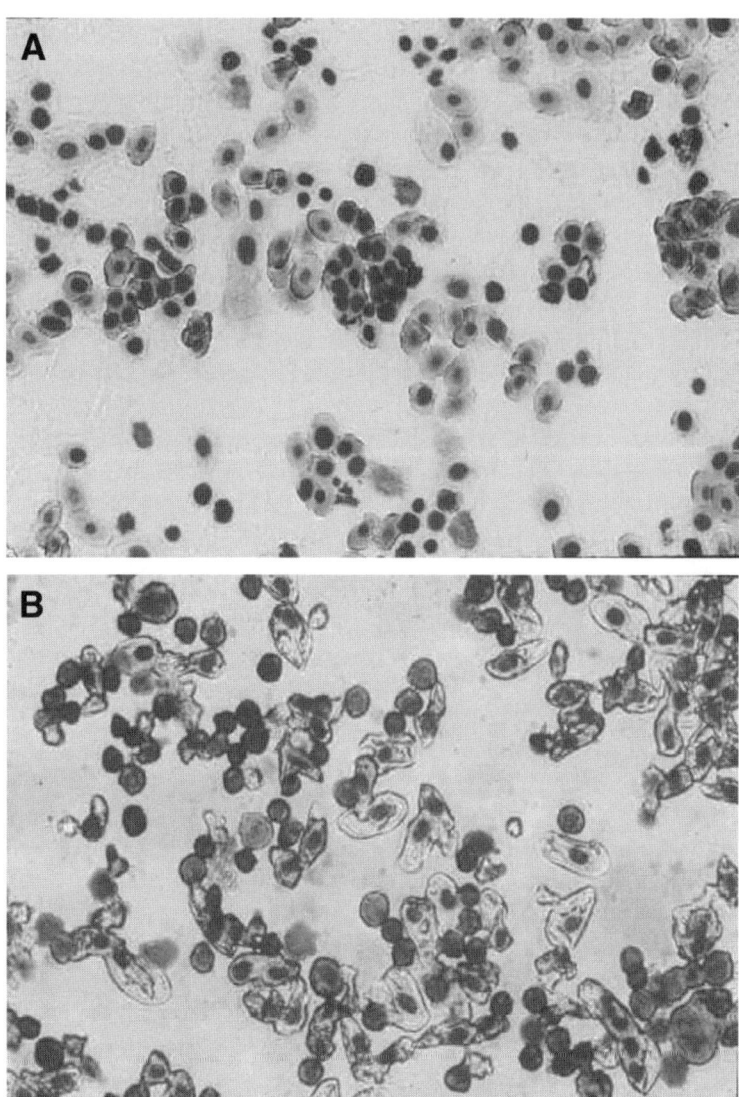

Fig. 1. *(continued)* (basophilic) cells are MEL (murine erythroleukemia) cells in which hemoglobin synthesis had been induced 4 d earlier by exposure to dimethyl sulfoxide *(23)*, and the oval/elliptical cells with a light (eosinophilic) cytoplasm and a centrally located, small nucleus are *Rana* larval (tadpole) erythrocytes. Hetero- and homopolykaryons of all combinations can be observed. In a number of the heterokaryons, it can be observed that the cytoplasm of the *Rana* larval erythrocytes has turned darker (more basophilic) after fusion with one or more MEL cells, indicating that the cytoplasms from the two cell types have mixed. In (B), approx 80% of the cells are participating in fusion. Both photographs were taken at 100× magnification.

amount of cell fusion differs with different cell types, their degree of differentiation, and how well the cells withstand the procedure, which has been revised over time to be as gentle as possible. To our knowledge, these levels of cell fusion are the highest that have been reported in the literature.

2. Materials

1. Sterile, premixed solutions and tissue culture media, as follows:
   a. Standard Ringer's solution *(33)* or a phosphate-buffered saline (PBS) of an osmolarity appropriate for the species and cells being used.
   b. Calcuim-free Ringer's solution and/or calcium-free PBS.
   c. Iscove's modified Dulbecco's tissue culture medium without serum and without antibiotics.
   d. Solution K (a high-potassium buffer solution): 5.35 mM NaCl, 135 mM KCl, 0.8 mM MgSO$_4$, 20 mM Tricine, and 2 mM MnCl$_2$, pH 7.8.
   e. Polyethylene glycol (PEG) in the 3000–3700 Da molecular range (Sigma Chemicals, St. Louis, MO) that has been melted in a water bath (37°C), then diluted 1/1 with an equal volume of sterile amphibian Ringer's solution (0.113 M NaCl, 2 mM KCl, 1.35 mM CaCl$_2$, 3.57 mM NaHCO$_3$), and the whole solution brought to pH 8.0 with 1 M NaOH. This solution should be equilibrated at room temperature (~21°C) before use.
   f. Postfusion medium: equal parts Iscove's medium (from c) and solution K (from d). For a single fusion sample, 10 mL postfusion medium (5 mL each of c and d) will suffice.
   g. Cell maintenance medium (CMMC *[34]*; per liter of solution): 2.0 g of NaCl, 7.0 g of G acid (disodium form), 4.0 g of N-2-hydroxyethylpiperazine-N-2-ethanesulfonic acid (HEPES; sodium form), 1.13 g of HEPES (hydrogen form), 0.25 g of KCl, 0.67 g glucose, 0.72 g of MgSO$_4$·7H$_2$O, and 10.0 g of bovine serum albumin (filtered). (G acid also is known as 2-naphthol-6,8-disulfonic acid, 7-hydroxy-1,3-naphthalenedisulfonic acid, or 2-hydroxynaphthalene-6,8-disulfonic acid; contact Aldrich Chemicals, Milwaukee, WI, for availability.) *See* **Note 1**.
   h. CMMC-Ca-free-Ringer's/PBS medium: 1 part CMMC and 1.5 parts calcium-free Ringer's or calcium-free PBS.
   i. Premixed Giemsa stain and Crystal/Mount medium for mounting cover slips to microscope slides. Both premixed products can be obtained from a general scientific supply house (e.g., Fisher Scientific).
2. A water bath equilibrated at 37°C and fitted with a test-tube rack suitable for 15-mL conical centrifuge tubes.
3. Two ice buckets with crushed ice.
4. Sterile, cotton-plugged glass, long-tip Pasteur pipets, with rubber suction bulbs.
5. Sterile pipet tips, pipets, and automatic pipetors capable of delivering volumes ranging from 10 μL to 2.0 mL.
6. IEC table-top clinical centrifuge (or equivalent) fitted with a rotor with cups for 15-mL conical centrifuge tubes.

7. Sterile, clear polypropylene 15-mL conical centrifuge tubes, graduated, with screw-caps (flat-top or plug-seal), such as Corning Orangeware tubes.
8. Clean microscope slides and cover slips.
9. Hemocytometer with cover slips, and a hand-operated counter.
10. Heparinized capillary tubes if cells are to be obtained from whole blood.
11. Compound microscope with phase-contrast optics and 10× and 20× phase objectives. An automatic camera (digital or film) attachment is helpful.
12. Suspensions of the two cell types to be fused, in Ringer's solution, PBS, or tissue culture medium.
13. A timing device that reads in seconds as well as minutes.

3. Methods

The methods outlined below describe (1) the order and procedures for completing a successful cell fusion using our techniques, and (2) the procedures that we have used to prepare microscope slides so that the progress of the fusion process can be followed visually.

3.1. Cell Fusion

1. All procedures are performed using sterile technique, and all manipulations are carried out at room temperature (~21°C) except as noted..
2. Begin with suspensions of the two cell types, in sterile medium (Ringer's, PBS, or tissue culture medium) in sterile 15-mL polypropylene centrifuge tubes in ice in an ice bucket (*see* **Note 2**).
3. Gently pellet the cells by centrifuging at approx 500*g* (speed 2 on the IEC tabletop clinical centrifuge) for a length of time sufficient to bring all the cells to the bottom of the tube.
4. Draw off the medium from the two cell pellets; wash the cells by resuspending the pellets in Ringer's solution or PBS, as appropriate for the cell type. Centrifuge briefly at 500*g*.
5. Repeat **step 3** and discard the supernatant fluid.
6. Resuspend the cells in both tubes in Ringer's solution or PBS.
7. Take a small sample (~200 μL) from the first tube of suspended, well-mixed cells using a sterile, long-tip Pasteur pipet. Apply a drop of cell suspension to opposite edges of the cover slip covering the hemocytometer grid and observe whether the suspension flows quickly and evenly under the cover slip on each end, such that each of the two copies of the counting grid is covered evenly with cells. For each grid, count the number of cells over the central 25 squares of the grid *(35)*. This figure equals the number of cells × 10^4 per milliliter of suspension. For example, if the count is 200, then the suspension contains 200×10^4 cells/mL, that is, 2×10^6 cells/mL. The counts over the two grids should agree within ±10%, and the raw values should be in the range of 100–300 cells in each count. If the values are much higher than 300, make a dilution of the suspension and recount; if the values are much lower, pellet the cells and resuspend in an appropriate volume of Ringer's solution or PBS.

8. Pellet the cells in each tube (as in **step 3**) and resuspend the pellets in a volume of Ringer's solution or PBS to give 1×10^8 cells per milliliter of suspension.
9. Into a fresh, sterile 15-mL centrifuge tube, pipet 500 µL (5×10^7 cells) from the suspensions of each of the cell types being used for fusion.
10. Gently pellet the cells (i.e, centrifuge at $500g$ for 3 min) to yield a mixed pellet of the cell types being fused. Carefully remove all the supernatant fluid from above the cell pellet using automatic pipetors (e.g., Pipetman) of appropriate volumes.
11. Caution: read **Note 3** before beginning this step. Once this step is begun, procedures must be performed in rapid succession, and there is no stopping point for at least 30 min. Add 200 µL of 1/1 PEG solution (from **step 1e** in **Subheading 2.**) and gently stir the pellet with the tip of a sterile, glass Pasteur pipet for 30 s.
12. Immediately but carefully add 2 mL of solution K so that the solution runs down the inside of the tube and dilutes the PEG solution without resuspending the pellet (*see* **Notes 4–7**).
13. Incubate the tube contents for 5 min at room temperature.
14. Add 2 mL of Iscove's medium to the tube and gently mix the Iscove's with the supernatant fluid in the tube, using a Pasteur pipet, without disturbing the cell pellet.
15. Incubate the tube contents for 5 min at room temperature.
16. Incubate the tube contents for 5 min in a 37°C waterbath (*see* **Note 8**).
17. Incubate the tube contents for 10 min at room temperature.
18. Carefully take small samples of cells from the cell pellet to make slides for wet mounts and histology as described in **Subheading 3.2., steps 1** and **2**. It is best if a second person makes the slides while the experimenter continues with the next step.
19. Centrifuge the contents of the tube at $500g$ for 2 min.
20. Remove the supernatant fluid without disturbing the pellet.
21. Very gently add 2 mL of postfusion medium.
22. Repeat **steps 19, 20**, and **21** without disturbing the pellet.
23. Gently remove the supernatant fluid and add 2 mL of Iscove's medium.
24. Incubate the tube contents (the fusion sample) at room temperature for times ranging from 30 min to 24 h.
25. The cell pellet may be gently sampled to make slides for microscopy at desired times after fusion, e.g., 1, 2, 4, 8, 12, and 24 h. We consider **step 11** as the beginning of fusion and **step 24** as 30 min; however, it takes practice to move swiftly so that **steps 11–23** can be accomplished in 30 to 40 min.
26. After the last time point, e.g., 12 or 24 h, the supernatant fluid is removed and the cell pellet processed for analysis, e.g., total RNA isolation or protein electrophoresis.

3.2. Preparing Slides for Microscopy

3.2.1. Preparation of Wet Mounts for Immediate Viewing With Phase Contrast Microscopy

1. Dilute a 10-µL sample taken from a cell pellet undergoing fusion with 10 µL of Ringer's solution or PBS (*see* **step 1a** in **Subheading 2.**).
2. Apply the 20 µL of diluted sample to the center of a standard, clean microscope slide and gently lay a circular, 18-mm diameter coverslip over the liquid.
3. View the specimen within 10 min, using phase contrast optics.

3.2.2. Preparation of Slides for Staining (36)

1. Dilute 10 μL of a cell pellet sample with 20 μL of CMMC-Ca-free-Ringer's solution/PBS composed of 10 mL CMMC medium and 15 μL of calcium-free Ringer's or calcium-free PBS.
2. Apply the 30 μL of this CMMC-diluted sample to a clean, standard microscope slide and gently spread the drop of liquid over a greater area with the edge of another microscope slide or cover slip. Allow the sample to air dry.
3. Apply liquid premixed Giemsa stain (Fisher Scientific) to the slide, covering the area of the dried sample. Allow the stain to stand 10 min.
4. Rinse the slide by dipping it in distilled water for 1 min. Stand the slide on end and let it air dry.
5. Add several drops of Crystal/Mount (Fisher Scientific), enough to cover the area of the stain, and lay a cover slip over the mounting solution. Place the slide on a horizontal, flat surface and allow the mounting medium to harden.
6. After the mounting medium has hardened such that the cover slip no longer moves, the slide may be viewed at any time using a compound microscope and bright-field illumination.

3.3. Summary

This set of procedures, when performed as described by practiced hands, is quite reproducible and routinely 20 to 80% of the cells in a mixture undergo fusion, the percentage depending on the cell types and specific pairings desired *(23)*. These are the highest percentages that we have found in the literature, indicating that this is the most efficient procedure to date for inducing cell fusion. Thus, this method may be the one of choice for a number of applications, including fusing stem cells with tissue-specific cell partners as part of a protocol for organ regeneration. Alternatively, the same fusion procedure could be used with erythrocyte ghosts or another membrane (a fusisome) or liposomes encapsulating a complex of gene-specific reprogramming molecules, with the aim of reprogramming the nuclei of recipient cells for specific purposes *(37)*.

4. Notes

1. CMMC also has been specifically tailored to be a cell protectant, specifically to support cells so there is minimal distortion in cell shape attributable to the handling required to make wet mounts and dried "smears" for histological staining. G acid *(34)* appears to be the key ingredient in this mixture (*see* **item 1g** in **Subheading 2.**).
2. Keeping the cells on ice for 15 to 30 min before initiating the fusions serves an important, fusion-related purpose in addition to prolonging viability. Colder temperatures are known to cause the dissociation of microtubules within cells, allowing them to round up and, thus, present more membrane surface area to a partner cell, thereby promoting cell fusion *(17)*.

3. **Step 11** of **Subheading 3.1.** is the most critical of the manipulations in the cell-fusion procedure. The PEG solution (the fusing agent) must be thoroughly mixed with the cells. However, this must be done carefully so that the cells are not dispersed. We have found that gentle stirring of the PEG with the narrow tip of a glass Pasteur pipet is the best way to gently bring the cells and the PEG together and still keep the cells in intimate contact with one another. Thus, there should be no tube "thumping" or tube inverting to mix the cells with PEG. And, above all, there is no vortexing in cell fusion!

4. In like manner, **steps 12–18** are designed to maintain the cell contacts created by centrifuging the cells into a mixed pellet. However, at the same time, the additions of other media are designed to gradually dilute the PEG without disturbing the inter-cell contacts.

5. Thus, the cells are exposed to the most concentrated PEG solution for a very short time (i.e., 30 s). This procedure is designed to coat as many cells as possible with PEG while keeping them close to other cells. At the same time, the PEG is also being diluted rapidly to avoid damage to the cells that would be caused by prolonged exposure to PEG at high concentrations.

6. Although the exposure to concentrated PEG is brief, cell fusion is a gradual process that probably takes much longer than the 30-min "quiet period" allowed. This is the reason for not disturbing the cell pellet for the first 30 min and only at intervals thereafter. Our results with this procedure indicate that the peak in numbers of heterokaryons formed probably occurs at sometime between 4 and 6 h after the addition of PEG *(17,22,23)*.

7. Another key to the success of this procedure is the composition of the first solution to bathe the cells after their contact with PEG. As defined in **item 1d** in **Subheading 2.**, solution K is a high-potassium solution. The purpose of its use in **Subheading 3.1.**, **step 12** is to protect the cells from an abrupt change in intracellular milieu as a result of the membrane "leaks" caused by interaction with PEG. Because the predominant intracellular cation is K^+, and the predominate extracellular cation is Na^+, use of the high-potassium buffer prevents the leakiness from turning the cells inside out, so to speak, and greatly reduces cell lysis.

8. Raising the temperature to 37°C for 5 min is an important step in promoting cell fusion. Raising the temperature promotes "melting" of the cell membrane phospholipid bilayer and increases membrane fluidity, thereby allowing membranes (in theory) to merge or fuse more easily.

References

1. Blau, H. M., Pavlath, G. K., Hardeman, E. C., Chiu, C. P., Silberstein, L., Webster S. G., Miller, S. C., and Webster, C. (1985) Plasticity of the differentiated state *Science* **230,** 758–766.

2. LaBarge, M. A., and Blau, H. M. (2002) Biological progression from adult bone marrow to mononucleate muscle stem cell to multinucleate muscle fiber in response to injury. *Cell* **111,** 589–601.

3. Petersen, B. E., Bowen, W. C., Patrene, K. D., Mars, W. M., Sullivan, A. K., Murase, N., et al. (1999) Bone marrow as a potential source of hepatic oval cells. *Science* **284,** 1168–1170.

4. Wang, X., Willenbring, H., Akkari, Y., Torimaru, Y., Foster, M., Al-Dhalimy, M., et al. (2003) Cell fusion is the principal source of bone-marrow-derived hepatocytes. *Nature* **422,** 897–901.

5. Vassilopoulos, G., Wang, P. R., and Russell, D. W. (2003) Transplanted bone marrow regenerates liver by cell fusion. *Nature* **422,** 901–904.

6. Alvarez-Dolado, M., Pardal, R., Garcia-Verdugo, J. M., Fike, J. R., Lee, H. O., Pfeffer, K., et al. (2003) Fusion of bone-marrow-derived cells with Purkinje neurons, cardiomyocytes and hepatocytes. *Nature* **425,** 968–973.

7. Camargo, F. D., Green, R., Capetenaki, Y., Jackson, K. A., and Goodell, M. A. (2003) Single hematopoietic stem cells generate skeletal muscle through myeloid intermediates. *Nat. Med.* **9,** 1520–1527.

8. Camargo, F. D., Chambers, S. M., and Goodell, M. A. (2004) Stem cell plasticity: from transdifferentiation to macrophage fusion. *Cell Prolif.* **37,** 55–65.

9. Camargo, F. D., Finegold, M., and Goodell, M. A. (2004) Hematopoietic myelomonocytic cells are the major source of hepatocyte fusion partners. *J. Clin. Invest.* **113,** 1266–1270.

10. Harris, H., Watkins, J. F., Ford, C. E., and Schoefl, G. I. (1966) Artificial heterokaryons of animal cells from different species. *J. Cell Sci.* **1,** 1–30.

11. Harris, H. (1965) Behaviour of differentiated nuclei in heterokaryons of animal cells from different species. *Nature* **206,** 583–588.

12. Ringertz, N. R. (1970) Activation of dormant cell nuclei. *Hoppe Seylers Z Physiol. Chem.* **351,** 779.

13. Ringertz, N. R. (1979) Analysis of differentiation by cell fusion techniques. *Differentiation* **13,** 63–64.

14. Harris, H. (1970) The use of cell fusion in the analysis of gene action. *Proc. R. Soc. Lond. B. Biol. Sci.* **176,** 315–317.

15. Harris, H. (1967) Hybrid cells from mouse and man: a study in genetic regulation. *Arzneimittelforschung* **17,** 1438–1439.

16. Ringertz, N. R., and Bolund, L. (1974) Reactivation of chick erythrocyte nuclei by somatic cell hybridization. *Int. Rev. Exp. Pathol.* **13,** 83–116.

17. Barker-Harrel, J., McBride, K. A., and Broyles, R. H. (1988) Formation of transient polykaryons by fusion of erythrocytes of different developmental programs. *Exp. Cell Res.* **178,** 435–48.

18. Baron, M. H., and Maniatis, T. (1987) Stage-specific reprogramming of globin gene expression. *Prog. Clin. Biol. Res.* **251,** 271 284.

19. Broyles, R. H. (1999) Use of somatic cell fusion to reprogram globin genes. *Semin. Cell Dev. Biol.* **10,** 259–265.

20. Broyles, R. H., Barker-Harrel, J., Ramseyer, L. T., McBride, K. A., and Sexton, D. L. (1989) Erythroid heterokaryons: a system for investigating the functional role of trans-acting factors in developmental hemoglobin switching. *Prog. Clin. Biol. Res.* **316B,** 83–96.

21. Broyles, R. H., Palmer, J. C., Ramseyer, L. T., Smith, D. J., Jarman, R. N., Do, T. H., and McBride, K. A. (1987) Hemoglobin switching across vertebrate classes: exchange of developmental signals by cell fusion. *Prog. Clin. Biol. Res.* **251**, 285–294.

22. Broyles, R. H., Ramseyer, L. T., Do, T. H., McBride, K. A., and Barker, J. C. (1994) Hemoglobin switching in Rana/Xenopus erythroid heterokaryons: factors mediating the metamorphic hemoglobin switch are conserved. *Dev. Genet.* **15**, 347–355.

23. Ramseyer, L. T., Barker-Harrel, J., Smith, D. J., McBride, K. A., Jarman, R. N., and Broyles, R. H. (1989) Intracellular signals for developmental hemoglobin switching. *Dev. Biol.* **133**, 262–271.

24. Stalder, J. (1988) Erythroid specific activation of the Xenopus laevis adult alpha-globin promoter in transient heterokaryons. *Nucleic Acids Res.* **16**, 11027–11045.

25. Freed, C. R. (2002) Will embryonic stem cells be a useful source of dopamine neurons for transplant into patients with Parkinson's disease? *Proc. Natl. Acad. Sci. USA* **99**, 1755–1757.

26. Baizabal, J. M., Furlan-Magaril, M., Santa-Olalla, J., and Covarrubias, L. (2003) Neural stem cells in development and regenerative medicine. *Arch. Med. Res.* **34**, 572–588.

27. Kerr, D. A., Llado, J., Shamblott, M. J., Maragakis, N. J., Irani, D. N., Crawford, T. O., et al. (2003) Human embryonic germ cell derivatives facilitate motor recovery of rats with diffuse motor neuron injury. *J. Neurosci.* **23**, 5131–5140.

28. Shamblott, M. J., Axelman, J., Littlefield, J. W., Blumenthal, P. D., Huggins, G. R., Cui, Y., et al. (2001) Human embryonic germ cell derivatives express a broad range of developmentally distinct markers and proliferate extensively in vitro. *Proc. Natl. Acad. Sci. USA* **98**, 113–118.

29. Jackson, K. A., Majka, S. M., Wulf, G. G., and Goodell, M. A. (2002) Stem cells: a minireview. *J. Cell Biochem. Suppl* **38**, 1–6.

30. Vassilopoulos, G., and Russell, D. W. (2003) Cell fusion: an alternative to stem cell plasticity and its therapeutic implications. *Curr. Opin. Genet. Dev.* **13**, 480–485.

31. von Degenfeld, G., Banfi, A., Springer, M. L., and Blau, H. M. (2003) Myoblast-mediated gene transfer for therapeutic angiogenesis and arteriogenesis. *Br. J. Pharmacol.* **140**, 620–626.

32. McLin, V. A., and Zorn, A. M. (2003) Organogenesis: making pancreas from liver. *Curr. Biol.* **13**, R96–R98.

33. Rugh, R. (1962) *Experimental Embryology,* 3rd ed. Burgess, Minneapolis.

34. Dorn, A. R., and Broyles, R. H. (1982) Erythrocyte differentiation during the metamorphic hemoglobin switch of Rana catesbeiana. *Proc. Natl. Acad. Sci. USA* **79**, 5592–5596.

35. Freshney, R. I. (1983) *Culture of Animal Cells, A Manual of Basic Technique,* Alan R. Liss, Inc., New York.

36. Humason, G. L. (1979) *Animal Tissue Techniques*, 4th ed. W.H. Freeman & Co., San Francisco.
37. Broyles, R. H. (2001) Hemoglobin Switching and developmental changes in erythropoietic sites and red blood cell populations of nonmammalian vertebrates, in *Hematopoiesis, A Developmental Approach* (Zon, L. I., ed.). Oxford University Press, New York, pp. 617–637.

6

Polyethylene Glycol-Mediated Cell Fusion

Jian Yang and Ming Hong Shen

Summary

Polyethylene glycol (PEG)-mediated cell fusion is a simple and efficient technique used widely for the production of somatic cell hybrids and for nuclear transfer in mammalian cloning. We describe a basic protocol of PEG-mediated cell fusion for the production of somatic cell hybrids. Fusion can be performed between adherent and suspension cells or between adherent cells or suspension cells. Either whole cells or microcells can be used as donors to fuse with recipient cells. Microcell fusion is particularly useful in transfer of a single or a limited number of chromosomes between various types of cells. Using this method, we have successfully introduced mammalian minichromosomes into a variety of vertebrate cells. The technique described here can be adapted for uses in other cell fusion involved research. This protocol, in principle, provides guidelines for further development of PEG mediated cell fusion technology.

Key Words: Polyethylene glycol; PEG; cell fusion; microcell fusion; nuclear transfer; chromosome transfer; somatic cell hybrids.

1. Introduction

Polyethylene glycol (PEG), a highly hydrated polymer, was shown to induce cell fusion in the 1970s *(1,2)*. Since then, this polymer has been widely used to mediate cell fusion for production of somatic cell hybrids that are invaluable tools in gene mapping, gene expression, the analysis of gene function, and the production of vaccines and antibodies *(3–5)*. PEG-mediated cell fusion has also been used for nuclear transfer in mammalian cloning *(6)*. PEG induces cell agglutination and cell-to-cell contact, leading to subsequent cell fusion. However, the detailed mechanisms underlying PEG-mediated cell fusion are not known. Model systems were used to define the molecular details of the fusion process *(7)*. These studies indicate that small perturbations in lipid packing within contacting bilayer leaflets are necessary and probably sufficient to

From: *Methods in Molecular Biology, vol. 325: Nuclear Reprogramming: Methods and Protocols*
Edited by: S. Pells © Humana Press Inc., Totowa, NJ

promote membrane fusion. In this chapter, we describe a basic protocol of PEG-mediated cell fusion for the production of somatic cell hybrids. The main procedures of the protocol include preparation of donor and recipient cells, cell fusion, hybrid selection and cloning, and characterization of hybrid cells as summarized in **Fig 1A**. Fusion can be performed between adherent and suspension cells or between adherent cells or suspension cells. Either whole cells or microcells can be used as donors to fuse with recipient cells. Microcell fusion is particularly useful for transfer of a single or a limited number of chromosomes between various types of cells. Using this method, we have successfully introduced mammalian minicchromosomes into a variety of vertebrate cells *(8,9)*. The technique described here can be adapted for uses in other cell fusion involved research. This protocol, in principle, provides guidelines for further development of PEG mediated cell fusion technology.

2. Materials

1. Cell lines: mouse LA9 cells containing a minichromosome, here designated LA9$^{hprt-hgy+}$cells, and chicken DT40 cells, here designated DT40$^{hprt+hyg-}$ cells. *See* **Note 1**.
2. Medium and supplements: DMEM (Life Technologies), β-mercaptoethanol (Sigma), fetal calf serum (FCS; Sigma), chicken serum (Gibco-BRL), hygromycin B (Roche), demecolcine (colcemid; Sigma), and hypoxanthine–aminopterin–thymine (HAT; Sigma).
3. Chemicals and biochemicals: phosphate-buffered saline: PBS, pH 7.3, 137 mM NaCl, 2.7 mM KCl, 10 mM Na$_2$HPO$_4$, 2 mM KH$_2$PO$_4$; 0.2% trypsin in Versene; cytochalasin B (Sigma); PEG 1500 (Roche); Percoll (Pharmacia); human Y alphoid satellite DNA either labelled with biotin or ^{32}P; primer pairs:

   P1 (5'-AACTTCATCAGTGTTACATCAAGG-3'
   5'-TGTGGCATTTTGTTATGTGG-3')
   P2 (5'-AGGAGATGTCAGGACTATCAGC-3'
   5'-TCCATCCAGCTGGTCATATT-3');

   streptavidin Alexa Fluor 488 (Molecular Probes); and biotinylated antistreptavidin (Vector).
4. Plastics and equipment: tissue culture flasks, dishes, multiple-well plates, test tubes, polycarbonate tubes (40 mL; Fisher), Beckman J-20 centrifuge and JA-20 rotor (Beckman), pulsed-field gel electrophoresis (PFGE) system Rotaphor (Biometra), and Peltier Thermal Cycler (MJ Research).

3. Methods

We use mouse LA9 cells that are *HPRT*⁻ and contain a mammalian minichromosome marked with the hygromycin B phosphotransferase (*Hyg*) gene (LA9$^{hprt-hyg+}$) as donor cells and chicken DT40 cells that are *hprt*⁺ (DT40$^{hprt+hyg-}$) as recipient cells to illustrate the protocol. The genotypes of the

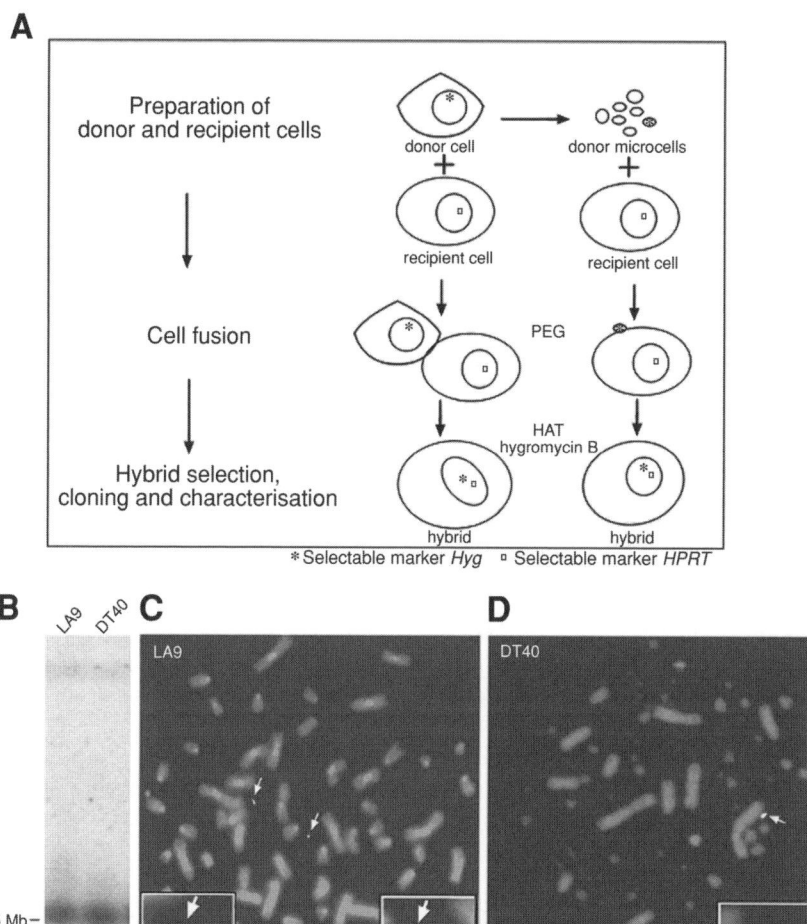

Fig. 1. Minichromosome transfer through PEG-mediated cell fusion. A minichromosome was transferred from mouse LA9 cells into chicken DT40 cells by PEG mediated cell fusion. (A) Procedures of PEG-mediated cell fusion. (B) PFGE analysis of the minichromosome in mouse LA9 and chicken DT40 cells. Human Y-specific alphoid DNA was labeled with ^{32}P and used as a probe to detect the 4.5-Mb minichromosome blotted onto a filter membrane. (C) FISH analysis of the minichromosome in a mouse LA9 cell. The minichromosome was detected with a biotin labelled human Y-specific alphoid probe. The arrows point to the minichromosome. The insets show the enlarged images of two copies of the minichromosome. (D) FISH analysis of the minichromosome in a chicken DT40 cell. The minichromosome was detected with a biotin labelled human Y-specific alphoid probe. The arrow points to the minichromosome. The inset shows the enlarged image of the minichromosome.

donor and recipient cells allow double positive selection with HAT and hygromycin B that kills both donor and recipient cells and selects for hybrid cells. LA9$^{hprt-hyg+}$ cells are adherent and DT40$^{hprt+hyg-}$ cells are grown in suspension. This difference facilitates the selection of hybrid types of interest. In this case, we are interested in the minichromosome transfer from LA9$^{hprt-hyg+}$ to DT40$^{hprt+hyg-}$ cells. Therefore, suspension cells are kept for subsequent selection, whereas attached cells are discarded. The main procedures of the protocol are delineated in **Fig. 1A**. Briefly, donor and recipient cells are prepared and fused with PEG. HAT and hygromycin B are used to select for hybrids. Hybrids are then cloned and characterised by molecular and cytological analysis.

3.1. Preparation of Donor and Recipient Cells (see Note 1)

3.1.1. Whole Donor Cells

1. Culture LA9$^{hprt-hyg+}$ cells in DMEM supplemented with 200 μM glutamine, 10% fetal calf serum (FCS), and 0.5 mg/mL hygromycin B. One 15-cm Petri dish of cells is prepared for whole cell fusion.
2. Feed cells with 30 mL of fresh medium 16 to 20 h before fusion.
3. Grow cells to approx 70% confluence.
4. Remove medium.
5 Rinse culture with 10 mL of PBS and remove PBS.
6. Add 2 mL of 0.2% trypsin and leave for approx 5 to 10 min.
7. Harvest cells and pellet cells at 400g for 5 min.
8. Wash pellets with 10 mL of PBS and pellet cells again.
9. Resuspend cells in 5 mL of prewarmed serum-free Dulbecco's modified Eagle's medium (DMEM).

3.1.2. Microcells

1. Culture LA9$^{hprt-hyg+}$ cells in DMEM supplemented with 200 μM glutamine, 10% FCS, and 0.5 mg/mL hygromycin B. Four 15-cm Petri dishes of cells are prepared for microcell fusion.
2. Feed cells with fresh medium 16 to 20 h before treating them with colcemid.
3. Add 0.1 μg/mL colcemid to the culture of approx 70% confluence.
4. Incubate cells for further 24 h at 37°C.
5. Remove medium.
6. Rinse culture with 10 mL of PBS and remove PBS.
7. Add 2 mL 0.2% trypsin and leave for 5 min.
8. Harvest micronucleated cells and pellet these cells at 400g for 5 min.
9. Wash pellets with 10 mL of prewarmed serum-free DMEM and pellet again.
10. Resuspend cells collected from each 15-cm Petri dish in a 40-mL solution containing prewarmed Percoll/serum-free DMEM (1/1) and 10 μg/mL cytochalasin B.
11. Transfer cells into a 40-mL polycarbonate tube.

12. Spin cell samples at 19,000 rpm at 37°C in a Beckman JA20 rotor for 75 min.
13. Collect approx 15 mL microcells in two visible bands by gentle aspiration or collect the whole supernatant without touching the gel pellet at the bottom if no visible bands are observed.
14. Mix 15 mL of microcells with 35 mL of prewarmed serum-free DMEM.
15. Pellet microcells at 500g for 10 min.
16. Pool microcells and divide into two 50-mL test tubes.
17. Wash microcells with 50 mL of prewarmed serum-free DMEM each tube and pellet microcells at 400g for 5 min.
18. Repeat **step 17**.
19. Resuspend microcells in 5 mL of prewarmed serum-free DMEM.

3.1.3. Recipient Cells

1. Culture DT40[hprt+hyg-] cells in DMEM supplemented with 200 μM glutamine, 10 μM β-mercaptoethanol, 10% FCS, and 1% chicken serum.
2. Add an equal volume of fresh medium to the cell culture 16 to 20 h before fusion.
3. Grow cells to a density of approx 10^6 cells per milliliter.
4. Collect 30 mL of cells for whole cell fusion or approx 60 to 80 mL cells for microcell fusion.
5. Pellet cells at 400g for 5 min.
6. Wash cells with 10 mL of prewarmed serum-free DMEM.
7. Pellet cells again.

3.2. Fusion of Donor and Recipient Cells (see Note 2)

1. Mix recipient cell pellets (DT40[hprt+hyg-] cells) with 5 mL donor cells (LA9[hprt-hyg+] cells) or microcells (LA9[hprt-hyg+] derived microcells).
2. Pellet the cell mixture at 400g for 5 min.
3. Resuspend cells in 1 mL of serum-free DMEM and incubate for 10 min at room temperature.
4. Pellet the mixture again.
5. Add 1 mL of 50% PEG 1500 to the pellet gently within 30 s and incubate for 2 min at room temperature.
6. Add 10 mL of serum-free DMEM slowly and leave the fusion at room temperature for 30 min.
7. Wash the fused sample twice in 50 mL of serum-free DMEM.
8. Resuspend cell pellets in 100 mL (for whole cell fusion) or 200 mL (for microcell fusion) of DMEM supplemented with 200 μM glutamine, 10 μM β-mercaptoethanol, 10% FCS, and 1% chicken serum in a 180-cm^2 flask.
9. Incubate cells for 2 h at 37°C.
10. Transfer suspension cells into a fresh 180-cm^2 flask and discard adherent cells.
11. Incubate suspension cells for another 24 h at 37°C.

3.3. Hybrid Selection, Cloning, and Characterization (see Note 3)

1. Apply double positive selection to the culture with a final concentration of 1X HAT and 2 mg/mL hygromycin B.
2. Split suspension cells into five 96-well plates for whole cell fusion or ten 96-well plates for microcell fusion, with 200 µL cells each well.
3. Incubate the culture for approx 10 to 15 d at 37°C.
4. Pick hybrid colonies from individual plate wells.
5. Divide each colony into two portions.
6. Use one portion of the cells for PCR analysis. The PCR primer pairs P1 and P2 are used to detect STS on the minichromosome in hybrids.
7. Transfer the second portion into a 24-well plate for culture expansion.
8. Transfer the culture into a 25-cm^2 flask after approx 3–5 d and incubate for approx 5 to 7 d.
9. Use approx 10^7 cells for PFGE and fluorescence *in situ* hybridization (FISH) analysis by standard methods. The ^{32}P-labeled human Y alphoid satellite DNA is used to detect the minichromosome in hybrids by PFGE (**Fig. 1B**). The biotin-labeled human Y alphoid satellite DNA is used to detect the minichromosome in hybrids by FISH (**Fig. 1C,D**). Streptavidin Alexa Fluor 488 and Biotinylated anti-streptavidin are used in the FISH detection system according to the manufacturer's instructions.

3.4. Other Considerations and Alternative Protocols

PEG is a key reagent in this protocol. PEG with a molecular weight of 1000 to 6000 can be used in cell fusion. Exposure of cells to 1 mL of 50% PEG 1500 solution for 2 min works well for most types of cells but the molecular weight, concentration, and treatment length of the polymer should be established to suit individual cell types. Some protocols use phytohemagglutinin to enhance cell fusion efficiency. In our experience, however, no significant enhancement of phytohemagglutinin in fusion efficiency was observed. In some protocols for microcell fusion, microcells are purified by filtration through 3- to 5-µm polycarbonate filters to remove whole cells and nuclei. This step is tedious and not necessary if proper double selection is applied. After selection applied to fusion culture, medium with proper selection is usually changed every 2 to 4 d, but in our protocol, medium changes are not required for the DT40 suspension cells. For the maintenance of hybrids, chemicals used for selection can be reduced to 30–50% of initial concentration for hybrid isolation that is enough to induce expression of selectable markers in most cases. In regard to FEG-mediated nuclear transfer from embryonic or somatic cells into enucleated oocytes, particular protocols should be considered (*6,10*).

4. Notes

1. Choice of cells for fusion and selectable markers: both donor and recipient cells must be exponentially growing before being harvested for fusion. In principle, cells between either different species or the same species can be fused with PEG, leading to the generation of interspecific or intraspecific hybrids. However, the fusibility varies depending on cell types. For example, it is much easier to fuse the mouse LA9 cells than it is the mouse E14 embryonic stem cells and the chicken DT40 cells. When fusion is performed between adherent and suspension cells, it is relatively easy to isolate hybrids of particular cell types if no efficient double positive selection is available. However, double positive selection usually is required to isolate hybrids if both donors and recipients in a fusion are suspension or adherent cells. Double selection requires donor and recipient cells to carry a different selectable marker respectively. A number of dominant markers are available for positive selection in vertebrate cells. The widely used markers include the *Hyg* gene, the neomycin resistance gene (*neo*), the bleomycin resistance gene (*ble*), the blasticidin resistance gene (*bsd*), the puromycin *N*-acetyl transferase gene (*pac*), and the histidinol dehydrogenase gene (*hisD*). These markers, driven by a suitable promoter, such as the cytomegalovirus promoter, can be introduced into donor or recipient cells by transfection or viral infection. However, these markers with a given promoter have to be tested in individual cell lines for function before being used in cell fusion. In addition, counter selection may be useful if cell fusion is used to transfer a single or a limited number of chromosomes between cells. Examples of selectable markers that can be used for counter selection are the *HPRT* gene and the *TK* gene. Cells with the *HPRT* gene are killed when grown in medium containing 6-thioguanine and cells with the *TK* gene are killed when grown in medium containing bromodeoxyuridine.

2. Whole cell fusion and microcell fusion. Whole cell fusion is simple and can be used in many areas involving production of somatic cell hybrids and nuclear transfer. Microcell fusion is particularly useful for transfer of a single or a limited number of chromosomes between cell lines. Microcells are prepared from micronucleated donor cells induced by colcemid. The colcemid concentration and treatment length are important factors for microcell preparation that vary with cell types. Treatment with 0.05 to 0.10 µg/mL colcemid for 10 to 48 h can be tried for most of vertebrate cell lines.

3. Characterization of hybrids: the characterization of hybrids may involve analysis of cell morphology, karyotypes, genome structure, gene expression, and gene function. Analysis of mitotic and structural stability of heterologous chromosomes can be extended to several months in the presence or absence of selection. Choice of techniques for hybrid characterisation is determined by particular purposes of a given cell fusion experiment.

Acknowledgments

This work was supported by the Medical Research Council and the Wellcome Trust in the UK.

References

1. Ahkong, Q. F., Fisher, D., Tampion, W., and Lucy, J. A. (1975) Mechanisms of cell fusion. *Nature* **253,** 194–195.
2. Davidson, R. L., and Gerald, P. S. (1976) Improved techniques for the induction of mammalian cell hybridization by polyethylene glycol. *Somatic Cell Genet.* **2,** 165–176.
3. Lane, R. D., Crissman, R. S., and Lachman, M. F. (1984) Comparison of polyethylene glycols as fusogens for producing lymphocyte-myeloma hybrids. *J. Immunol. Methods* **72,** 71–76.
4. Kamarck, M. E., Barker, P. E., Miller, R. L., and Ruddle, F. H. (1984) Somatic cell hybrid mapping panels. *Exp. Cell Res.* **152,** 1–14.
5. Cho, M. S., Yee, H., and Chan, S. (2002) Establishment of a human somatic hybrid cell line for recombinant protein production. *J. Biomed. Sci.* **9,** 631–638.
6. Sims, M., and First, N. L. (1994) Production of calves by transfer of nuclei from cultured inner cell mass cells. *Proc. Natl. Acad. Sci. USA* **91,** 6143–6147.
7. Lee, J., and Lentz, B. R. (1997) Outer leaflet-packing defects promote poly(ethylene glycol)-mediated fusion of large unilamellar vesicles. *Biochemistry* **36,** 421–431.
8. Shen, M. H., Mee, P. J., Nichols, J., Yang, J., Brook, F., Gardner, R. L., et al. (2000) A structurally defined mini-chromosome vector for the mouse germ line. *Curr. Biol.* **10,** 31–34.
9. Shen, M. H., Yang, J. W., Yang, J., Pendon, C., and Brown, W. R. (2001) The accuracy of segregation of human mini-chromosomes varies in different vertebrate cell lines, correlates with the extent of centromere formation and provides evidence for a trans-acting centromere maintenance activity. *Chromosoma* **109,** 524–535.
10. Tesarik, J., Nagy, Z. P., Mendoza, C., and Greco, E. (2000) Chemically and mechanically induced membrane fusion: non-activating methods for nuclear transfer in mature human oocytes. *Hum. Reprod.* **15,** 1149–1154.

7

Epigenetic Reprogramming of Somatic Genomes by Electrofusion With Embryonic Stem Cells

Masako Tada and Takashi Tada

Summary

Cell fusion is an approach for combining genetic and epigenetic information between two different types of cells. Electrofusion for generating hybrid cells between mouse embryonic stem cells and somatic cells, which is a type of nonchemically induced and nonvirus-mediated cell fusion, is introduced here as a highly effective, reproducible, and biomedically safe in vitro system. Under optimized electrofusion conditions, cells are aligned and form pearl chains between electrodes in response to AC pulse stimulation, and subsequently adjacent cytoplasmic membranes are fused by DC pulse stimulation. Hybrid cells survive as drug-resistant colonies in selection medium. Cell fusion is a technique that is applied widely in the life sciences. A recent topic of great interest in the field of stem cell research is the successful production of cloned animals via epigenetic reprogramming of somatic nuclei. Interestingly, nuclear reprogramming for conferring pluripotency on somatic nuclei also occurs via cell fusion between pluripotential stem cells and somatic cells. Furthermore, it has been shown that spontaneous cell fusion contributes to generating the intrinsic plasticity of tissue stem cells. Cell fusion technology may make important contributions to the fields of regenerative medicine and epigenetic reprogramming.

Key Words: Cell fusion; hybrid cell; cell alignment; electrofusion; stem cell; ES cell; EG cell; reprogramming; epigenetics; cloning.

1. Introduction

Fertilization is one of the well-known cell fusion phenomena occurring in vivo. The union of two pronuclei from an MII oocyte and a sperm creates the next generation and endows it with genetic and epigenetic diversity. Genetically programmed spontaneous cell fusion also is known to occur in the formation of polykaryones, such as myotubes, osteoclasts, and syntrophoblasts in vivo. Remarkably, recent developments in the field of stem cell research have

From: *Methods in Molecular Biology, vol. 325: Nuclear Reprogramming: Methods and Protocols*
Edited by: S. Pells © Humana Press Inc., Totowa, NJ

revealed that spontaneous cell fusion plays an important role in maintaining homeostasis of many tissues and organs via degeneration during defined processes of self-renewal and after tissue damage. The capability of spontaneous cell fusion has been shown by the generation of hybrid cells via in vitro co-culturing of mouse bone marrow cells and embryonic stem (ES) cells *(1)* and of mouse brain cells and ES cells *(2,3)*. The in vivo contribution of spontaneous cell fusion to regeneration of tissues has been proven by the transdifferentiation of bone-marrow-derived hybrid cells into Purkinje neurons, cardiomyocytes, and hepatocytes *(4–7)*.

A fascinating breakthrough in the area of stem cell research is the successful production of cloned animals via transplantation of committed somatic cell nuclei into enucleated unfertilized oocytes *(8)*. This technique recently has been applied to generate human ES cells derived from cloned blastocysts *(9)*. Nuclear reprogramming from the "somatic type" to the "pluripotential type" is induced by genome-wide epigenetic changes via the activity of *trans*-acting factors in unfertilized oocytes. Interestingly, using electrofusion between ES cells and adult somatic cells, Tada et al. *(10)* demonstrated that ES cells have an intrinsic capacity for epigenetic reprogramming of somatic genomes. In hybrid cells between ES cells and thymocytes, nuclear reprogramming of somatic genomes has been shown by (1) the successful contribution of the ES hybrid cells to the normal embryogenesis of chimeric embryos, (2) reactivation of the inactivated X chromosome derived from female thymocytes, (3) reactivation of pluripotential cell-specific genes (*Oct4* and *Tsix*) derived from somatic cells, (4) redifferentiation of a variety of cell types independent of the origin of thymocytes in teratomas, (5) tissue-specific gene expression from the reprogrammed somatic genomes after in vivo and in vitro differentiation, and (6) decondensed chromatin formation in the reprogrammed somatic nuclei as marked by histone-tail modifications of H3 and H4 hyperacetylation and H3 lysine 4 hypermethylation *(10–13)*. Moreover, embryonic germ (EG) cells derived from gonadal primordial germ cells of mouse E11.5 and 12.5 embryos, in which most parental imprints are erased *(14)*, possess additional potential for inducing the reprogramming of somatic cell-derived parental imprints accompanied by the demethylation of allele-specific DNA methylation in the EG hybrid cells *(15)*. Therefore, electrofusion between pluripotential stem cells and somatic cells will contribute to elucidating the mechanisms of epigenetic reprogramming involved in DNA and chromatin modifications.

The technique of cell fusion pioneered in 1965 by Harris *(16)* has proven to be a powerful approach for analyzing biological interactions between differentiated cell types and for studying genomic plasticity *(17–19)*. It has been shown that cell fusion caused by membrane fusion between two different cells is induced by treatment with various chemical agents, such as calcium ions, lysolecithin, and polyethylene glycol and also by mediation with viruses such as paramyxoviruses, including Sendai virus (i.e., HVJ), oncornavirus, coronavirus, herpesvirus, and poxvirus. However, chemical and virus-mediated cell fusion has difficulties with respect to the efficiency of inducing cell fusion and also with respect to the reproducibility and biomedical safety when used for clinical applications. Since the 1980s, efforts have been made to optimize many parameters of electrofusion, resulting in the improvement of the fusion efficiency, which is higher than that of cell fusion with polyethylene glycol *(20)*. However, the electrofusion parameters were optimized for making hybridomas for the purpose of stable production of antibodies.

In this chapter, we introduce an experimental procedure to create hybrid cells by electrofusion between pluripotential stem cells and somatic cells. Electrofusion is nonchemically induced and nonvirus-mediated cell fusion that presents many advantages in making hybrid cells: (1) the optimized conditions are reproducible in independent cell fusion experiments, (2) the electrofused cells are biomedically safer than the chemically or virus-induced fused cells, and (3) the microscale production of hybrid cells can be manipulated by using a Microslide chamber under a microscope without possible contamination with other cells. A description of the electrofusion procedure can be subdivided into several sections: (1) The setup of the Electro Cell Manipulator (ECM) 2001 (BTX), AC/DC Pulse Generator with a Microslide chamber; (2) cell culture conditions of ES and fused cells; (3) pretreatment of somatic cells; (4) operation of the electrofusion generator; (5) cell selection of cells after fusion; and (6) the isolation and cloning of fused cells. In the process of electrofusion, an alternative current (AC) induces alignment and compression of the cells, and a direct current (DC) transiently makes reversible pores in cytoplasmic membranes to initiate the process of fusing adjacent cytoplasmic membranes. The alignment voltage, pulse length, and electroporation voltage and number of DC pulses should be precisely controlled. The overall procedures are summarized in **Fig. 1**. The following optimized conditions are suitable for electrofusion experiments between ES cells and thymic lymphocytes with the AC/DC pulse generator ECM 2001 (BTX).

2. Materials
2.1. Instruments
2.1.1. Cell Fusion

1. Electro Cell Manipulator ECM 2001 (BTX).
2. Microslide chambers with 1-mm electrode gap (BTX P/N450-1).
3. Micrograbber cables (BTX 464).
4. Inverted microscope with 10× and 20× objectives.

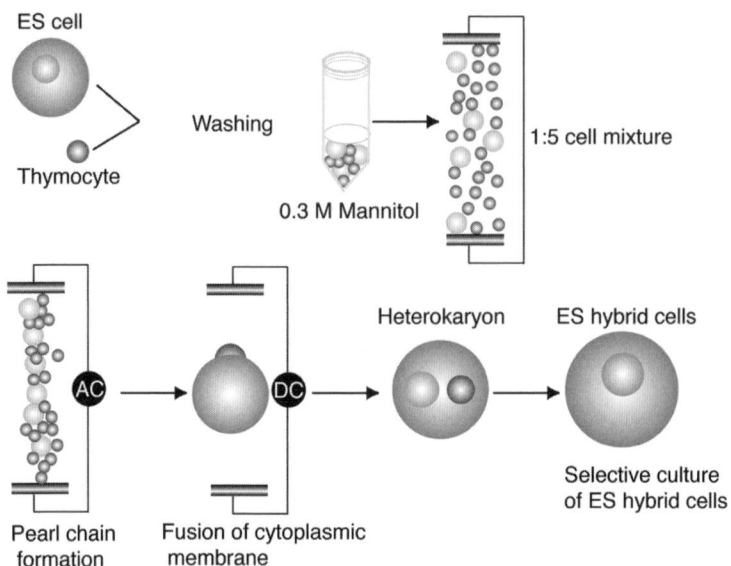

Fig 1. Scheme of electrofusion system. The cell mixture of ES cells and somatic cells suspended in nonelectrolyte 0.3 M mannitol is placed into the 1-mm gap between two electrodes fixed on a Microslide. Presettings of optimized AC voltage, AC duration, DC pulse voltage, and DC pulse number are required. The function of AUTOMATIC START in ECM 2001 initiates cell alignment through AC application and reversible breakage of cytoplasmic membranes by sequential DC electroporation pulse, leading to the fusion of cell membranes between two different cells. After the cell fusion procedure, ES cells and ES–ES hybrid cells are selectively killed using selection medium, while thymocytes are nonadherent. ES–somatic hybrid cells are selectively grown in the selection medium and isolated in independent culture dishes.

2.1.2. Cell Culture

1. Humidified incubator at 37°C, 5% CO_2 and 95% air.
2. 60-mm Plastic tissue culture dishes
3. 60-mm and 100-mm Bacterial dishes.
4. 10-mm and 30-mm Well plastic tissue culture plates.
5. 15-mL and 50-mL Conical tubes.
6. 0.2-μm Microfilter.
7. 200-μL and 1000-μL Dispensable pipets with autoclaved tips.

2.1.3. Single-Cell Isolation of Adult Thymocytic Lymphocytes

For this procedure, a 2.5-mL syringe, 18-gage needle, and 50-mL conical tube are required.

2.2. Reagents

1. Dulbecco's modified Eagle's medium/nutrient mixture F12 HAM (DMEM/F12; cat. no. D6421, Sigma).
2. Dulbecco's modified Eagle's medium (DMEM; cat . no. D5796, Sigma)
3. Fetal bovine serum (FBS; cat. no. 12003-78P, JRH Biosciences).
4. Recombinant leukemia inhibitory factor (LIF; cat. no. ESG1107, Chemicon).
5. 200 mM Glutamine (cat. no. 320-5030A G, Gibco).
6. 2-Mercaptoethanol (cat. no. M7520, Sigma).
7. 10,000 U/mL Penicillin and 10 mg/mL streptomycin (penicillin–streptomycin 100X; cat. no. P-0781, Sigma).
8. 100 mM Sodium pyruvate (cat. no. S8636, Sigma).
9. 7.5% Sodium bicarbonate (cat. no. S8761, Sigma).
10. Ca^{2+}/Mg^{2+}-free phosphate-buffered saline (PBS; cat. no. 10010-023, Gibco).
11. 0.25% Trypsin/1 mM ethylene diaminetetraacetic acid · 4Na (cat. no. 25200-056, Gibco).
12. Mitomysin C (cat. no. M0503, Sigma).
13. Gelatin from porcine skin, Type A (cat. no. G-1890, Sigma).
14. D-mannitol (cat. no. M-9546, Sigma).
15. G418 (Geneticin; cat. no. G-9516, Sigma).
16. Hypoxanthine–aminopterin–thymine media supplement, 50X (HAT; cat. no. H0262, Sigma).
17. 70% Ethanol.

2.3. Solutions

1. PEFs medium: mix 500 mL of DMEM, 50 mL of FBS, 5 mL of 200 mM glutamine, 5 mL of 10,000 U/mL penicillin, and 10 mg/mL streptomycin. Store at 4°C.
2. ES medium: mix 500 mL of DMEM/F12, 75 mL of FBS, 5 mL of 200 mM glutamine, 5 mL of 100X penicillin–streptomycin, 5 mL of 100 mM sodium pyruvate, 8 mL of 7.5% sodium bicarbonate, 4 μL of 10^{-4} M 2-mercaptoethanol, and 50 μL of 10^7 U/mL LIF (final 1000 U/mL). Store at 4°C.
3. 0.25% Trypsin/1 mM ethylenediaminetetraacetic acid · 4Na. Dispense into aliquots and store at –20°C.
4. Mitomycin C (MMC) Prepare solution at 0.2 mg/mL in PBS, dispense into aliquots and store at –20°C.
5. 0.1% Gelatin: dissolve 0.1 g of gelatin in 100 mL of distilled water. Sterilize by autoclaving and store at 4°C.
6. Fresh nonelectrolyte solution: 0.3 M mannitol. Dissolve 2.74 g of mannitol in 50 mL of distilled water. Filter through a 0.2-μm filter. Keep at 4°C.
7. ES medium with G418: dissolve antibiotic G418 in distilled water at 50 mg/mL. Sterilize through a 0.2-μm filter and store at 4°C. Add 50 μL of the G418 solution to 10 mL of ES medium to obtain a final concentration of 250 μg/mL.

8. ES medium with HAT: dissolve HAT media supplement supplied in a vial in 10 mL of 50X DMEM stock solution and store at –20°C. Each vial contains 5×10^{-3} M hypoxanthine, 2×10^{-5} M aminopterin, 8×10^{-4} M thymidine. Add 200 μL of the stock solution to 10 mL of ES medium.

2.4. Cells and Animals

Adult mice, ES cells, and *neo*[r] primary embryonic fibroblasts (PEFs) produced from E12.5 embryos of ROSA26 transgenic mice carrying the ubiquitously expressed *neo/lacZ* gene *(21)*.

3. Methods
3.1. Setup of the Electro Cell Manipulator ECM 2001
3.1.1. Setup of AC/DC Pulse Generator, ECM 2001, and the Accessories

1. Place the ECM 2001 beside the inverted microscope (**Fig. 2A**).
2. Connect the Micrograbber cable to the output jacks on the back of the ECM2001.
3. Switch power on at the back.

3.1.2. Automatic Operating Parameters

Set the optimized electrical parameters to fuse ES cells and thymocytes (**Fig. 2A,G**).

1. AC: 10 V.
2. AC duration: approx 60 to 99 s.
3. DC: approx 250 to 300 V (280 V). Adjust DC voltage according to gap distance between electrodes. The appropriate electric field strength is approx 2.5 to 3.0 kV/cm. When 2-mm gap Microslides are used, the DC should be around 600 V.
4. DC pulse length: 10 μs.
5. Number of DC pulses: 1.
6. Post-Fusion AC: 8 s.

3.1.3. Setup of Microslide Chambers

1. Sterilize the Microslides by immersion in 70% ethanol followed by flaming.
2. Set a Microslide in a 100-mm plastic dish chamber made from a bacterial dish.
3. For each electrofusion, apply 40 μL of cell mixture into the 1-mm electrode gap on the Microslide.
4. Connect the Microslide with the Micrograbber cable to the ECM 2001 (**Fig. 2B**).
5. Place the chamber on an inverted microscope to check cell alignment and compression. It is important to determine the optimal fusion conditions under the microscope each time. The fusion conditions may vary depending on cell density and cell size.

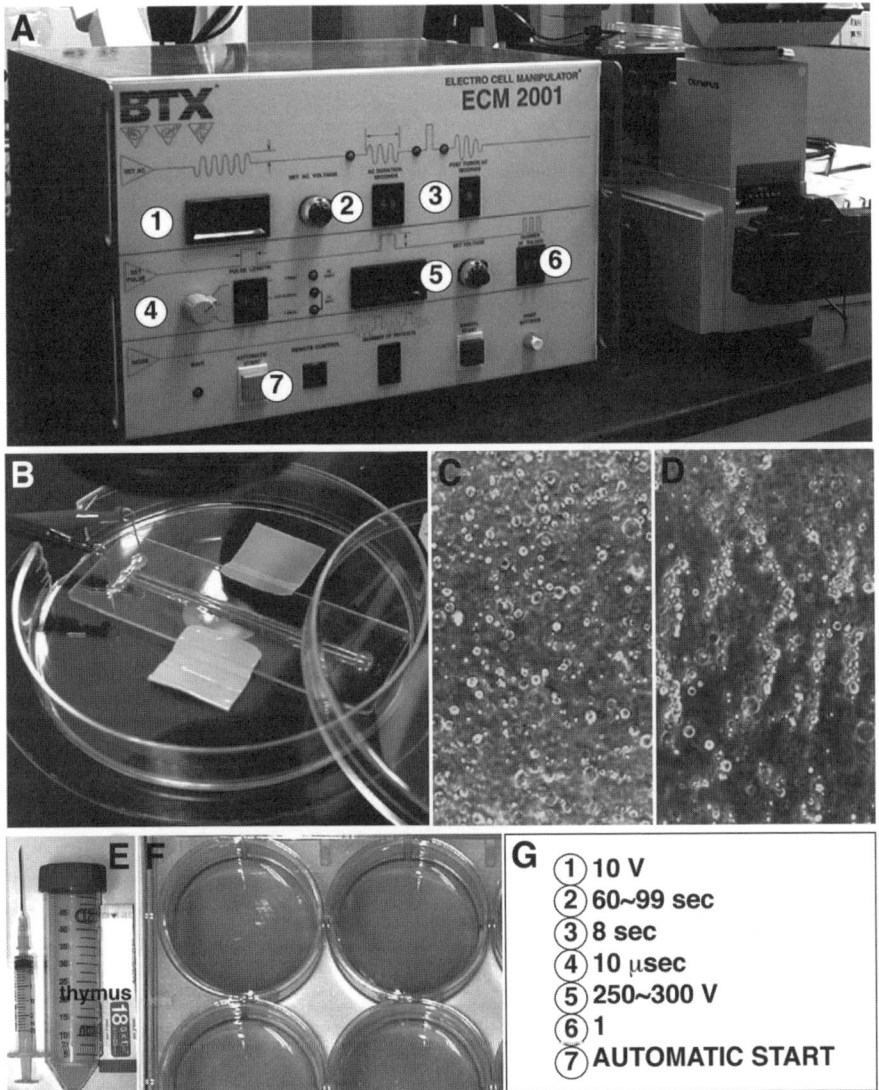

Fig 2. **(A)** Setup of AC/DC pulse generator ECM2001 (BTX) **(B)** Microslide chamber in a 100-mm plastic dish made using a bacterial dish and Micrograbber cable. **(C)** Cell mixture of ES cells and thymocytes applied between electrodes. **(D)** Pearl chain formation during AC application. **(E)** Instruments for dissociating mouse thymus into single cells. **(F)** Culture dishes with inactivated PEFs prepared 1 d before cell fusion. **(G)** The recommended fusion parameters.

3.2. Electrofusion Protocol

3.2.1. Feeder Cells for ES Cells

1. Coat 60-mm culture dishes with 0.1% gelatin for at least 30 min at 37°C.
2. Incubate *neo^r* PEFs with 10 μg/mL MMC for 2 h at 37°C in a CO_2 incubator to produce mitotically inactivated feeder cells. Frozen stocks of MMC-treated PEFs are prepared at a cell concentration of 5×10^6 cells/mL/cryotube and are stored in liquid nitrogen. Inactivated *neo^r* PEFs are routinely used as feeder cells (1×10^6 cells/60-mm culture dish and 2.5×10^6 cells/100-mm culture dish) for culture of ES and hybrid cells, and also for selection of hybrid cell colonies with G418.

3.2.2 PEFs for Fused Cells

1. One day before cell fusion, coat 30-mm culture wells (6-well culture plate) with 0.1% gelatin for at least 30 min at 37°C.
2. Prepare inactivated PEFs (4×10^5 cells/ 30-mm well) in 3 mL of ES medium (**Fig. 2F**).

3.2.3. ES Cells

One of the most important requirements for cell fusion experiments is that the culture conditions have to be optimized for maintaining the pluripotential competence and full set of chromosomes (80 chromosomes) derived from mouse ES cells and somatic cells through numerous cell divisions. It is strongly recommended that one identify a satisfactory production lot of FBS that can supply supplements to support effective cell growth without differentiation induction.

1. Prepare exponentially growing ES cells cultured on the inactivated PEFs with changes of culture medium once or twice a day.
2. Ascertain that complete sets of chromosomes are maintained in the ES cells before cell fusion.

3.2.4. Somatic Cells

1. Sterilize all dissection instruments (scissors and forceps) by immersion in 70% ethanol, followed by flaming.
2. Sacrifice a 6- to 8-wk-old adult mouse humanely and dissect out the thymus in a clean room if a clean bench is not available.
3. Wash the tissues with sterilized PBS twice in 60-mm Petri dishes.
4. Place a half lobe of a thymus in the barrel of a sterile 2.5-mL syringe fitted with a sterile 18-gage needle (**Fig. 2E**).
5. Expel and draw up the thymus gently through the needle via the tip of the needle several times into a 50-mL conical tube with 2 mL of DMEM to dissociate the thymus into a single-cell suspension.
6. Place for several minutes at room temperature.
7. Transfer the supernatant excluding cell clumps to a 15-mL conical tube and add 10 mL of DMEM.

8. Spin down the thymocytes in 15-mL conical tubes at > 400g (1500 rpm) for 5 min.
9. Resuspend them in 10 mL of DMEM.

3.2.5 Purification of ES Cells

1. Coat a 60-mm culture dish with 0.1% gelatin for at least 30 min at 37°C.
2. Trypsinize the ES cells and remove excess trypsin quickly. Add 3.0 mL of ES medium to inactivate the trypsin and dissociate the cells into a single-cell suspension by gentle pipetting.
3. Plate them on a fresh gelatin-coated 60-mm culture dish.
4. Incubate the ES cells in a CO_2 incubator for 30 min to separate feeder cells from ES cells.
5. Collect unattached ES cells and harvest them by centrifugation at > 400g (1500 rpm) for 5 min.
6. Resuspend the cell pellet in 10 mL of DMEM and transfer the cell suspension into a 15-mL conical tube.

3.2.6. ES–Somatic Cell Mixture in 0.3 M Mannitol

1. Spin down the ES cells and the thymocytes in 15-mL conical tubes at 1500 rpm for 5 min, separately.
2. Wash them with 10 mL of DMEM and spin down at 1500 rpm for 5 min and repeat again to remove FBS completely.
3. Add 5 to 10 mL of DMEM and adjust the density of ES cells and thymocytes to 1 × 10⁶ cells/mL each.
4. Pellet a 1:5 mixture of ES cells and thymocytes (1 mL of ES cell suspension and 5 mL of thymocyte suspension). Keep the remaining cells for control experiments.
5. Spin down and resuspend the cell pellet in 0.3 M mannitol to the appropriate cell density of 6 × 10⁶ cells/mL. Usually, 1 mL of the mixture of ES cells and thymocytes is sufficient for the following fusion experiment.
6. Use the cells for electrofusion immediately.

3.2.7. Operation of ECM2001 With Microslides

The automatic operation switch is used to initiate AC followed by DC. AC is used to induce a nonhomogeneous (or divergent) electric field, resulting in cell alignment and pearl chain formation. DC is used to produce reversible temporary pores in the cytoplasmic membranes. When juxtaposed pores in the physically associated cells reseal, cells have a chance to be hybridized by cytoplasmic membrane fusion. AC application after the DC pulse induces compression of the cells, which helps the process of fusion between the cell membranes.

1. Apply 40 µL of cell mixture between 1-mm gap electrodes on the Microslide at room temperature (**Fig. 2C**).
2. Place the Microslide in the 100-mm plastic dish chamber and connect the cable.

3. Place the chamber on an inverted microscope to allow for observation of cell alignment (**Fig. 2D**).
4. Press the automatic operation switch of the ECM 2001.
5. Add 40 µL of DMEM to the fusion mixture between the electrodes to induce membrane reformation immediately after electroporation.
6. Leave the cell mixture for 10 min at room temperature.
7. Transfer the cell mixture directly to a 30-mm culture dish containing inactivated PEFs with 3 mL of ES medium supplemented with LIF (**Fig. 2F**).
8. Repeat the cell fusion procedure sequentially by using several Microslides. Usually, cell suspensions recovered from three Microslides (80 µL ×3) are plated into one 30-mm culture dish. As a control, plate the untreated cell mixture and culture under the same conditions.
9. Incubate the cells at 37°C in a CO_2 incubator for 24 h.
10. Change the medium to ES medium with the proper supplement to select for ES hybrid cells 24 h after cell fusion.
11. Change the selection medium once a day.

As a result of the 7-d drug treatment, nonfused ES cells and hybrid cells derived from ES cells are killed, and hybrid cells derived from thymocytes are nonadherent. Thus, the hybrid cells derived from an ES cell and a somatic cell survive and form colonies. Several colonies of hybrid cells per 10^4 host ES cells appear using the aforementioned procedures for electrofusion. *See* **Subheading 4**.

3.3. Isolation and Cloning of Hybrid Cells

We describe two independent chemical selection systems to select for hybrid cells between ES cells and somatic cells. Normal ES cells are hybridized with thymocytes containing the bacterial neomycin resistance (*neo*[r]) gene *(10,15)*. Thymocytes are derived from ROSA26 transgenic mice, which carry the ubiquitously expressed *neo/lacZ* transgene *(21)*. Only ES–thymocyte hybrid cells can survive and grow in ES medium supplemented with the protein synthesis inhibitor G418. In this case, the ES hybrid cells and their derivatives are visualized by positive reaction with X-gal through β-galactosidase activity, which allows us to analyze their contribution to the development of chimeric embryos and tissues *(10,11,15)*. In another selection system, male ES cells deficient for the *Hprt* gene on the X chromosome have been used for selecting hybrids with wild-type somatic cells. Electrofusion-treated cells are cultured in ES medium supplemented with HAT. The HAT medium is fatal to the *Hprt*-deficient ES cells, whereas ES hybrid cells, which are rescued by the thymocyte-derived wild-type *Hprt* gene, are able to survive and grow *(11–13)*. In the synthesis of DNA, purine nucleotides can be synthesized by the *de novo* pathway and recycled by the salvage pathway. *Hprt* is a purine salvage enzyme responsible for converting the purine degradation product, hypoxanthine, to inosine monophosphate, a precursor of ATP and GTP. In the presence of aminopterin, the *de novo* pathway is inhibited and only the salvage pathway functions. Consequently, dysfunction of *Hprt* induces cell death in culture with the HAT medium.

3.3.1. Selection With G418

1. Perform the automatic procedure for electrofusion of the mixture of normal ES cells and thymocytes collected from 6- to 8-wk-old ROSA26 transgenic mice carrying the *neo/lacZ* transgene.
2. Culture the electrofusion-treated cells in ES medium for 24 h.
3. Change the medium to ES medium supplemented with G418. ES hybrid cell colonies can be detected by 7 to 10 d.
4. Prepare a 24-well culture plate containing 1 × 10⁵ inactivated *neo*ʳ PEFs per 10-mm well and 0.8 mL of ES medium supplemented with G418 for selection.
5. Pick up the colonies with a micropipette and transfer each colony into a 10-mm well of the 24-well culture plate on inactivated PEFs.
6. Subculture the colonies every 2 or 3 d and gradually expand the number of cells in 30-mm and 60-mm culture dishes with inactivated PEFs and ES medium. When the cells become nearly confluent in a 60-mm culture dish, we determine that a hybrid cell line is established at passage 1.
7. Analyze the karyotypes of the hybrid cells before using for further studies.

3.3.2. Selection With HAT

1. Use ES cells (XY) deficient for *Hprt* and normal thymocytes collected from 6- to 8-wk-old female (XX) or male (XY) mice.
2. Culture the electrofusion-treated cells in ES medium for 24 h.
3. Change the medium to ES medium supplemented with HAT. ES hybrid cell colonies can be detected by 7 to 10 d.
4. Analyze the karyotypes of the ES hybrid cells, which should be 80, XXXY or 80, XXYY.

4. Note

1. Successful formation of pearl chains during AC duration is extremely important for the efficiency of cell fusion. The pearl chain formation is influenced mainly by the following factors: (1) cell density in mannitol, (2) contamination by cell debris, and (3) contamination by serum or salts from cultured medium. To improve the conditions:
   - Pellet the mixed cells by centrifugation and resuspend the cells in a suitable amount of fresh 0.3 *M* mannitol.
   - Increase the cell density if pearl chains are formed poorly.
   - Decrease the cell density if cell movement is disturbed.

- The total cell volumes in 0.3 M mannitol have to be carefully controlled to obtain a smooth electric current. When other somatic cells larger than thymocytes and similar-sized with ES cells are used as a fusion partner, prepare a 1/1 cell mixture at 2×10^6 cells/mL.

References

1. Terada, N., Hamazaki, T., Oka, M., Hoki, M., Mastalerz, D. M., Nakano, Y., et al. (2002) Bone marrow cells adopt the phenotype of other cells by spontaneous cell fusion. *Nature* **416**, 542–545.
2. Ying, Q. L., Nichols, J., Evans, E. P., and Smith, A. G. (2002) Changing potency by spontaneous fusion. *Nature* **416**, 545–548.
3. Pells, S., Di Domenico, A. I., Callagher, E. J., and McWhir, J. (2002) Multipotentiality of neuronal cells after spontaneous fusion with embryonic stem cells and nuclear reprogramming in vitro. *Cloning Stem Cells* **4**, 331–338.
4. Vassilopoulos, G., Wang, P. R., and Russell, D. W. (2003) Transplanted bone marrow regenerates liver by cell fusion. *Nature* **422**, 901–904.
5. Wang, X., Willenbring, H., Akkari, Y., Torimaru, Y., Foster, M., Al-Dhalimy, M., et al. (2003) Cell fusion is the principal source of bone-marrow-derived hepatocytes. *Nature* **422**, 897–901.
6. Alvarez-Dolado, M., Pardal, R., Garcia-Verdugo, J. M., Fike, J. R., Lee, H. O., Pfeffer, K., et al. (2003) Fusion of bone-marrow-derived cells with Purkinje neurons, cardiomyocytes and hepatocytes. *Nature* **425**, 968–973.
7. Weimann, J. M., Johansson, C. B., Trejo, A., and Blau, H. M. (2003) Stable reprogrammed heterokaryons form spontaneously in Purkinje neurons after bone marrow transplant. *Nat. Cell Biol.* **5**, 959–966.
8. Wilmut, I., Schnieke, A. E., McWhir, J., Kind, A. J., and Campbell, K. H. S. (1997) Viable offspring derived from fetal and adult mammalian cells. *Nature* **385**, 810–813.
9. Hwang, W. S., Ryu, Y. J., Park, J. H., Park, E. S., Lee, E. G., Koo, J. M., et al. (2004) Evidence of a pluripotent human embryonic stem cell line derived from a cloned blastocyst. *Science* **303**, 1669–1674.
10. Tada, M., Takahama, Y., Abe, K., Nakatsuji, N., and Tada, T. (2001) Nuclear reprogramming of somatic cells by in vitro hybridization with ES cells. *Curr. Biol.* **11**, 1553–1558.
11. Tada, M., Morizane, A., Kimura, H., Kawasaki, H., Ainscough, J. F., Sasai, Y., et al. (2003) Pluripotency of reprogrammed somatic genomes in embryonic stem hybrid cells. *Dev. Dyn.* **227**, 504–510.
12. Kimura, H., Tada, M., Hatano, S., Yamazaki, M., Nakatsuji, N., and Tada, T. (2002) Chromatin reprogramming of male somatic cell-derived XIST and TSIX in ES hybrid cells. *Cytogenet. Genome Res.* **99**, 106–114.
13. Kimura, H., Tada, M., Nakatsuji, N., and Tada, T. (2004) Histone code modifications on pluripotential nuclei of reprogrammed somatic cells. *Mol. Cell. Biol.* **24**, 5710–5720.

14. Tada, T., Tada, M., Hilton, K., Barton, S. C., Sado, T., Takagi, N., et al. (1998) Epigenotype switching of imprintable loci in embryonic germ cells. *Dev. Gene Evol.* **207,** 551–561.

15. Tada, M., Tada, T., Lefebvre, L., Barton, S. C., and Surani, M. A. (1997) Embryonic germ cells induce epigenetic reprogramming of somatic nucleus in hybrid cells. *EMBO J.* **16,** 6510–6520.

16. Harris, H. (1965) Behaviour of differentiated nuclei in heterokaryons of animal cells from different species. *Nature* **206,** 583–588.

17. Blau, H. M., Pavlath, G. K., Hardeman, E. C., Chiu, C. P., Silberstein, L., Webster, S. G., et al. (1985) Plasticity of the differentiated state. *Science* **230,** 758–766.

18. Baron, M. H., and Maniatis, T. (1986) Rapid reprogramming of globin gene expression in transient heterokaryons. *Cell* **46,** 591–602.

19. Blau, H. M., and Baltimore, D. (1991) Differentiation requires continuous regulation. *J. Cell Biol.* **112,** 781–783.

20. Neil, G. A., and Zimmermann, U. (1993) Electrofusion. *Methods Enzymol.* **220,** 174–196.

21. Friedrich, G., and Soriano, P. (1991) Promoter traps in embryonic stem cells: a genetic screen to identify and mutate developmental genes in mice. *Genes Dev.* **5,** 1513–1523.

8

Quantification of Cell Fusion by Flow Cytometry

Stephen Sullivan, Martin Waterfall, Ed J. Gallagher, Jim McWhir, and Steve Pells

Summary

Cells of different types can be induced to fuse by electroshock. Cells of one type are typically dominant and are able to reprogram the nuclei derived from cells of the other type, in fusion hybrids derived from one cell of each type. Flow cytometry provides a quick and objective technique to assess cell fusion for nuclear reprogramming studies. Two cell types are each stained with a different fluorescent dye and then induced to fuse to form fusion products called heterokaryons. Heterokaryons can be identified and quantified by flow cytometry as double-stained events. Protocols are provided for the optimization of cell staining under conditions that minimize cell clumping and dye leakage. If spectral overlap occurs between emission spectra of the two stained cell types, the data will need to be electronically compensated.

Key Words: Cell fusion; heterokaryon; homokaryon; hybrid; flow cytometry; nuclear reprogramming.

1. Introduction: Cell Fusion as a Nuclear Reprogramming System

As a model system for studying the process of nuclear reprogramming (NR), nuclear transfer (NT) is imperfect. Only a single cell is reprogrammed with each technically demanding manipulation, and the surviving embryo must then be transferred back into the uterus to develop. As measured by the frequency of live births, NT has a low efficiency of NR (typically 1–2% of NT embryos) and thus the signal-to-noise ratio of a reprogramming experiment using NT is low. However, artificially induced reprogramming can also be detected in systems that do not require embryos or embryo transfer.

Different cell types can fuse spontaneously at a low rate when cultured together (1–4). Under appropriate conditions, the rate of such fusion can be dramatically increased. Fusion generates a binucleate cell called a *heterokaryon*,

From: *Methods in Molecular Biology, vol. 325: Nuclear Reprogramming: Methods and Protocols*
Edited by: S. Pells © Humana Press Inc., Totowa, NJ

A NO FUSION

B HOMOKARYON FORMATION

C HETEROKARYON FORMATION

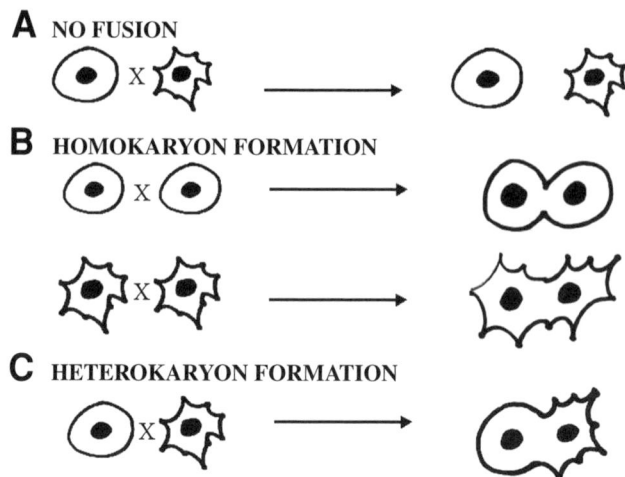

Fig. 1. Inducing a mixture of two cell types to fuse commonly results in a mixture of three possible outcomes: (**A**) cells fail to fuse with other cells; (**B**) cells fuse with other cells of the same cell type: these fusion products are called homokaryons; or (**C**) cells fuse with other cells of the other cell type: these are termed heterokaryons.

and at a low rate the two nuclei may fuse to generate a hybrid cell with a single, tetraploid nucleus (*see* **Fig. 1**). Cell lines may be easily established from such hybrids. The resulting characteristics of the hybrids are not those of an intermediate phenotype midway between that of the two parental cells; rather, the embryonic, or "less-differentiated", cell (usually) appears to reprogram the somatic, or "more-differentiated" partner (*5,6*). Because this reprogramming event usually involved the loss of specific differentiated functions in the somatic cell, the phenomenon traditionally was called *extinction*. Ample evidence exists from a variety of fusion systems that this reprogramming of one partner in the fusion by the other is not merely a passive, "cytoplasmic dominance" effect, but is an active reprogramming process. In fusions of embryonic stem (ES) cells or embryonic germ (EG) cells and somatic cells, the activation of ES-specific gene expression from alleles derived from the somatic partner has been demonstrated at either the ribonucleic acid (RNA) or protein levels in several instances (*2,4,7–12*). Examples of embryo-specific genes that may be activated by an NR event after fusion include *Oct-4* and related transgenes such as *OctNeo* or *Oct-4* promoter-driven green fluorescent protein (GFP), *Foxd3*, *Rex-1*, *Nanog*, and telomerase.

This chapter describes a method of generating and quantifying fusion hybrids from ES cells and somatic cells developed in our laboratory from the EG cell–thymocyte system originally described by Tada et al. (*[9]* and Chapter 7,

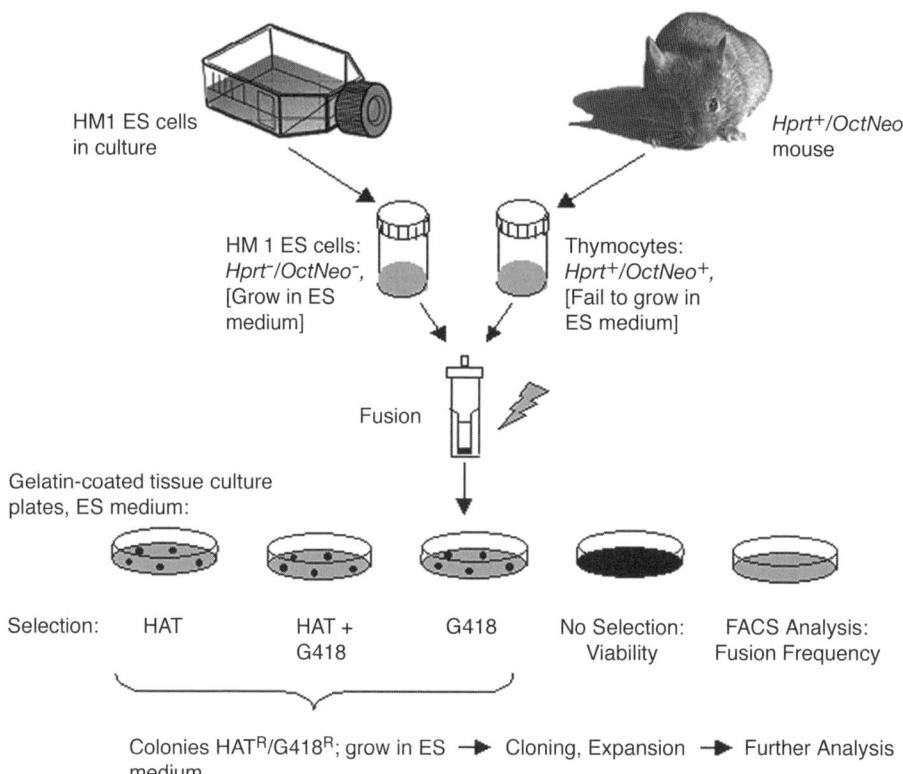

HM1 ES cells in culture

Hprt+/OctNeo mouse

HM 1 ES cells: *Hprt⁻/OctNeo⁻*, [Grow in ES medium]

Thymocytes: *Hprt+/OctNeo+*, [Fail to grow in ES medium]

Fusion

Gelatin-coated tissue culture plates, ES medium:

Selection: HAT   HAT + G418   G418   No Selection: Viability   FACS Analysis: Fusion Frequency

Colonies HATR/G418R; grow in ES → Cloning, Expansion → Further Analysis medium

Fig. 2. An overview of the process to generate fusion hybrids from ES cells and adult murine thymocytes.

this volume). The protocol is summarized in **Fig. 2** and in essence consists of the fusion by electroshock of a reprogramming cell type (the HM-1 ES cell) and a somatic cell type (the thymocyte), followed by a selection regime, which requires fusion to form a hybrid for cell survival. This protocol reliably generates approx 100 fusion hybrid colonies from HM-1 murine ES cells and murine thymocytes after hypoxanthine–aminopterin–thymidine (HAT) selection, which selects for a wild-type *hprt* gene that the ES cell lacks. All the colonies are phenotypically ES (**Fig. 3B**).

Such fusion hybrids generated in vitro provide a very tractable and replicable system for the generation and study of NR events. This protocol may therefore be used for generating fusion hybrids to study NR events, as the starting point for developing a system to fuse other cell types, and also as a positive control in such fusion experiments. However, it is important to remember that a change to the protocol can result in a different number of fusion hybrid colo-

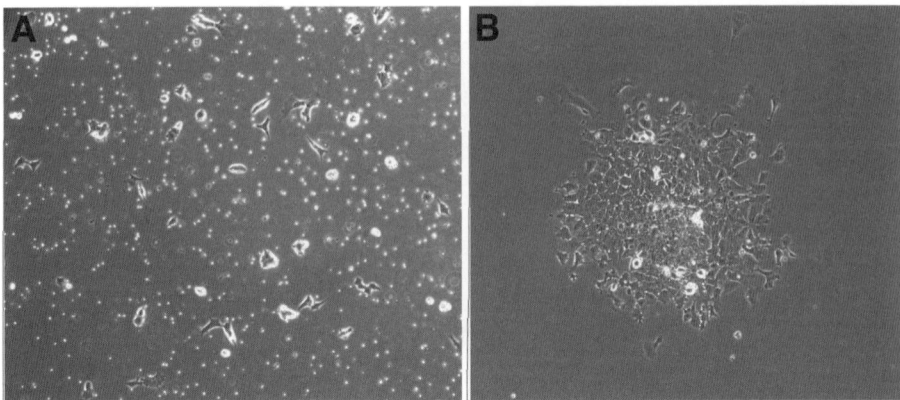

Fig. 3. **(A)** A field of a murine ES cell–thymocyte fusion experiment tissue culture plate at 24 h after fusion. Two main cell types are visible. The small round, bright cells are dead thymocytes, which have failed to attach themselves to the gelatin/tissue culture plastic substrate. The cells that have adhered to the gelatin substrate are ES cells. Most of these are diploid ES cells and will be killed over subsequent days in selection. A small number represent ES–thymocyte fusions and will survive in selection to generate a (clonal) colony such as that shown in **(B)**, which can then be picked, expanded, and studied in more detail. (B) A murine ES (HM1)–thymocyte fusion colony, after 8 d growing in HAT medium. The cells are phenotypically ES, with a modest level of differentiation at the periphery of the colony as is typical for colonies of ES cells. In a few more days, this colony would have expanded further and be ready for picking.

nies being recovered for reasons other than a change in the rate of nuclear reprogramming, for example, a change in the percentage of cells surviving the procedure, or a change in the *fusion* rate rather than NR rate. For this reason, monitoring the rate of fusion during such experiments is essential. To identify and isolate cell fusion products, we used two-color flow cytometry *(see* **Note 1** *[13])*. As with microscopic analysis, this technique allows the assessment of cell fusion on a "one-cell-at-a-time" basis *(14)*, but it is much quicker, less subjective *(see* **Note 2**), and allows analysis of a far larger number of cells.

Flow cytometry exploits the fact that light is scattered by biological particles and that they autofluoresce when they are exposed to laser light of a specific wavelength *(15)*. Such light scattering and autofluorescence is dependent on physical characteristics of the cells, such as size, shape, and granularity. Thus, one may use differences in how fluorescently labeled cells are excited by laser light to distinguish physical differences between individual cells in a sample.

Two-color flow cytometry for the purpose of assessing cell fusion involves two steps: (1) staining two cell populations, each with a different fluorescent dye before fusion induction and (2) the analysis of cell fluorescence after fusion induction. When cell staining and flow cytometric analysis are conducted, unstained, single-stained, and double-stained cells can be detected and quantified. Fusion products called heterokaryons are formed when one cell type fuses to another cell type. These appear as double-stained events. In addition to heterokaryon formation other outcomes are also common (**Fig. 1**), such as the failure of cells to fuse and homokaryon formation. A homokaryon is a fusion hybrid formed by the fusion of two cells of the same type. The outcomes of cells failing to fuse and homokaryon formation are easily distinguished from heterokaryon formation as they result in single-stained events. Typical data from an experiment of this kind are shown in **Fig. 4**.

2. Materials

2.1. Animals

The thymi from three wild-type female mice for the *Hprt* gene are required for each fusion. We have used mice of CBA strain and also have used *OctNeo* mice *(16)*, which are 94% CBA and also carry the *OctNeo* transgene. The suitability of this strain for fusion experiments studying NR is considered in some more detail in **Note 3**.

2.2. Cells

1. HM1 murine ES cells (*see* **Notes 4** and **5**).
2. Thymocytes derived from the adult mice in **Subheading 2.1.** (*see* **Notes 5** and **6**), prepared as described in **Subheading 3.1.**

2.3. ES Cell Culture

1. Standard ES Medium: Murine ES cells are routinely cultured in Glasgow's Modified Eagle's Medium (GMEM, Life Technologies), supplemented with 5% fetal calf serum, 5% newborn calf serum, 2 mM L-glutamine, nonessential amino acids (0.1 mM each of glycine, L-alanine, L-aspartic acid, L-asparagine and L-glutamic acid; 0.2 mM each of L-proline and L-serine), 1 mM sodium pyruvate, 0.1 mM β-mercaptoethanol, and 500 U/mL leukemia inhibitory factor.
2. Grow ES cells on tissue culture plastic coated with gelatin. Prepare 0.1% gelatin in water by adding 400 mL of Analar water to 0.4 g of gelatin in a 0.5-liter Duran bottle. Autoclave, allow to cool, and autoclave a second time. Apply the gelatin to the tissue culture plastic a few minutes before the cells are to be added and aspirate immediately before adding medium and cells.
3. ES cells are passaged typically every 3 d 1:6 by washing in phosphate-buffered saline (PBS) and treating with trypsin–EGTA (TEG) for approx 2 min to release the cells. Neutralize the TEG with standard ES medium before recovering the cells by centrifugation for 5 min at 200g and plating out as desired.
4. PBS: dissolve 8 g of NaCl, 0.2 g of KCl, 1.44 g of Na_2HPO_4, and 0.24 g of KH_2PO_4 in 800 mL of double-distilled water and adjusting the pH to 7.4 with HCl. Make the volume to 1 L with double-distilled water and autoclave.

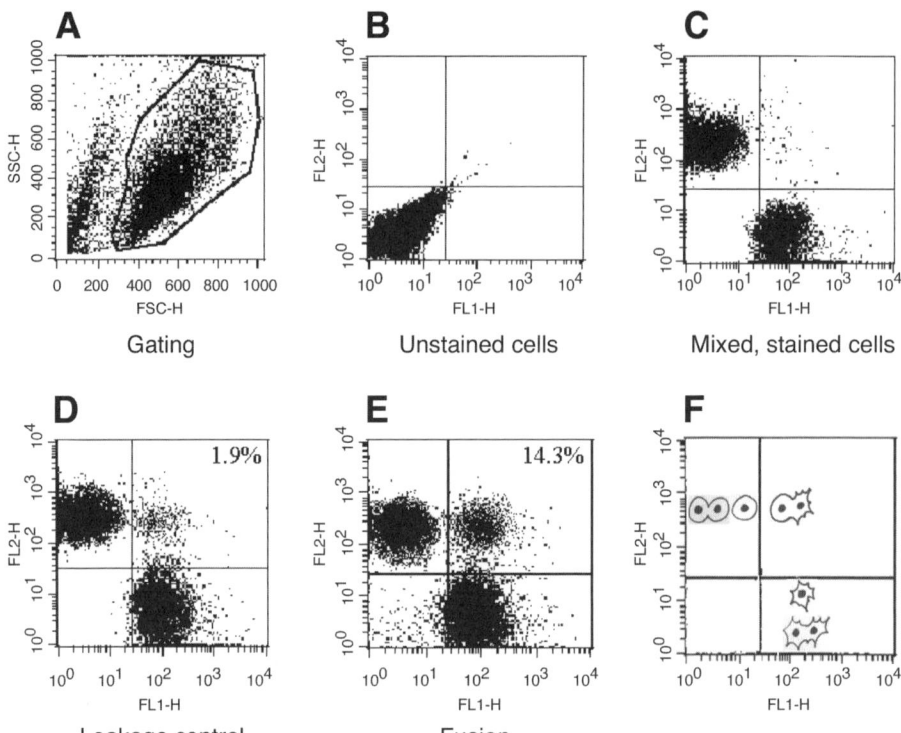

Fig. 4. Typical data from a two-color flow cytometry experiment to measure cell fusion. One cell population has been stained with Celltracker Green (CMFDA) and the other with Celltracker Orange (CMTMR) **(A)** A forward scatter/side scatter (FSC/SSC) plot showing the "live" gate used to exclude cellular debris and dead cells; this gate is then applied to all subsequent plots. **(B)** Voltage settings for the fluorescence photomultiplier tubes are set so that unstained cells lie close to the origin of an FL1/FL2 dot plot as shown. In this example, FL1 refers to recorded fluorescence after light emitted from the sample has passed through a FITC filter (i.e., green fluorescence) FL2 refers to fluorescence recorded after the light has been passed through a PE filter (i.e., orange fluorescence) Stained (and consequently more fluorescent) cells appear further along the axes than do unstained cells. **(C)** Celltracker Green (CMFDA)-stained cells and Celltracker Orange (CMTMR)-stained cells when mixed and analyzed appear as two distinct populations. **(D)** A control for dye leakage (where stained cells are electropulsed separately and then mixed together) shows some dye leakage has taken place. **(E)** An actual fusion sample (in which the two stained cell populations were mixed and then induced to fuse) is also shown. The same quadrant gate has been applied to all dot plots so that the percentage of events that are double-stained can be measured. Here the percentage of viable cells fusing is 12.4% (Fusion [14.3%] – Leakage Control [1.9%]) **(F)** A schematic shows unfused cells and homokaryons register in the upper left and lower right quadrants and heterokaryons will appear in the upper right quadrant.

5. TEG: dissolve 6.3 g of NaCl, 0.12 g of Na$_2$HPO$_4$, 0.216 g of KH$_2$PO$_4$, 0.333 g of KCl, 0.9 g of D-glucose, 2.7 g of Tris, and 0.9 mL of 1% phenol red in 800 mL of Analar water, then add 100 mL of 2.5% trypsin, 0.4 g of ethylenebis(oxyethylenenitrilo)tetra-acetic acid, and 0.1 g of polyvinyl alcohol. Adjust the pH to 7.6 and make to 1 liter. Filter through a sterile 0.2-µm filter, aliquot, and store at −20°C.

6. 5 × 10^{-3} *M* aminopterin: weigh out approx 100 mg of aminopterin; and dissolve at a concentration of 2.2 mg/mL in double-distilled water. One mL of 1 *M* NaOH may have to be added to make the aminopterin dissolve. Filter sterilize and store at −20°C in a foil-wrapped tube.

7. 100X HAT: to prepare HAT selective medium, standard ES medium is supplemented with hypoxanthine, aminopterin, and thymidine at final concentrations of 0.1 m*M* hypoxanthine, 1 µ*M* aminopterin, and 10 µ*M* thymidine. 100X HAT is prepared as follows:

   a. Dissolve 272.2 mg of hypoxanthine in 2 mL of 1 *M* NaOH.
   b. Add 8 mL of double-distilled water and mix well.
   c. Add a further 10 mL of double-distilled water and mix until most of the hypoxanthine is dissolved.
   d. Add another 1 mL of 1 *M* NaOH and mix until all the hypoxanthine is dissolved.
   e. Add 75 mL of double-distilled water.
   f. Add 48.4 mg of thymidine and mix well.
   g. Add a further 100 mL of double-distilled water.
   h. Add 4.0 mL of 5 × 10^{-3} *M* aminopterin (*see* **step 6** in this subheading) and mix well.
   i. Filter sterilize and store as 20-mL aliquots at −20°C in foil-wrapped tubes.

8. 100X HT, for weaning cells off HAT selection, is prepared and stored as in **step 7** but omitting the aminopterin. Because it is the aminopterin that is light-sensitive, HT aliquots need not be kept wrapped in foil.

9. To prepare G418 medium for selection of *OctNeo*-expressing cells, supplement standard ES medium with either 100 µg/mL or 150 µg/mL G418 depending on whether the thymocytes are heterozygous or homozygous for the *OctNeo* transgene respectively. Prepare stock G418 by dissolving G418 in double-distilled water and making to a final concentration of 100 mg/mL. Filter sterilize and store at −20°C in 5-mL aliquots.

2.4. Fusion Reagents

1. 0.3 *M* D-mannitol, osmolarity of 280 mOsmoles (*see* **Note 7**).
2. Standard PBS, sterile.
3. Cell marker dyes (Cambridge Bioscience), 10-µL aliquots at 10 m*M* concentration in DMSO, stored at −70°C in foil-wrapped tubes. Green: 5-chloromethylfluorescein diacetate (CMFDA) and Orange: (5-(and-6)-(((4-chloromethyl)benzoyl)amino) tetramethylrhodamine) (CMTMR; *see* **Note 8**).

2.5. Fusion Equipment

1. Tissue culture plates, 10 cm.
2. 18-gage Needles.

3. 10-mL Syringes.
4. Sterile 7-mL bijou tubes.
5. Microscope, inverted phase contrast.
6. Hemocytometer, Neubauer Improved.
7. Cell incubator. Humidified air, 37°C, 5% CO_2 in air.
8. Bench-top centrifuge and sterile, disposable universal tubes (14 mL and 50 mL).
9. Bio-Rad "Gene Pulser" electroporation apparatus.
10. 2-mm Gene Pulser cuvets.
11. Flow cytometer (e.g., Beckton-Dickinson FACScan). The flow cytometer should be equipped with a 488 nm emission laser and filters for FITC (530 ± 15 nm) and PE (585 ± 21 nm) for measuring the green and orange fluorescences of Celltracker Green (CMFDA) and Celltracker Orange (CMTMR), respectively.
12. Osmometer (Advanced Micro-Osmometer model 3MO+, Advanced Instruments, Inc., Norwood, MA).

3. Methods

3.1. Fusion Protocol

If cell fusion is to be quantified by flow cytometry, identification of optimal dye concentrations (*see* **Subheading 3.2.1.**) must be performed before proceeding with the following fusion protocol.

1. Kill three 6- to 7-wk-old female mice, strain CBA, *OctNeo* transgenics or equivalent, for each fusion experiment.
2. Isolate the thymi from all three mice and put them in a bijou tube with 5 mL of PBS, supplemented with 100 U/mL penicillin and 100 μg/mL streptomycin.
3. Chop up the thymi with dissecting scissors and then triturate the PBS/tissue mixture with an 18-gage needle and syringe to generate a cloudy suspension.
4. Allow the crude cell suspension to settle for 5 min and then remove the approximately middle 4 mL of suspension (*see* **Note 9**).
5. While the thymus suspension is settling, isolate ES cells from their flask as normal by aspirating the medium, washing with PBS, and then incubating at 37°C for 1 to 2 min with TEG. Resuspend in ES medium.
6. Count both cell types. Recover the cells by centrifugation at 200*g* for 5 min.
7. Resuspend the cells in the CellTracker/ES medium solutions at concentrations as defined in **Subheading 3.2.** and prepared as in **Note 8** at a final concentration of 10^7 cells/mL, thymocytes in CMTMR (orange) and ES cells in CMFDA (green).
8. Incubate the cells in the incubator at 37°C, 5% CO_2 in air for 30 min (*see* **Note 10**).
9. Wash the cells three times by centrifugation for 5 min at 200*g* and resuspending in PBS.
10. Count the cells.
11. Mix 10^7 ES cells and 5×10^7 thymocytes, centrifuge at 200*g* for 5 min as described previously, and resuspend in 400 μL of 0.3 *M* D-mannitol, 280 mOsmoles solution. For the fusion dye leakage controls (*see* **Subheading 3.2.3.**), take identical numbers of ES cells or thymocytes *alone* and resuspend them in the same

volume of mannitol and process them in parallel with the fusion samples in subsequent steps.

12. Transfer the ES cell/thymocyte/mannitol suspension to a 0.2 cm Gene Pulser cuvet.
13. Centrifuge the cuvet in a 50 mL tube padded with tissue paper for 5 min at 200g to pellet the mixture of cells at the bottom of the cuvet (*see* **Note 11**).
14. Fuse the cells by inserting the cuvet into the Gene Pulser electroporation apparatus and applying a single pulse of 300 volts, 25 μF (*see* **Note 12**).
15. Allow the cells to stand at room temperature in the cuvet for 15 to 20 min after pulsing. To prepare dye leakage controls, mix an "ES cell only" and a "thymocyte only" cuvet (both cuvets pulsed individually) at this stage.
16. Gently resuspend the cells and add to 50 mL of ES cell medium. Mix. Add 10 mL to each of five gelatinized 10-cm tissue culture plates and place in the incubator.

The next day, 24 h after fusion:

17. Refeed three plates with selective medium (*see* **Note 13** and **Fig. 3A**). One further plate remains in nonselective medium as a growth control. The last plate is used to monitor fusion efficiency by flow cytometry (*see* **Subheading 3.2.**).
18. Flow cytometric analysis for the detection of cell fusion requires the following samples, which should be at similar cell concentrations (*see* **Subheading 3.2.**):
    a. Unstained ES cells and thymocytes to permit the detection voltages for FL1 and FL2 to be set in the first log decade of appropriate fluorescence plots.
    b. Single-stained ES cells and thymocytes to perform electronic compensation for spectral emission overlap between the fluorescent dyes.
    c. A mixture of both single-stained cell populations to determine dye leakage (mixed together after electropulsing as single populations).
    d. Electrofused ES cell–thymocyte sample for quantification of cell fusion. Fused cells are positive for both the green and orange fluorescence, whereas unfused cells are only positive for one color (*see* **Note 14**).

The following days to 2 wk:

19. Refeed plates with 10 mL of selective or nonselective medium as appropriate to allow clones to grow (*see* **Note 15** and **Fig. 3B**).
20. Resistant colonies are cloned as normal (*see* **Note 16**), and then expanded (*see* **Note 17**).

3.2. Cell Staining

Laser excitation of fluorescent dyes results in the emission of a spectrum of light, not a single wavelength. The optics in the flow cytometer are designed to collect light within a specific bandwidth for each fluorescence detector, but emitted light outside this bandwidth may overlap with a secondary detector's bandwidth, resulting in the appearance of false double-positive cells. Spectral overlap can be compensated for electronically by the subtraction of the overlapping signals from the inappropriate fluorescence channel, which requires single-stained cell populations as compensation controls. Overstaining cells with fluorescent dyes will increase spectral overlap and, because electronic compensation reduces the sensitivity of the cytometer, dye concentrations for staining should be optimized in advance by titration to avoid wasting time, energy, and expensive reagents. The levels of compensation may vary between experiments and therefore compensation controls should always be included in an experiment.

3.2.1. Identification of Optimal Dye Concentrations

1. Make up seven concentrations of the Celltracker dye solutions (0, 1.25, 2.5, 5, 10, 25, and 50 μM) using standard cell growth medium. Use these to stain aliquots of 10^6 cells for 30 min (10^6 cells/mL dye-supplemented medium). The aliquots should be foil wrapped and placed in an incubator using standard conditions (e.g., 37°C, 5% CO_2 for most mammalian cells) during incubation (*see* **Note 18**). Shaking the cells during incubation is not required as the dye freely permeates the cell. Wash the cells three times in PBS to remove excess dye before using them in fusion experiments.

2. Set up an acquisition page on the flow cytometer, including a forward scatter (FSC, cell size)/side scatter (SSC, cell "complexity") dot plot, histogram plots for FL1 (fluorescence channel 1; usually FITC/green) and FL2 (fluorescence channel 2; usually PE/orange), and an FL1/FL2 dot plot. Take at least 2×10^5 unstained cells of the type intended for use as a fusion partner, resuspend them in PBS, and transfer them into a tube before loading it into the flow cytometer. Start acquiring data. As the cells are analyzed, dots will appear on the plot. FSC and SSC voltages should be altered until all events are just within the limits of the dot plot. As **Fig. 4A** shows, there should be two clusters of events: one near the origin of the dot plot (representing cellular debris and dead cells) and another cluster further along both axes (representing the live cells). Draw an electronic gate around the cluster of dots representing the live cells. Apply the live gate to histograms and the FL1/FL2 fluorescent dot plot so that only gated events are shown.

3. The fluorescence detectors should now be adjusted. To ensure that the flow cytometer can register fluorescently stained cells, it is important that the autofluorescence peaks of unstained controls lie within the first log decade of the histogram. This is accomplished by adjusting the voltage setting for fluorescence channels FL1 and FL2. Change settings for both FL1 and FL2 to generate histograms as shown in **Fig. 5A**.

4. The FL1/FL2 dot plot should now resemble **Fig. 4B**, in which unstained cells appear near the origin. Record data for at least 10,000 "live" gated events.

5. Load aliquots of single-stained cells for each dye concentration series into the flow cytometer and record data for at least 10,000 "live" gated events. To identify the appropriate concentration of each Celltracker dye, overlay the fluorescence histograms from each series; the results should look similar to those in **Fig. 5**. The flow cytometer can display fluorescence up to 1000-fold that of unstained negative control cells. Ideally stained cells should be in the third log decade of fluorescence detection (**Fig. 5C**).

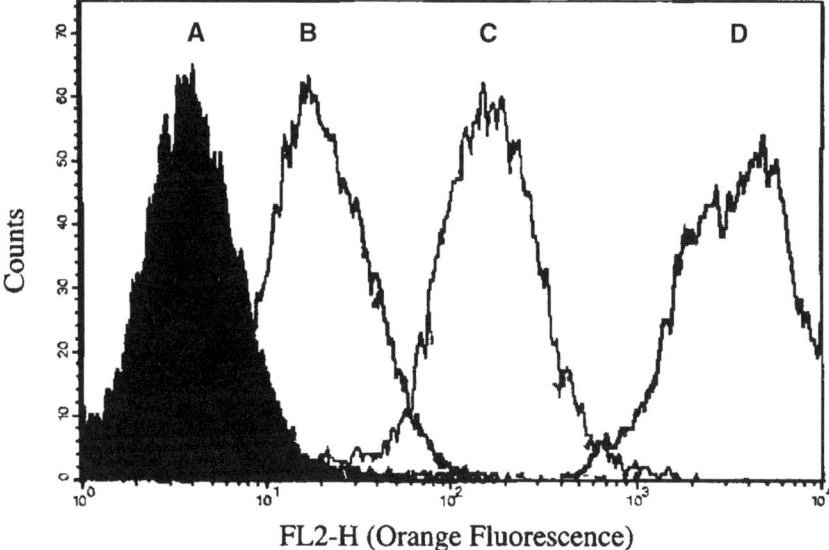

FL2-H (Orange Fluorescence)

Fig. 5. Identifying the optimal dye concentration. This figure shows cells stained with Celltracker Orange (CMTMR) (**A**) Auto-fluorescence of unstained cells. (**B**) The fluorescence distribution of cells not stained strongly enough to properly distinguish them from unstained cells. (**C**) Fluorescence of cells that have been optimally stained. (**D**) The fluorescence distribution of cells that have been overstained (part of the emission spectrum lies outside the range of detection).

6. Having identified the optimal concentration for each Celltracker dye for their appropriate cell type, these samples should now be reacquired utilizing the FL1/FL2 dot plot to adjust the spectral overlap of the dyes by electronic compensation. Following compensation the mean FL2 fluorescence for FL1 (CMFDA)-stained cells should be the same as that of unstained cells, and the mean FL1 fluorescence for FL2 (CMTMR)-stained cells should be the same as that of un-stained cells. Once this has been achieved record data for at least 10,000 "live" gated events for each single-stained cell type and for a mixture of the two samples. The resulting dot plot should look like that in **Fig. 4C**.
7. After set-up and compensation, the instrument settings for optimal concentrations of dyes should be saved and used unchanged where possible for all future experiments to allow direct comparison between experiments.

3.2.2. Controlling for Dye Leakage

Inducers of cell fusion (either electrical, chemical, viral, or laser-based) de-stabilize cell membranes and such disruption results in dye leakage (S. Sullivan; unpublished data). Controls should be set up to assess how much dye leakage is occurring between cell types, so that it may be taken into account when measuring cell fusion. Perform the following procedure using the optimal concentration of dye determined in **Subheading 3.2.1.**

1. Stain duplicate aliquots of each cell type with CMFDA or CMTMR. For the fusion experiment, mix together one aliquot of each stained cell type and electropulse. For dye leakage controls the remaining aliquot of each stained cell type should be electropulsed separately and then immediately mixed together (*see* **Note 19**).
2. The dye leakage control is analyzed concurrently with the actual fusion experiment (*see* **Subheading 3.3.** for details of analysis). The number of double-stained events should be lower than in the fusion experiment (*see* **Fig. 4**).
3. The percentage of double-stained events is calculated by applying quadrant gates to distinguish unstained, single-stained, and double-stained events. The same gate should be applied to all samples in an experiment. To assess the cell fusion, the percentage of double-stained cells in the leakage control is subtracted from that of the fusion sample (*see* **Fig. 4**).

3.3. Quantifying Fusion Products

After the cells have been induced to fuse, the cell suspension should be repelleted at 200*g* for 5 min and disaggregated to a single-cell suspension before being analyzed using the flow cytometer. It is important that cells are analyzed as single-cell suspensions so that clumped cells are not registered by the flow cytometer as fusion products. The best way to avoid this problem is to disaggregate the sample (usually using TEG) to a single-cell suspension immediately before loading the sample into the flow cytometer. If there is a sample-agitation option on the flow cytometer, use it (this resuspends the sample in the tube during sample acquisition, so reducing cell aggregation).

1. If electrofusion is used to induce cell fusion, the fluorescence of electropulsed cells is erratic for several hours after electropulsing (S. Sullivan, unpublished observations *[13]*). During this time, cells should be reseeded in appropriate conditions using complete growth medium. The cells should be incubated in the dark for the required time and then disaggregated again into a single-cell suspension using TEG before flow cytometric analysis.
2. Mix an aliquot of the Celltracker Green (CMFDA)-stained cell type with an aliquot of the Celltracker Orange (CMTMR)-stained cell type. Load and analyze the mixture. On the FL1/FL2 dot plot, you should see two distinct populations (*see* **Fig. 4C**), each further along one axis than where unstained cells appeared. If this is not the case and instead the two populations are not distinct, repeat the setup procedure outlined in **Subheading 3.2.1.** (*see* **Note 20**).
3. Analyze the fusion and dye leakage samples. Double-stained events will appear in the upper right quadrant of FL1/FL2 dot plots (*see* **Fig. 4**). Apply the same quadrant gates to the data and use the statistics option in the software to determine the percentage of events in each quadrant. Subtract the percentage of events

in the upper right quadrant for the dye leakage sample from the corresponding percentage for the experimental fusion sample. This will give the percentage of live cells fusing to form heterokaryons (*see* **Note 21**).

4. If the flow cytometer has a sorting capability the heterokaryons can be isolated by collecting the double positive cells in the upper right quadrant. Consult the flow cytometer user manual for details.

4. Notes

1. We have chosen to assume that most readers of this chapter will be fusing two different cell types to study nuclear reprogramming. There is no physical reason why cells of the same cell type cannot also be fused.

2. For example, it often is hard to differentiate overlaid cells on a slide from fused cells by microscopic analysis.

3. The *OctNeo* transgene in *OctNeo* mice *(16)* consists of the early embryo-specific *Oct-4* promoter *(17–19)* driving expression of the neomycin resistance gene, the product of which provides resistance to the nucleoside analog G418. Because this transgene is not expressed in the thymus of *OctNeo* mice (S. C. Pells, unpublished observations), reactivation of *OctNeo* expression and concomitant resistance to G418 provides a functional assay for nuclear reprogramming at this transgene locus. The difference in colony frequency between plates subjected to HAT selection, and those subjected to both HAT and G418 selection indicates how many phenotypic hybrids are actually reprogrammed, at least at the transgenic *OctNeo* locus.

4. HM-1 is a male ES cell line *(20)* with a single copy of the *Hprt* gene carrying the $Hprt^{b-m3}$ mutation *(21)*. This mutation is a deletion of exons 1 and 2 of the gene and, as such, reversion events are not observed. Therefore, in HAT selection, *no* colonies should be observed in an ES cell-only control plate as colony formation is entirely dependent upon fusing an ES cell with a somatic cell containing a wild-type X chromosome.

5. The ES cells and the somatic cells should be free of microbial contamination such as mycoplasma. Such contamination can affect cell viability and cell-cell fusion frequencies.

6. Thymocytes are mostly arrested in the G0 phase of the cell cycle, and are dependent upon T-cell receptor activation and interleukin stimulation for growth. They do not adhere to the gelatin substrate recommended in **Subheading 2.3., item 2**. As such, ES cell culture conditions do not support thymocyte growth and to survive a thymocyte needs to fuse with an ES cell.

7. The osmolarity of the mannitol solution in which electrofusion is conducted is important to the frequency at which hybrids are obtained. In our laboratory, we make the 0.3 *M* D-mannitol solution and then correct osmolarity to 280 mOsmoles, ±10 mOsmoles, by adding 0.1 m*M* MgSO₄ dropwise, mixing and monitoring the osmolarity of the solution with an Osmometer. The mannitol should then be used immediately or stored at –20°C.

8. Just before use, dilute the CellTracker dyes at least 1:100 in ES tissue culture medium. Concentrations of either 0.25 μ*M* if the cells are to be analyzed the same day as the fusion, or 2.5 μ*M* if the cells will be examined by flow cytometry on the day following fusion work well for murine ES cells and thymocytes, but for other cell types, these concentrations should be optimized as described in **Subheading 3.2.** The dyes are light-sensitive and should be kept in the dark as much as possible, even during the incubation of cells. This is most easily done by wrapping them in foil. If GFP expression is to be used to investigate the reprogramming of a specific gene after fusion, then a green fluorescent probe such as CMFDA cannot be used because both GFP and CMFDA are detected in the same fluorescence channel (FL1): use CellTracker Red (CMTPX) in its place. Red fluorescence can be detected in the FL3 (Red) fluorescence channel using a 650-nm long-pass filter.

9. Avoid any remaining clumps of tissue at the bottom of the tube and any fatty tissue floating at the top.

10. The CellTracker dyes are vital stains, that is, they stain the cell but do not harm it. They are taken up by the cells during the 30-min incubation at 37°C at **step 8** of **Subheading 3.1.** If flow cytometry is not to be used to monitor fusion efficiency, **steps 6** to **8** inclusive of the protocol may be omitted.

11. This step concentrates all the cells at the bottom of the cuvet, placing them all in close proximity for fusion.

12. When performing experiments requiring multiple electroshocks, we feel it is better to leave an interval of a couple of minutes between pulses. Otherwise, the electroporation equipment appears to deliver less accurate later pulses. This can be seen as a drift in the measured value of the time constant of the pulse with increasing number of pulses. A 300-V, 25-μF pulse typically gives a time constant of 2 to 2.5 s under the aforementioned conditions, and the time constant increases to more than 10 s if several pulses are performed in relatively quick succession.

13. Selective media vary depending on the exact combination of cells in use. For thymocytes from wild-type mice and HM1 ES cells, use HAT medium. For thymocytes from *OctNeo* mice and HM1 cells either HAT, a combination of HAT and G418, or G418 alone may all be used. In a successful experiment, the control plate in nonselective medium quickly reaches confluence as there should be sufficient viable ES cells to grow over the plate. In selective media, most cells die, but over approximately the next 2 wk, resistant colonies arise at a rate of approx 20 per plate. This represents a fusion hybrid frequency of approx 100 fusions in a complete fusion experiment consisting of 10^7 ES and 5×10^7 somatic cells.

14. Flow cytometry to monitor fusion frequency is "optional" but is important when quantitative measures of reprogramming events are required. Some modifications to the protocol will give more or fewer hybrid colonies not because of a change in the rate of NR, but because the rate of fusion changes, and to detect this, the assay of fusion efficiency described in **Subheading 3.3.** is required.

15. Early in selection when there are many dying cells in the plate, cultures should be fed every day, possibly with a gentle wash in PBS first to remove loose dead

matter that is toxic to live cells. Once most of the cells are dead, the plate has cleared and colonies are expanding, feeding every 2 or 3 d is sufficient.

16. Clonal colonies are picked with a sterile 20-μL pipet, transferred into a drop of TEG for 2 min to disaggregate the cells, and then plated individually into wells of a 24-well tissue culture plate.

17. After single-cell-cloning, clones need not be maintained in selective medium. If the cells are in HAT medium, note that aminopterin inhibition of DHFR, and hence the *de novo* purine pathways, is only slowly reversed after removal of the aminopterin from the medium because the aminopterin–polyglutamate derivative actually responsible for enzyme inhibition must be degraded by γ-glutamyl hydrolase first. Therefore, if HAT selection is removed, cells will die even if they are wild-type for *Hprt* unless they are supplemented with hypoxanthine and thymidine (at the same concentrations as HAT medium) for 3 d after the removal of HAT selection. This transition medium is referred to as HT. After having spent 3 d in HT medium, previously HAT-selected clones may be grown in standard ES medium. G418 may be removed immediately with no ill effects. However, it should be noted that the hybrids are tetraploid cells and as such may be genetically unstable. In our laboratory, we therefore typically maintain fusion hybrids in selective media permanently to select for at least those chromosomes carrying the selectable markers.

18. CellTracker reagents are fluorescent chloromethyl derivatives that freely diffuse through the membranes of live cells. Once inside the cell, these mildly thiol-reactive probes react with intracellular components to produce cells that are both fluorescent and viable for at least 24 h after loading.

19. It could be argued that in dye leakage controls, cells may still be fusing after fusion induction has ceased. It has been shown that electropulsed cells have a higher fusion rate when centrifuged up to 40 min after the application of a single high-intensity pulse *(22)*, which is perhaps surprising, given lipid bilayers in an aqueous environment reform in a matter of micro- or milliseconds when electric pulses are applied *(23)*. It is the authors' experience that fusion only occurs at all in dye leakage controls very infrequently and that the incidence of dye leakage is far greater. Thus, using leakage controls makes the measurement of cell fusion more accurate.

20. If the manuals or the technical support for your flow cytometer are unable to provide assistance with regard to compensating data, a very useful internet resource is cytometry@flowcyt.cyto.purdue.edu, where flow cytometer experts provide help to fellow flow cytometer users.

21. If fusion rates between cell populations are consistently low (<1%), try evaluating other dissociation procedures for the generation of single-cell suspensions. Extracellular matrix components of the plasma membrane may inhibit cell fusion. One of us (S. Sullivan) observed that treating H9 human embryonic stem cells with hyaluronidase prior to fusion induction increased heterokaryon formation fivefold (Unpublished data). Ohno-Shosaku and Okada *(24)* have found previously that pronase treatment increased fusion of mouse cells.

Acknowledgments

The work described in this chapter was supported by the Geron Corporation and the BBSRC and was carried out in Dr. Jim McWhir's laboratory at the Roslin Institute (Edinburgh, UK).

References

1. Weiss, M., and Green, H. (1967) Human-mouse hybrid cell lines containing partial complements of human chromosomes and functioning human genes. *Proc. Natl. Acad. Sci. USA* **58,** 1104–1111.
2. Ying, Q.-L., Nichols, J., Evans, E. P., and Smith, A. G. (2002) Changing potency by spontaneous fusion. *Nature* **416,** 545–547.
3. Tarada, N., Hamazaki, T., Oka, M., Hoki, M., Mastalerz, D. M., Nakano, Y., et al. (2002) Bone marrow cells adopt the phenotype of other cells by spontaneous fusion. *Nature* **416,** 542–545.
4. Pells, S., Di Domenico, A. I., Gallagher, E. J., and McWhir, J. (2002) Multipotentiality of neuronal cells following spontaneous fusion with es cells and nuclear reprogramming *in vitro. Cloning Stem Cells* **4,** 331–338.
5. Kennett. R. H. (1979) Cell fusion. *Methods Enzymol.* **58,** 345–359.
6. Kikyo, N., and Wolffe, A.P. (2000) Reprogramming nuclei: insights from cloning, nuclear transfer and heterokaryons. *J. Cell Sci.* **113,** 11–20.
7. Chiu C. P., and Blau. H. M. (1984) Reprogramming cell differentiation in the absence of DNA synthesis. *Cell* **37,** 879–887.
8. Ben-Shushan E., Pikarsky, E., Klar A., and Bergman Y. (1993) Extinction of Oct-3/4 gene expression in embryonal carcinoma x fibroblast somatic cell hybrids is accompanied by changes in the methylation status, chromatin structure, and transcriptional activity of the Oct-3/4 upstream region. *Mol. Cell Biol.* **13,** 891–901.
9. Tada, M,. Tada, T., Lefebvre, L., Barton, S. C., and Surani, M. A. (1997) Embryonic germ cells induce epigenetic reprogramming of somatic nucleus in hybrid cells. *EMBO J.* **16,** 6510–6520.
10. Tada, M., Takahama, Abe, K., Nakatsuji, N., and Tada, T. (2001) Nuclear reprogramming of somatic cells by in vitro hybridization with ES cells. *Curr. Biol.* **11,** 1553–1558.
11. Flasza, M. S. A., Smith, K., Andrews, P. W., Talley, P., and Johnson, P. A. (2003) Reprogramming in inter-species embryonal carcinoma-somatic cell hybrids induces expression of pluripotency and differentiation markers. *Cloning Stem Cells* **5,** 339–354.
12. Do, J. T., and Scholer, H. R. (2004) Nuclei of embryonic stem cells reprogram somatic cells. *Stem Cells* **22,** 942–949.
13. Scott-Taylor, T. H., Pettengell, R., Clarke, I., Stuhler, G., La Barthe, M. C., Walden, P., and Dalgleish, A. G. (2000) Human tumour and dendritic cell hybrids generated by electrofusion: potential for cancer vaccines. *Biochim. Biophys. Acta.* **1500,** 265–279.

14. Alkan, S. S., Mestel, F., Jiricka, J., and Blaser, K. (1987) Estimation of heterokaryon formation and hybridoma growth in murine and human cell fusions. *Hybridoma* **6,** 371–379.

15. Jaroszeski, M. J., Gilbert, R., and Heller, R. (1998) Flow cytometric detection and quantitation of cell–cell electrofusion products, in *Methods in Molecular Biology. Flow Cytometry Protocols* (Jaroszeski, M. J., Heller, R., and Gilbert, R., eds.), Humana Press, Totowa, NJ, pp. 149–156.

16. McWhir, J., Schnieke, A., Ansell, R., Wallace, H., Colman, A., Scott, A., et al. (1996) Selective ablation of differentiated cells permits isolation of embryonic stem cell lines from murine embryos with a non-permissive genetic background. *Nat. Genet.* **14,** 223–236.

17. Scholer, H., Ruppert, S., Suzuki, N., Chowdury, K., and Gruss, P. (1990) New type of POU domain in germ line-specific protein Oct-4. *Nature* **344,** 435–439.

18. Okamoto, K., Okazawa, H., Okuda, A., Sakai, M., Muramatsu, M., and Hamada, H. (1990) A novel octamer binding transcription factor is differentially expressed in mouse embryonic cells. *Cell* **60,** 461–472.

19. Rosner, M., Vigano, M., Ozato, K., Timmons, P., Poirer, F., Rigby, P., and Staudt, L. (1990) A POU-domain transcription factor in early stem cells and germ cells of the mammalian embryo. *Nature* **345,** 686–692.

20. Magin, T., McWhir, J., and Melton, D. (1992) A new mouse embryonic stem cell line with good germ line contribution and gene targeting frequency. *Nucl. Acids Res.* **20,** 3795–3796.

21. Thompson, S., Clarke, A., Pow, A., Hooper, M., and Melton, D. (1989) Germ line transmission and expression of a corrected HPRT gene produced by gene targeting in embryonic stem cells. *Cell* **56,** 313–321.

22. Sowers, A. E. (1987) The long-lived fusogenic state induced in erythrocyte ghosts by electric pulses is not laterally mobile. *Biophys. J.* **52,** 1015–1020.

23. Benz R., and Zimmermann, U. (1981) The resealing process of lipid bilayers after reversible electrical breakdown. *Biochim. Biophys. Acta.* **640,** 169–178.

24. Ohno-Shosaku T., and Okada, O. Y. (1984) Facilitation of electrofusion of mouse lymphoma cells by the proteolytic action of proteases. *Biochem. Biophys. Res. Commun.* **120,** 138–143.

9

Modulation of Cell Fate Using Nuclear and Cytoplasmic Extracts

Anne-Mari Håkelien, Kristine G. Gaustad, and Philippe Collas

Summary

The direct transformation of one somatic cell type into another somatic cell type would be beneficial for producing isogenic replacement cells for therapeutic use. Various approaches for altering cell fate are being developed, including methods for differentiating stem cells isolated from somatic tissues. This chapter describes a procedure for turning one somatic cell type (the "donor" cell) into another somatic "target" cell type using cellular extracts. The method also can be used to promote differentiation of a somatic stem cell along a specific pathway. The procedure involves permeabilization of the donor cell, incubation of the permeabilized cell in a nuclear and cytoplasmic extract derived from the target cell type, the resealing of donor cell membrane, and culture. Cells can be analyzed for induction of new gene and protein expression, as well as for the establishment of cellular functions specific to the target cell type. We also describe a slight modification of the procedure to allow analysis of extract-induced chromatin remodeling in nuclei purified from somatic cells. Because large numbers of cells and nuclei can be treated in cell extracts and because extracts can be fractionated or supplemented with various agents, this system constitutes a powerful tool to examine the molecular mechanisms of nuclear reprogramming and of cell differentiation, at least as they take place in vitro.

Key Words: Cell extract; cell culture; chromatin; differentiation; fibroblast; nucleus; nuclear reprogramming; permeabilization; plasma membrane; stem cell; streptolysin O; transdifferentiation.

1. Introduction

The direct conversion of a differentiated cell type into another differentiated cell type or into a pluripotent cell would present technological, medical, and societal benefits. Altering cell fate would be beneficial for producing isogenic replacement cells for treatment of a variety of life-threatening diseases. It

From: *Methods in Molecular Biology, vol. 325: Nuclear Reprogramming: Methods and Protocols*
Edited by: S. Pells © Humana Press Inc., Totowa, NJ

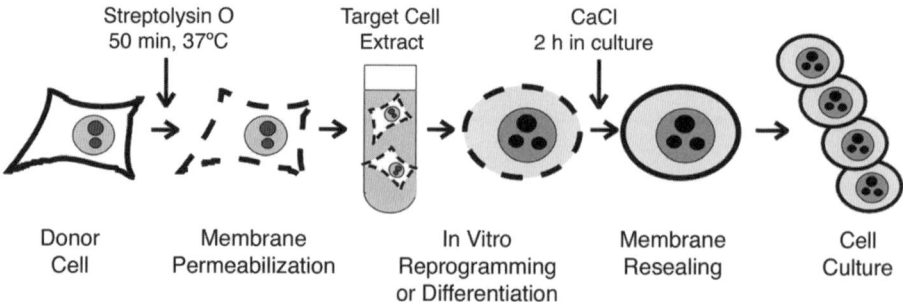

Fig. 1. In vitro cell reprogramming and differentiation approach. The plasma membrane of "donor" cells (e.g., fibroblasts) is reversibly permeabilized with streptolysin *O*. The permeabilized cells are exposed for 1 h to a nuclear and cytoplasmic extract derived from a target cell type (e.g., Jurkat cells, insulin-producing INS-1E cells, cardiomyocytes). After exposure to extract, cells are resealed for 2 h in culture medium containing 2 m*M* CaCl$_2$, and cultured. Differentiation of somatic stem cells such as adipose tissue stem cells into cells expressing cardiomyocyte-specific markers can also be achieved using this system.

would also alleviate limitations of current cell therapeutic approaches resulting from graft rejection, use of immunosuppressants, and safety and ethical issues raised by the use of cells of embryonic, fetal, or animal origin. A variety of approaches for manipulating cell fate are being investigated *(1)*. These include transgenesis, coculture of two different cell types, reprogramming nuclear function by transplantation of somatic nuclei into eggs (cloning) *(2–5)*, or fusion of a somatic cell with a pluripotent cell *(6,7)*.

We also are developing an in vitro approach for directly turning one somatic cell type into another (**Fig. 1**). The system relies on the transient uptake of regulatory components from a nuclear and cytoplasmic extract derived from a "target cell" by the nucleus of a reversibly permeabilized cell (referred to as the "donor cell" *[8,9]*). Reprogramming extracts are prepared by sonication of a target cell type and sedimentation of the coarse material. The supernatant, or extract, provides nuclear regulatory factors that mediate alterations in the gene expression profile of the donor cell and cytoplasmic components required to promote import of the nuclear factors into the donor cell nucleus *(9)*. At the end of incubation in the extract, the donor cell is resealed and cultured to assess expression of target cell-specific markers and the establishment of target cell-specific functions. Additionally, for molecular assessments of nuclear reprogramming, purified nuclei, as opposed to cells, can be treated with the extract.

Using this system, we have reported a partial reprogramming of 293T cells with a Jurkat T-cell extract to take on properties of T cells. Reprogramming was illustrated by induction of 293T-cell chromatin remodeling (hyperacetylation of histone H4 at the *IL2* promoter), induction of expression of an array of hematopoietic cell-specific genes, downregulation or repression of genes expressed in fibroblasts, expression of T cell-specific surface antigens, and the establishment of T-cell regulatory functions *(8)*. The new phenotypes were shown to persist for several months in culture, proving that some reprogramming of nuclear function had taken place. More recently, extracts of rat cardiomyocytes were shown to elicit the differentiation of human adipose tissue stem cells (ATSCs) toward a cardiomyocyte phenotype *(10)*. Because cell extracts can be manipulated easily, they constitute a powerful tool for investigating the mechanisms of nuclear reprogramming, at least as they take place in vitro.

This chapter describes procedures required to direct the program of a model human cell line, human adult stem cells, and primary rat fibroblasts towards defined target cell types. The methods describe (1) the preparation of cells to be reprogrammed or differentiated (the "donor" cells), (2) the preparation of nuclear and cytoplasmic extracts, (3) the permeabilization of the donor cells, (4) the setup of the reprogramming reaction, (5) the resealing of cells, (6) examples of assessment of cell reprogramming and differentiation, (7) the purification of somatic cell nuclei, and (8) the handling of purified nuclei in extracts for use in, for instance, chromatin remodeling experiments. The procedures are described with a focus on several cell types, including, as donor cells, 293T cells, rat fibroblasts, and ATSCs, and as source of extract, INS-1E cells, Jurkat cells, and primary rat cardiomyocytes. Notwithstanding minor adjustments, the procedures can in principle be applied to many other donor and target cell types.

2. Materials
2.1. In Vitro Reprogramming or Differentiation of Cells

1. Cells: donor cell type to be used is the investigator's choice. We have used 293T cells, NIH3T3 cells, cultured primary rat fibroblasts, and ATSCs. Procedures described in this chapter are essentially the same for each cell type.
2. Poly-L-lysine (cat. no. P8920; Sigma-Aldrich Co.; St. Louis, MO).
3. Phosphate-buffered saline (PBS).
4. Antibiotics: penicillin–streptomycin mix (100X solution; cat. no. 15140-122; Gibco-BRL; Paisley, UK).
5. 200 mM L-Glutamine (cat. no. 25030-024; Gibco-BRL).
6. Nonessential amino acids (NEAA, 100X solution, cat. no. 11140-036M; Gibco-BRL).
7. 100 mM Sodium pyruvate (cat. no. 11360-039; Gibco-BRL).
8. RPMI-1640 medium (cat. no. R0883; Sigma).
9. Dulbecco's modified Eagle's medium (cat. no. D6546; Sigma).
10. Hanks' balanced salt solution (HBSS; cat. no. 14170-088; Gibco-BRL).
11. Protease inhibitor cocktail (cat. no. P2714; Sigma). This is a 100X stock solution.

12. Cell lysis buffer: 50 mM NaCl, 5 mM MgCl$_2$, 20 mM HEPES, pH 8.2, 1 mM dithiothreitol (DTT), 0.1 mM phenylmethylsulfonyl fluoride, and protease inhibitor cocktail. Prepare fresh and maintain on ice until use.
13. Streptolysin O (SLO; cat. no. S5265; Sigma): 100 µg/mL in sterile-filtered H$_2$O, dispensed in 10-µL aliquots and frozen at –20°C.
14. 1 M CaCl$_2$ (cat. no. C4901; Sigma) in sterile H$_2$O.
15. ATP (cat. no. A3377; Sigma): 200 mM in H$_2$O, aliquoted and stored at –20°C.
16. Creatine kinase (cat. no. C3755; Sigma): 5 mg/mL in H$_2$O, aliquoted and stored at –20°C.
17. Phosphocreatine (cat. no. P7936; Sigma): 2 M in H$_2$O, aliquoted and stored at –20°C.
18. GTP (cat. no. G8752; Sigma): 10 mM in H$_2$O, aliquoted and stored at –20°C.
19. Nucleotide triphosphate (NTP) set (cat. no. 1277057; Roche; Basel, Switzerland). Prepare a stock solution by mixing 20 µL of each NTP in the set at a 1/1/1/1 ratio on ice. Aliquot in 10 µL and store at –20°C. This makes an NTP mix at 25 mM of each NTP. Prepare more stock solution as needed.
20. Texas Red-conjugated 70,000 M_r dextran (cat. no. D1830; Sigma).
21. Pulse sonicator fitted with a round 2-mm diameter probe (Model Labsonic M, B. Braun Biotech International, Melsungen, Germany).
22. Cell incubator set at 100% humidity, 37°C and 5% CO$_2$ in air.
23. 50-mL and 15-mL plastic conical tubes (Corning, Corning, NY).
24. 1.5-mL centrifuge tubes.
25. 48-well cell culture plates (Corning).

2.2. Chromatin Remodeling Experiments

In addition to the aforementioned materials, the following are needed to prepare somatic cell nuclei and handle them in cell extracts. We have worked with nuclei from the PC12 cell line, derived from a pheochromocytoma of rat adrenal medulla. However, the procedures are similar for nuclei of other cell lines (nevertheless, *see* **Note 1** for variations in the method for isolating nuclei).

1. PC12 cells, grown in RPMI-1640 containing 10% fetal calf serum (FCS; source and lot number should be tested by the investigator), 5% horse serum (source and lot number should be tested by the investigator), 2 mM L-glutamine, 1 mM sodium pyruvate, 1/100 dilution of NEAA and 1% antibiotics.
2. Glass homogenizer (2 mL) with tight-fitting pestle type B (Kontes; cat. no. 885300-0002; Vineland, NJ).
3. Glycerol (cat. no. G789-3; Sigma).
4. Buffer N: 250 mM sucrose; 10 mM HEPES, pH 7.5, 2 mM MgCl$_2$, 25 mM KCl, 1 mM DTT, protease inhibitor cocktail. Prepare fresh and maintain on ice until use. Add DTT and protease inhibitors immediately before use.
5. 2 M sucrose solution made in H$_2$O. This can be stored frozen.
6. Freezing medium: glycerol diluted to 70% in buffer N.
7. 1 M Sucrose solution made in buffer N.

3. Methods

3.1. Seeding Donor Cells

On the day before the reprogramming reaction, plate donor cells in their regular culture medium. 293T cells are cultured in RPMI-1640 supplemented with 10% FCS, 2 mM L-glutamine, 1 mM sodium pyruvate, 1/100 dilution of NEAA, and 1% antibiotics. Rat and mouse fibroblasts and ATSCs are cultured in Dulbecco's modified Eagle's medium supplemented as described previously. Cells should be seeded to reach a confluency of approx 50% on day of use for reprogramming (293T cells, fibroblasts) or differentiation (ATSCs). Because doubling time varies between cell types, seeding concentration also is dependent on cell type.

3.2. Preparation of the Reprogramming Extract

Unless primary cells freshly isolated from a given organ (e.g., cardio-myocytes *[10]*) are used, we prepare extracts from cells in exponential growth phase to benefit from maximum transcriptional activity. Cells are harvested as per standard procedure. If trypsin is used, it should immediately be inactivated by addition of 10 volumes of complete culture medium to the cell suspension. Cells growing in suspension can be immediately processed as described in **Subheading 3.2.1., step 1**.

3.2.1. Cell Collection

1. Transfer the cell suspension culture or the trypsinized cells into 50-mL conical tubes and sediment the cells at 800g for 10 min at 4°C.
2. Wash the cells twice in ice-cold PBS by suspension and sedimentation at 800g for 10 min at 4°C. The cells can be pooled into a single tube after the first wash in PBS.

3.2.2. Swelling of the Cells

1. Resuspend the cells in 10 mL of ice-cold cell lysis buffer. It is preferable to use a graduated 15-mL conical tube to evaluate the volume of the cell pellet after sedi-mentation.
2. Centrifuge at 800g for 10 min at 4°C.
3. Estimate cell pellet volume. Resuspend cells into, as a rule, one volume of ice-cold cell lysis buffer. For extracts of rat cardiomyocytes cells are resuspended in 1.5 volumes of cell lysis buffer.
4. Hold cells on ice for 45 min to allow swelling. Swelling facilitates cell lysis during sonication. Keep cells suspended by occasional tapping of the tube.

3.2.3. Extract Preparation

The procedure described to prepare nuclear and cytoplasmic extracts is similar for any cell type we have used. Only adjustments in sonication time and power might be necessary (*see* **step 1**).

1. Aliquot the suspension of swollen cells into 200 µL in 1.5-mL centrifuge tubes chilled on ice. Sonicate each tube on ice until all cells and nuclei are lysed. Lysis is assessed by complete disruption of cells and nuclei as judged by phase contrast microscopy. Once lysis is achieved, keep the tube on ice and proceed with the other tubes. The power and duration of sonication vary with each cell type. For Jurkat cells, sonication at 25% amplitude and 0.5-s pulse cycle over the course of 1 min, 40 s is recommended when using the Labsonic M sonicator. For INS-1E cells, we use 40% power, 0.5-s pulses over the course of 1 min, 30 s. Occasionally, it may be necessary to sonicate for another 20 to 30 s at 50 to 70% amplitude. For rat cardiomyocytes, we use 80% amplitude, 0.7-s pulses over the course of 1 min, 30 s. *See* **Note 2**.
2. Pool all cell lysates into one or more prechilled 1.5-mL centrifuge tubes. Sediment the lysate at 15,000g for 15 min at 4°C in a fixed-angled or swing-out rotor.
3. Carefully collect the supernatant (the extract) with a 200-µL pipet and transfer into a new 1.5-mL tube chilled on ice. The pellet is discarded.
4. Aliquot the extract into 200-µL tubes such as those used for polymerase chain reaction, with 100 µL of extract per tube. Snap-freeze each tube in liquid N$_2$ and store at –80°C. We recommend, however, carrying out reprogramming reactions with freshly made extract because the stability of the extract at –80°C may vary with cell types and batches. This variability has not been examined thoroughly to date and needs to be evaluated by the investigator.
5. After sedimentation in **step 2**, remove 20 µL of extract to determine protein concentration and pH. Protein concentration should be approx 30 mg/mL. pH should be between 6.7 and 7.0 (*see* **Note 3**). Note, however, that extracts prepared from rat cardiomyocytes are more diluted (5–7 mg/mL protein) and of higher pH (7.7) than extracts of Jurkat or INS-1E cells.

3.2.4. Assessment of Extract Toxicity

Each new batch of extract requires a cell toxicity test. The assay is based on the microscopic evaluation of cell lysis after incubation of intact cells in the extract.

1. Add 50,000 293T cells (or the cell type used for reprogramming or differentiation) to 30 µL of extract on ice in a 1.5-mL centrifuge tube. The extract does not need to contain any additives (unlike for a reprogramming or differentiation reaction).
2. Incubate for 1 h at 37°C in a water bath. Maintain the cells in suspension by occasionally tapping the tube.
3. Remove a 3-µL aliquot and assess cell morphology by phase contrast microscopy. **Figure 2** illustrates rat fetal fibroblasts after a 30-min exposure to INS-1E cell

A Reprogramming Extract (pass) **B** Apoptotic Extract (fail) **C** Cell lysis buffer

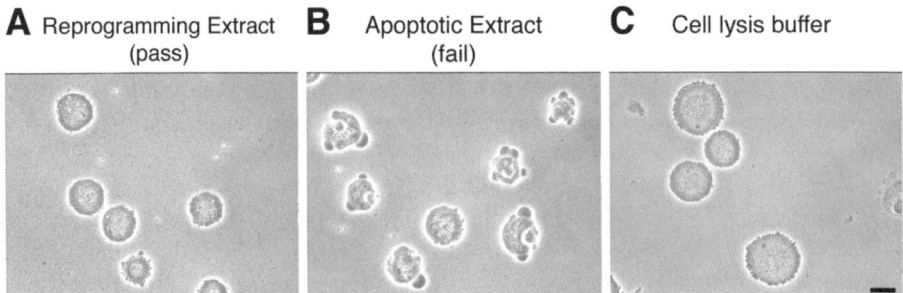

Fig. 2. Assessment of extract toxicity. Intact cells (here, HeLa) are incubated in each new batch of extract for 30 to 60 min. An aliquot is removed to a slide and examined by phase contrast microscopy. Cells shown in (**A**) survived exposure to the extract (here, INS-1E extract), whereas cells in (**B**) did not survive extract exposure and will die in subsequent culture. Batches of extract giving rise to such cells should be discarded. Cells shown in (**C**) were exposed to cell lysis buffer alone. Bar, 20 μm.

extracts. Cell morphology immediately after exposure to the extract reflects survival in culture. Cells shown in **Fig. 2A** survived exposure to extract whereas cells in **Fig. 2B** were exposed to an extract causing cell death and did not survive in culture. Batches of extract producing such cells should be discarded. Cells in **Fig. 2C** were exposed to cell lysis buffer and show swelling expected from a hypotonic buffer.

4. If desired, replate the cells directly from the extract in complete culture medium for an overnight evaluation of cell viability. There is no need to remove the extract before replating.

3.3. Permeabilization of Donor Cells

Although some cell types, such as ATSCs, endocytose readily (*10*), most cell types require efficient permeabilization for components from the extract to be taken up. Permeabilization should be reversible as cells are expected to be cultured after exposure to the extract. Permeabilization is accomplished with the *Streptococcus pyogenes* toxin, SLO, a cholesterol-binding toxin that forms large pores in the plasma membrane of mammalian cells (*11*).

3.3.1. SLO Stock Solution

1. On ice, dissolve the SLO powder in sterile-filtered MilliQ H_2O to 100 μg/mL.
2. Aliquot 10 μL in 200-μL tubes on ice and store at –20°C.
3. Discard all tubes after one month of storage at –20°C and prepare a new stock of SLO. Stock aliquots should be thawed only once. More details on SLO are given in **Note 4**.

3.3.2. Reversible Donor Cell Permeabilization

1. Dilute the SLO stock 1/100 in ice-cold HBSS. This constitutes a working solution that will provide a final SLO concentration of 230 ng/mL (**step 6**). We use this concentration for 293T cells, rat and mouse fibroblasts, and ATSCs. For other cell types, the concentration may be adjusted using the cell permeabilization assay described in **Subheading 3.3.3**. Adjustments should be made to achieve the highest efficiencies of permeabilization and subsequent resealing.
2. Keep the SLO working solution on ice until adding it to the cells.
3. Harvest donor cells from culture as per standard procedures. Wash cells twice in ice-cold PBS and once in ice-cold cell lysis buffer, each by sedimentation at 400g for 10 min at 4°C. This is essential to remove all Ca^{2+} from the culture medium as Ca^{2+} inhibits SLO activity.
4. Resuspend the cell pellets in aliquots of 20,000 cells in 100 µL of HBSS in 1.5-mL tubes placed on ice, centrifuge at 400g for 5 min at 4°C, and discard the supernatant. Note that as proportionally augmenting reaction volumes and cell numbers has to date not produced successful results (for unclear reasons), we recommend using as many 20,000-cell samples as necessary.
5. Carefully resuspend the cells in 16.4 µL of ice-cold HBSS using a 200-µL pipet whose tip has been cut with clean, sterile scissors.
6. Place the tubes in a water bath at 37°C for 2 min and add 4.6 µL of the SLO working solution prepared in **step 1**.
7. Incubate the cells at 37°C for another 50 min; maintain cells in suspension by occasional tapping of the tube.
8. Place the tubes on ice and add 200 µL of ice-cold HBSS.
9. Sediment the cells at 400g for 3 min at 4°C using a swing-out rotor. Remove the supernatant and proceed to **Subheading 3.4.2**. At this stage, the use of a fixed-angle rotor damages the cells as they streak along the tube wall during centrifugation.

3.3.3. Cell Permeabilization Assay (see **Note 5**)

The assay allows the evaluation of the efficiency of SLO treatment. It is based on the uptake of a Texas Red- (or Alexia Red)-conjugated 70,000 M_r dextran by the permeabilized cells (**Fig. 3**).

1. Harvest the donor cells from culture as per standard procedure. Wash the cells twice in ice-cold PBS and once in ice-cold cell lysis buffer, each by suspension and sedimentation at 400g for 10 min at 4°C.
2. Place the resuspended cell pellets in aliquots of 20,000 cells in 100 µL of HBSS in 1.5-mL tubes on ice, centrifuge at 400g for 5 min at 4°C, and discard the supernatant.
3. Resuspend the cells in 16.4 µL of ice-cold HBSS containing 50 µg/mL Texas Red-conjugated dextran.
4. Place the tubes in a water bath at 37°C for 2 min and add 4.6 µL of SLO working solution prepared as described in **Subheading 3.3.2., step 1**. Mix by gentle pipetting.

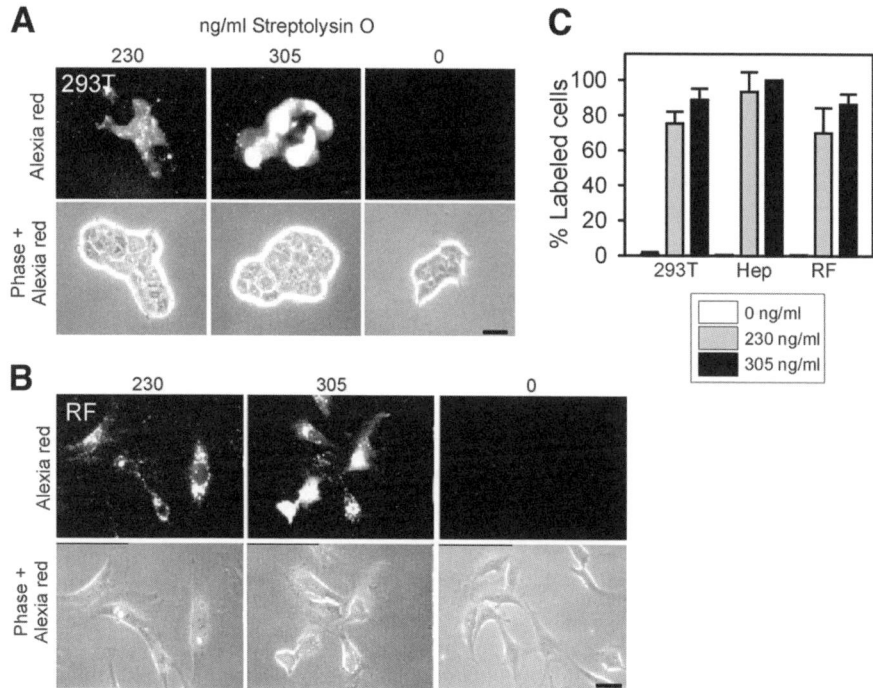

Fig. 3. Reversible cell permeabilization. (A) 293T cells and (B) primary RFs were incubated for 50 min in 0, 230, or 305 ng/mL SLO containing 50 µg/mL Alexia Red-conjugated dextran (here, a 10,000 M_r dextran was used, but similar results are obtained with a 70,000 M_r dextran). Cells were resealed with 2 mM CaCl$_2$ and cultured for approx 24 h before observation by phase contrast and epifluorescence microscopy. Bars, 20 µm. (C) Proportions (±SD) of 293T cells, RFs and rat hepatoma cells (Hep) displaying Alexia Red fluorescence 24 h after exposure to SLO and dye. Results from three experiments (>200 cells/treatment).

5. Incubate at 37°C for 50 min and maintain the cells in suspension by occasional tapping of the tube.
6. Place the tubes on ice and add 200 µL of ice-cold HBSS.
7. Sediment the cells at 400g for 3 min at 4°C using a swing-out rotor and add 200 µL of complete culture medium (use the medium recommended for the cell type being used) containing 2 mM CaCl$_2$ added from the 1 M stock (*see* **Subheading 2.1.**).
8. Transfer the cell suspension from one tube into one well of a 48-well culture plate and culture for 2 to 4 h to allow resealing and replating of the living cells. Replating of the cell can take place on a glass cover slip placed in the well if necessary.
9. Remove nonplated (dead) cells by careful aspiration and add fresh complete culture medium.

10. Observe cells by epifluorescence microscopy. Cells displaying Texas Red fluorescence have been permeabilized, have taken up the dye, were resealed and reseeded successfully. When dose-titrating a new batch of SLO, or when testing a new cell type, we recommend carrying out a second observation approx 24 h after resealing to confirm cell survival. **Figure 3** displays 293T cells (**Fig. 3A**) and primary rat fibroblasts (RFs; **Fig. 3B**) 24 h after exposure to two SLO concentrations (230 and 305 ng/mL) containing an Alexia Red-conjugated dextran. Depending on the cell type, 70 to 90% of replated cells display dye uptake (**Fig. 3C**). Note that in our hands, cell lysis after SLO treatment averages approx 50%. These cells do not replate and thus are not represented in **Fig. 3**.

3.4. In Vitro Reprogramming/Differentiation Reaction

3.4.1. Preparation of Extract

During the SLO treatment, the extract should be prepared for the reprogramming or differentiation reaction. The extract should be ready to use and contain all ingredients at the time the SLO treatment described in **Subheading 3.3.2.**, **step 8** is completed. (*See* **Note 6**.)

1. Prepare the ATP-regenerating system: mix on ice ATP:GTP:creatine kinase:phosphocreatine in a 1/1/1/1 ratio from each separate stock (*see* **Subheading 2.1.**) and keep on ice.
2. Aliquot the freshly prepared extract into as many 20-µL samples as needed and hold on ice. If the extract is frozen, promptly thaw the extract between the fingers, place on ice and prepare 20-µL aliquots.
3. Add 1 µL of the ATP-regenerating system mix to 20 µL of extract on ice.
4. Add 0.75 µL of the 25 mM NTP mix (*see* **Subheading 2.1.**) to 20 µL of extract on ice.
5. Vortex briefly and replace the extract on ice.

3.4.2. Reprogramming/Differentiation Reaction

These steps are to be taken only after the donor cells have been permeabilized with SLO (**Subheading 3.3.2.**, **step 8**) and once the extract is ready for use (**Subheading 3.4.1.**, **step 5**).

1. To each tube containing sedimented permeabilized donor cells, add 20 µL of extract and gently suspend the cells by tapping of the tube.
2. Incubate the tubes in a water bath at 37°C for 1 h. Maintain the cells in suspension by occasional tapping.

3.5. Resealing the Cells

1. At the end of incubation, to each tube add 0.5 mL of preheated (37°C) complete culture medium (needed for the target cell type) containing 2 mM CaCl$_2$ added from the 1 M stock. The extract does not need to be removed before adding the Ca^{2+}-containing medium.

2. Transfer the contents of three to five tubes into one well of a 48-well culture plate.
3. Culture for 2 to 4 h in a 5% CO_2 incubator at 37°C.
4. Remove dead (floating) cells and the Ca^{2+}-containing medium by gentle aspiration and replace with 250 µL of complete culture medium (without added $CaCl_2$).
5. Place the cells back into the 5% CO_2 incubator and culture until they are ised to assess the reprogramming.

3.6. Assessment of Cell Reprogramming

Various assessments of nuclear and cell reprogramming can be performed, according to the purpose of the experiment, donor, and target cell types. Using the methods described here, we have reported changes in the gene expression profile of reprogrammed 293T cells, using complementary deoxyribonucleic acid arrays from R&D Systems (Abington, UK *[8]*) (*see* **Note 7**). Expression of new proteins can be monitored at regular intervals after the reprogramming or differentiation reaction by immunofluorescence or flow cytometry. We have shown expression of several antigens that are specific to hematopoietic cells on the surface of the reprogrammed cells *(8)*. A variety of functional assays can also be performed, such as cytokine secretion in response to stimulation of pathways downstream of the T-cell receptor/CD3 complex in reprogrammed cells or expression of additional cytokine receptors on the cell surface *(8)*. We also have reported the expression of cardiomyocyte-specific proteins in ATSCs induced to differentiate by exposure to rat cardiomyocyte extracts *(10)*.

3.7. Isolation of Somatic Nuclei for Chromatin Remodeling Assays

This section describes procedures for isolating nuclei from a variety of somatic cell lines. Variations from the methods reported below are indicated in **Note 1**. Similar procedures for purifying intact, membrane-enclosed nuclei from cultured cells have been reported earlier *(9,12)*.

3.7.1. Purification of Nuclei From Somatic Cells

1. Harvest the cells from culture and wash in PBS as per standard protocol.
2. Sediment the cells in a 50-mL conical tube at 400*g* for 10 min at 4°C.
3. Discard the PBS and resuspend the cells in 20 mL of ice-cold hypotonic buffer N (buffer N made without sucrose) and centrifuge at 400*g* for 10 min at 4°C.
4. Discard the buffer and resuspend the cells in 10 to 20 volumes ice-cold hypotonic buffer N.
5. Allow the cells to swell on ice for 10 to 60 min depending on cell type (10 min is sufficient for 293T cells; >30 min is recommended for other cell types).
6. Lyse the cells using the glass Dounce homogenizer recommended in **Subheading 2.1.** Place the mortar on ice while homogenizing. 293T cells usually require 10 to 20 strokes, HeLa cells, >100 strokes, and PC12 cells approx 50 strokes.

7. Monitor cell lysis under a phase contrast microscope using a 40× objective. In contrast to intact cells, isolated nuclei are highly refringent with a clearly defined nuclear envelope. All cells should be lysed, but nuclei should remain intact. Keep homogenizing carefully if not all the nuclei are isolated.

8. When cell lysis is achieved and nuclei are released from the cytoplasm, transfer the contents of the homogenizer to a 15-mL conical tube placed on ice.

9. Add 125 μL of the 2 M sucrose solution (**Subheading 2.1.**) per milliliter of lysate. Mix well by inversion and place the tube on ice.

10. Sediment the lysate at 400g in 15-mL conical tubes for 15 to 20 min at 4°C in a swing-out rotor. A fixed angle rotor will cause nuclei to streak along the tube wall and lyse.

11. Resuspend the pellet of nuclei into 10 mL of ice-cold buffer N and sediment as in **step 10**.

12. Decant the supernatant and resuspend the nuclei in ice-cold buffer N at 10^7 nuclei/mL and keep on ice. Alternatively, resuspend the nuclei in freezing medium (*see* **Subheading 2.2.**).

3.7.2. Freezing and Thawing of Nuclei

1. Resuspend the nuclei obtained in **Subheading 3.7.1.**, **step 11**, into the desired volume of freezing buffer to obtain a concentration of 10^6 nuclei/mL. Replace the tube on ice as soon as possible.

2. Promptly prepare 500-μL aliquots of the nuclear suspension into 1.5-mL tubes. This will give 5×10^5 nuclei per aliquot, which is more than sufficient for most of our applications.

3. Immediately freeze the aliquots in a methanol/dry ice bath and store at –80°C. Freezing nuclei in liquid nitrogen should be avoided.

4. To thaw an aliquot of nuclei, remove one tube of frozen nuclei and thaw at room temperature. Place the tube on ice.

5. Promptly dilute with 1 mL of ice-cold buffer N, mixing well but gently with a 1-mL pipet tip.

6. Centrifuge at 1000g for 15 min at 4°C using a swing-out rotor.

7. Discard the supernatant and resuspend the pelleted nuclei in 500 μL of ice-cold buffer N.

8. Centrifuge as in **step 6** of this subheading. Resuspend the nuclei in ice-cold buffer N to 10^7 nuclei per milliliter and hold on ice until use.

3.8. Exposure of Purified Nuclei to Extract (see Fig. 4)

1. Prepare a reprogramming extract exactly as described for incubation of permeabilized cells in **Subheading 3.4.1.**

2. Using a 10-μL pipet cut at the tip, add no more than 1/10 volume of the nuclei suspension to the reprogramming extract, to reach a final concentration of 5000 nuclei/μL of extract. Gently suspend the nuclei by pipetting. Note that for chromatin immunoprecipitation experiments, we use 50,000 nuclei/μL of extract.

3. Incubate for 1 to 2 h in a 37°C water bath. Maintain the nuclei in suspension by occasionally tapping the tube.

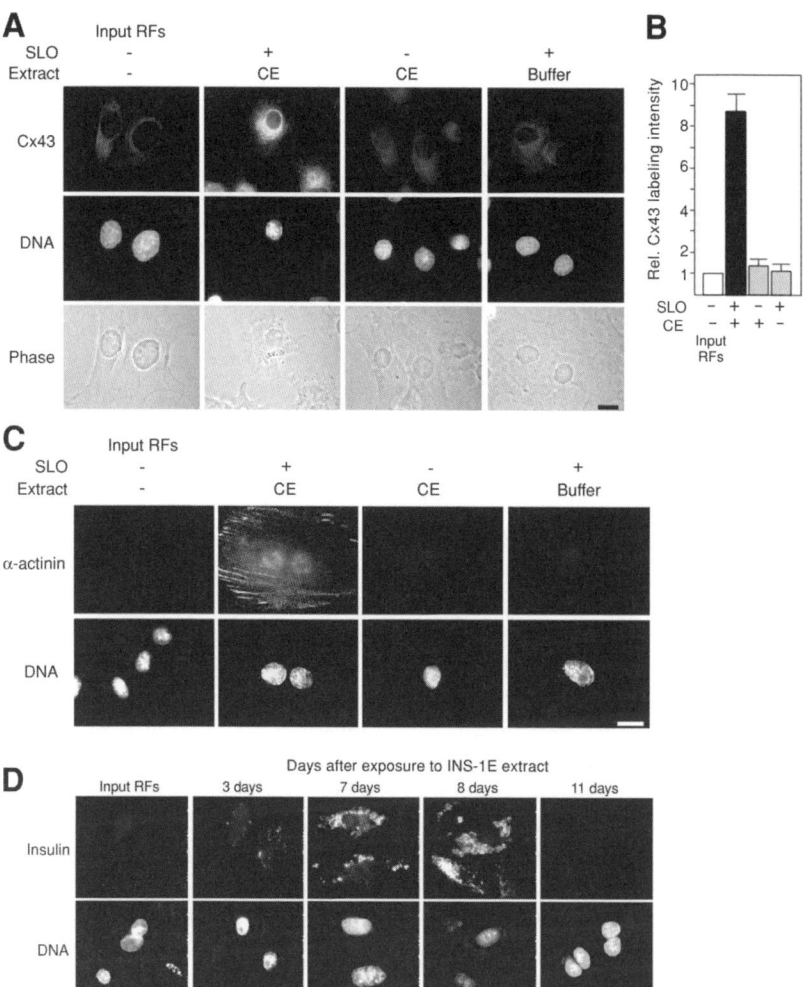

Fig. 4. Expression of proteins specific for target cell types as a result of exposure to extracts. **(A)** RFs, either intact (SLO–) or permeabilized (SLO+), were exposed for 1 h to a rat cardiomyocyte extract (CE) or PBS (buffer), resealed, cultured, and examined for overexpression of connexin43 (Cx43), a gap junction protein. DNA was counterstained with 10 μg/mL Hoechst 33342 (Sigma). **(B)** Relative anti-Cx43 immunolabeling intensity in RFs exposed to cardiomyocyte extract and in control cells treated as in (A). Note the induction of overexpression of Cx43 after exposure to extract. **(C)** Immunofluorescence analysis of expression of cardiac sarcomeric α-actinin in RFs exposed to cardiomyocyte extract and controls as described in (A). Cells in (A–C) were analyzed for as long as 3 months after incubation in extract. **(D)** RFs were exposed to an extract of the INS-1E rat insulinoma cell line, resealed and cultured. Onset of detection of intracellular (pro)insulin was monitored by immunofluorescence analysis. (A–D) Bars, 10 μm.

4. At the end of the incubation, dilute the extract 10-fold with hypotonic buffer N and sediment the nuclei at 110g for 5 min at 4°C using a swing-out rotor.
5. Resuspended the nuclei in a desired volume of ice-cold buffer N. At this stage, the nuclei are ready for analysis. Nevertheless, should contamination with extract remnants be judged excessive (as evaluated by phase contrast microscopy), the nuclei can at this stage be resuspended in buffer N and sedimented at 1000g. Recovered nuclei can for example be sedimented onto cover slips for immunofluorescence analysis. For chromatin immunoprecipitation experiments *(13)*, nuclei are recovered from the extract by sedimentation at 1000g for 10 min at 4°C through an equal volume of a 1 M sucrose cushion made in buffer N.

4. Notes

1. Nuclei can be purified from lymphoid cell lines (such as Jurkat, Bjab, Reh) and from peripheral blood T cells simply by suspending the cells in PBS containing 0.5% NP-40 (cat. no. N6507; Sigma) for 5 min at room temperature, sedimenting the nuclei at 1000g for 10 min at 4°C in a swing-out rotor and washing the nuclei in buffer N. No Dounce homogenization is required for these cell types.
2. It is currently difficult to objectively assess the extent of sonication of any of the cell type we have used. It is important to sonicate until all cells and nuclei are completely lysed. This results in the appearance of "debris" under a phase contrast microscope. The effect of sonication time and power on reprogramming or differentiation efficiency has not been tested; thus, whether extended sonication after cell lysis is complete is detrimental or beneficial is, at present, unknown.
3. pH of the extract: we usually observe a drop of 1 to 1.5 pH units upon extract preparation, which explains the pH 8.2 of the cell lysis buffer. Notably, raising the pH of the cell lysis buffer to 8.7 with a HEPES buffer does not increase the pH of the final extract. Other buffers with greater buffering capacity have not been tested. The pH of primary rat cardiomyocyte extracts is higher (pH 7.7) because of the greater dilution of the cells with pH 8.2 cell lysis buffer before lysis *(10)*.
4. Commercially available SLO batches may vary in specific activity. Efficiency of SLO-mediated permeabilization also varies between cell types. Thus, it is recommended to test a range of SLO concentrations (100–1000 ng/mL) on a given cell type or after preparing a new batch of SLO prior to initiating reprogramming/differentiation reactions. Optimum concentration should thereafter be fine-tuned. Other laboratories have investigated SLO-mediated cell permeabilization in detail *(14,15)*.
5. It is recommended that one conduct a cell permeabilization assay in parallel to each reprogramming or differentiation reaction as daily variations of permeabilization efficiency can be seen.
6. Variability in batches of extract exists, even among extracts that have been rated as "nontoxic" in the toxicity assay described here. Variability is evident by the absence of target cell type-specific markers of reprogramming or differentiation.

7. With current techniques, expression of a reprogrammed phenotype occurs for at least two months for 293T cells reprogramming in Jurkat extract. The reprogrammed phenotype may also last for shorter periods depending on the cell type reprogrammed (primary as opposed to transformed), the target cell type or the markers analyzed.

Acknowledgments

Our work is supported by the Research Council of Norway, the Norwegian Cancer Society Nucleotech LLC, and the University of Oslo.

References

1. Collas, P. and Håkelien, A. M. (2003) Teaching cells new tricks. *Trends Biotechnol.* **21**, 354–361.
2. Cibelli, J. B., Stice, S. L., Golueke, P. J., Kane, J. J., Jerry, J., Blackwell, C., et al. (1998) Cloned transgenic calves produced from nonquiescent fetal fibroblasts. *Science* **280**, 1256–1258.
3. Gurdon, J. B., Laskey, R. A., De Robertis, E. M., and Partington, G. A. (1979) Reprogramming of transplanted nuclei in Amphibia. *Int. Rev. Cytol. Suppl.* **9**, 161–178.
4. Munsie, M. J., Michalska, A. E., O'Brien, C. M., Trounson, A. O., Pera, M. F., and Mountford, P. S. (2000) Isolation of pluripotent embryonic stem cells from reprogrammed adult mouse somatic cell nuclei. *Curr. Biol.* **10**, 989–992.
5. Wilmut, I., Schnieke, A. E., McWhir, J., Kind, A. J., and Campbell, K. H. S. (1997) Viable offspring derived from fetal and adult mammalian cells. *Nature* **385**, 810–813.
6. Blau, H. M. and Blakely, B. T. (1999) Plasticity of cell fate: insights from heterokaryons. *Semin. Cell Diff.* **10**, 267–272.
7. Tada, M., Takahama, Y., Abe, K., Nakastuji, N., and Tada, T. (2001) Nuclear reprogramming of somatic cells by in vitro hybridization with ES cells. *Curr. Biol.* **11**, 1553–1558.
8. Håkelien, A. M., Landsverk, H. B., Robl, J. M., Skålhegg, B. S., and Collas, P. (2002) Reprogramming fibroblasts to express T-cell functions using cell extracts. *Nat. Biotechnol.* **20**, 460–466.
9. Landsverk, H. B., Håkelien, A. M., Küntziger, T., Robl, J. M., Skålhegg, B. S., and Collas, P. (2002) Reprogrammed gene expression in a somatic cell-free extract. *EMBO Rep.* **3**, 384–389.
10. Gaustad, K. G, Boquest, A. C., Anderson, B. E., Gerdes, A. M., and Collas, P. (2004) Differentiation of human adipose tissue stem cells using extracts of rat cardiomyocytes. *Biochem. Biophys. Res. Commun.* **314**, 420–427.
11. Walev, I., Hombach, M., Bobkiewicz, W., Fenske, D., Bhakdi, S., and Husmann, M. (2002) Resealing of large transmembrane pores produced by streptolysin O in nucleated cells is accompanied by NF-kappaB activation and downstream events. *FASEB J.* **16**, 237–239.

12. Collas, P., Le Guellec, K., and Tasken, K. (1999) The A-kinase anchoring protein, AKAP95, is a multivalent protein with a key role in chromatin condensation at mitosis. *J. Cell Biol.* **147,** 1167–1180.

13. O'Neill, L. P. and Turner, B. M. (1996) Immunoprecipitation of chromatin. *Methods Enzymol.* **274,** 189–197.

14. Walev, I., Bhakdi, S. C., Hofmann, F., Djonder, N., Valeva, A., Aktories, K., et al. (2001) Delivery of proteins into living cells by reversible membrane permeabilization with streptolysin-O. *Proc. Natl. Acad. Sci. USA* **98,** 3185–3190.

15. Weller, U., Muller, L., Messner, M., Palmer, M., Valeva, A., Tranum-Jensen, J., et al. (1996) Expression of active streptolysin O in *Escherichia coli* as a maltose-binding-protein-streptolysin-O fusion protein. The N-terminal 70 amino acids are not required for hemolytic activity. *Eur. J. Biochem.* **236,** 34–39.

10

Transgenic Systems in Nuclear Reprogramming

Megan Munsie, Peter Mountford, and Jennifer Nichols

Summary

Transgenic reporters have proved to be invaluable in the study of nuclear reprogramming, from demonstrating revival or silencing of gene expression in fusion hybrids to providing a means to display levels and distribution of specific gene products after nuclear transfer. Here, the method of piezo-assisted direct injection, which has been used previously to generate blastocysts and subsequently embryonic stem cell lines by transfer of nuclei from transgenic reporter mice, is described. This protocol differs from previously described techniques in that the donor nucleus is placed in the recipient oocyte before removal of the host metaphase plate.

Key Words: *LacZ*; β*geo*; GFP; nuclear transfer; transgene; piezo; blastocyst; reactivation.

1. Introduction

Transgenic systems have been used very successfully to advance our understanding of the mechanisms involved in the process of nuclear reprogramming. In nuclear transfer experiments, a random gene trap line, ZIN40, that ubiquitously expresses nuclear *lacZ (1)*, has been used effectively to prove that reprogrammed somatic nuclei can support conversion into pluripotent embryonic stem (ES) cells *(2)*. In addition, re-expression of developmentally regulated genes has been demonstrated using transgenes incorporating the *lacZ* reporter *(3)*. Reawakening of pluripotent marker genes in thymocytes fused with ES cells was observed by means of an Oct4-GFP transgene carried by the mice from which the thymocytes were purified *(4)*. Transgenic reporters also have been invaluable in demonstrating that apparent transdifferentiation of cells from one lineage to another may actually be explained by spontaneous cell fusion and subsequent reprogramming of certain genes in the hybrid cell thus formed *(5,6)*. The ingenious use of an X-linked fluorescent transgenic reporter incorporating a selectable marker has contributed to our understanding of the process of reactivation and silencing of the chosen X chromosome and the

From: *Methods in Molecular Biology, vol. 325: Nuclear Reprogramming: Methods and Protocols*
Edited by: S. Pells © Humana Press Inc., Totowa, NJ

Table 1
Composition of Embryo Culture Media

Component	Concentration in various media			
	mCZB	Sr-mCZB	H-mCZB[a]	G2
NaCl (mM)	81.62	81.62	81.62	85.16
KCl (mM)	4.83	4.83	4.83	5.50
CaCl$_2$·2H$_2$O (mM)	1.28	None	1.28	1.80
NaH$_2$PO$_4$·2H$_2$O (mM)	None	None	None	0.50
MgSO$_4$·7H$_2$O (mM)	1.18	1.18	1.18	1.00
KH$_2$PO$_4$ (mM)	1.18	1.18	1.18	None
EDTA·2Na (mM)	0.11	None	0.11	None
NaHCO$_3$ (mM)	25.12	25.12	5.00	25.00
HEPES.Na (basic) (mM)	None	None	20.00	None
Sodium strontium (mM)	None	10.00	None	None
Sodium lactate (60% syrup, mM)	23.00	23.00	23.00	11.74
Sodium pyruvate (mM)	0.27	0.27	0.27	0.10
D-Glucose (mM)	5.55	5.55	5.55	3.15
L-Glutamine (mM)	1.00	1.00	1.00	1.00
Penicillin G potassium salt (g/L)	0.05	0.05	0.05	0.06
Streptomycin sulfate (g/L)	0.07	0.07	0.07	0.05
Phenol red (g/L)	0.10	0.10	0.10	0.01
Essential AA[b]	None	None	None	1X
Nonessential AA[c]	None	None	None	1X
BSA (g/L)	5.00	5.00	None	3.00
PVA (g/L)	None	None	0.1	None
Osmolarity (± 3 mosmoles)	265	271	275	264

(continued)

behavior of its imprint during nuclear reprogramming *(7)*. The importance of accurate reactivation of embryonic genes such as Oct-4 to allow establishment of the pluripotent lineage in cloned embryos has also been demonstrated using Oct4-GFP transgenic mice *(8)*.

In this chapter, we describe piezo-assisted nuclear transfer using nuclei of cumulus cells from mice carrying a ubiquitously expressed transgene, ZIN40 *(1)*. The injection step is conducted before enucleation, which we postulated could improve the efficiency of nuclear reprogramming by maximizing the exposure of the nucleus to the oocyte cytoplasm before cell division and the onset of zygotic transcription. Reactivation of the transgene of choice is visualized in this protocol by X-gal staining, indicating expression of Lac Z.

2. Materials

Table 1 lists all components and composition of the media required for this protocol. Most of the reagents are available from Sigma, BDH, or other popu-

Table 1 *(Continued)*
Composition of Embryo Culture Media

[a]pH adjusted to 7.4.

[b]Essential amino acids:

L-Arginine HCl	126.4 mg/L
L-Cystine disodium monohydrate	30.2 mg/L
L-Histidine HCl monohydrate	41.9 mg/L
L-Isoleucine	52.5 mg/L
L-Leucine	52.5 mg/L
L-Lysine HCl	73.1 mg/L
L-Methionine	14.9 mg/L
L-Phenylalanine	33.0 mg/L
L-Threonine	47.6 mg/L
L-Tryptophan	10.2 mg/L
L-Tyrosine disodium dihydrate	51.9 mg/L
L-Valine	46.9 mg/L

[c]Nonessential amino acids:

L-Alanine	8.9 mg/L
L-Asparagine monohydrate	15.0 mg/L
L-Aspartic acid	13.3 mg/L
L-Glutamic acid	14.7 mg/L
Glycine	7.5 mg/L
L-Proline	11.5 mg/L
L-Serine	10.5 mg/L

All media components were Sigma or BDH products, excluding BSA (Life Technologies) and essential and nonessential amino acids (ICN Pharmaceuticals, Costa Mesa, CA).

lar suppliers. The sources of the more specialized materials are included where appropriate.

1. mCZB (*see* **Table 1** *[9,10]*).
2. H-mCZB (*see* **Table 1**).
3. Sr-mCZB (*see* **Table 1**; add Sr solution to 10% and 2.5 µg/mL cytochalasin B immediately before use).
4. Sr solution: 100 m*M* strontium chloride in ultrapure water.
5. Cytochalasin B solution: 10 mg/mL in dimethyl sulfoxide stored frozen in aliquots.
6. PVP solution: 5% polyvinylpyrolidone in H-mCZB.
7. G2 (*see* **Table 1** *[11]*).
8. Osmometer (*see* **Note 1**).
9. D2 female mice, 4- to 12-wk-old (progeny of C57Bl/6 x DBA2 mating; *see* **Note 2**).
10. Transgenic nuclear donor females (in this case ZIN40, *see* **Fig. 1**).
11. Hormones for superovulation: pregnant mare's serum and human chorionic gonadotrophin (hCG). Both are made up to 50 U/mL in phosphate-buffered saline (PBS) and stored frozen in aliquots (*see* **Note 3**).
12. Hyaluronidase: 0.1% in H-mCZB.
13. Tissue culture-grade Petri dishes, 10 cm in diameter.

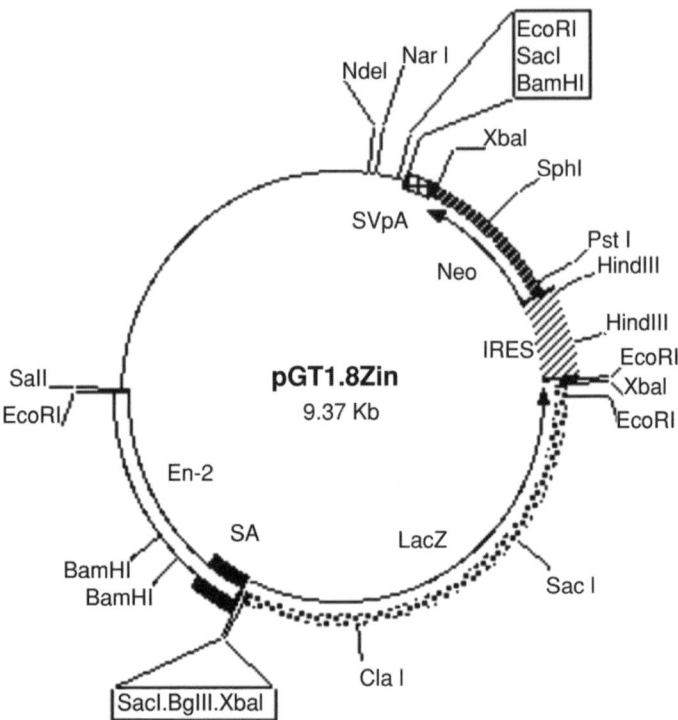

Fig. 1. Map of ZIN40 plasmid used to generate mice with ubiquitous nuclear expression of LacZ by random insertion into ES cells. Plasmid name, pGT1.8Zin; plasmid size, 9.37 kb; constructed by P. S. Mountford; construction date, July, 1992; comments/references: (1) *Hind*III digest pGT1.8Bgeo, Klenow fill + ligate on *Sal*I adaptors (blunt/sticky). T4 kinase + recirularize. (2) *Cla*I/*Sph*I excise LacZ/Neo hybrid sequence + replace with *Cla*I/*Sph*I fragment from 6P-LacZ(3)-TINBS-MO. Fragment includes 3'Z-T3-Ires-Neo 5'.

14. Glass capillary tubing for holding, injection, and enucleation pipet: 1.0-mm outer, 0.58-mm inner diameter for holding pipet, 1.0-mm outer, 0.78-mm inner diameter for injection and enucleation pipet (Clark Electromedical Instruments).
15. Electrode puller.
16. Microforge.
17. Fine and course platinum filaments.
18. Hydrofluoric acid (*see* **Note 4**).
19. Ultrapure distilled water.
20. Isopropanol.
21. Mineral oil (embryo tested).
22. Hoechst or 4',6-diamidino-2-phenylindole dihydrochloride (DAPI).
23. Microinjection assembly.
24. Mercury (*see* **Note 5**).

25. Hamilton syringe.
26. Piezo impact drive unit.
27. Fluorescence microscope.
28. X-gal fix: 1% (v/v) formaldehyde, 0.2% (v/v) glutaraldehyde in PBS supplemented with 1% fetal calf serum.
29. X-gal wash: PBS with 1% (v/v) fetal calf serum or bovine serum albumin.
30. X-gal stain: 4 mM potassium ferricyanide, 4 mM potassium ferrocyanide, 2 mM magnesium chloride, and 1 mg/mL 5-bromo-4-chloro-3-indolyl-beta-D-galactosylpyranoside (X-gal) in PBS.

3. Methods
3.1. Preparation of Pipets for Micromanipulation
3.1.1. Holding Pipets

1. Pull holding pipet by hand over a fine-flamed Bunsen burner or spirit lamp.
2. Break the tip off cleanly at an internal diameter of approx 70 µm using a microforge by fusing the shank of the pipet to a glass bead attached to the platinum filament and then turning the heat off suddenly.
3. Reduce the diameter of the tip of the holding pipet to approx 10 µm by holding the heated glass bead close to the opening.
4. Place a bend of about 30° approx 4 mm from the tip by applying heat from a fine filament to one side of the pipet.

3.1.2. Injection and Enucleation Pipets

Injection and enucleation pipets are made using an electrode puller to produce a thin tapered point extending for approx 1 cm (*see* **Note 6**).

1. Using the microforge, make a clean, perpendicular opening by gently fusing the pipet at the desired break point, without allowing thickening or distortion of the wall, to the heated filament, then turning off the heat suddenly (*see* **Note 7**). For injection pipets, the desired external diameter is 7 to 8 µm; for enucleation pipets, it is 8 to 10 µm.
2. Place a 30° bend approx 4 mm from the tip by applying heat from the fine filament to one side of the pipet.
3. Etch the internal and external surfaces of the injection or enucleation pipet (*see* **Note 8**). This is achieved by dipping the tip in 25% hydrofluoric acid for 10 to 20 s while maintaining a current of air through it by means of a 10-mL syringe connected to a piece of rubber tubing modified to connect the capillary to the syringe nozzle. A capillary cap may be used for this purpose.
4. Rinse the etched pipet extensively in ultrapure water and isopropanol before disconnecting it from the syringe apparatus.

3.1.3 Setting Up the Micromanipulation Assembly

Holding, injection, and enucleation pipets are attached to the micromanipulation apparatus and controlled using oil-filled injectors. A piezo impact drive system is used for operating injection and enucleation pipets. Before attachment of injection or enucleation pipets to the apparatus, insert a few microliters of mercury (enough to occupy approx 5 mm of the barrel of the pipet) with the aid of a Hamilton syringe. The Hamilton syringe can be filled conveniently from a 1-mL plastic syringe. Fill the proximal end of the pipet with mineral oil, leaving a small gap between mercury and oil.

3.2. Recovery of MII Oocytes

3.2.1. Superovulation

Superovulate female mice by sequential intraperitoneal (ip) injection of 5 U each of pregnant mare's serum and hCG made up in PBS 46 to 48 h apart. Aliquots of hormone should be stored frozen (for a maximum of a few months) and thawed immediately before use. Oocyte and cumulus cell donors can be superovulated simultaneously.

3.2.2. Recovery of MII Oocytes and Cumulus Cells

Oocytes and cumulus cells should be prepared 12.5 to 13 h after hCG injection (*see* **Fig. 2**).

1. Release the cumulus masses from the oviduct by tearing the ampulla in a small dish of H-mCZB.
2. Remove cumulus cells by incubation for 5 to 10 min in 0.1% hyaluronidase.
3. Wash recipient oocytes free from adherent cumulus cells and store in pre-equilibrated mCZB at 37°C, in a humidified atmosphere containing 5% carbon dioxide.
4. Rinse the cumulus cells for injection in H-mCZB to remove hyaluronidase then store in 5% PVP at room temperature.

3.3. Micromanipulation

3.3.1. Setting Up Chambers for Micromanipulation

The lids of 10-cm tissue culture grade plastic Petri dishes are used as chambers.

1. For the injection chamber make two parallel columns of drops (~10 μL each), one of H-mCZB, the other of PVP solution containing the donor cumulus cells, and a large rinse drop of H-mCZB.
2. Pour mineral oil gently into the lid to cover the drops.
3. For enucleation chambers place small drops (~5 μL) of H-mCZB containing 2.5 μg/mL cytochalasin B in a grid formation with an additional large rinse drop of H-mCZB also containing cytochalasin B and cover with oil.

3.3.2. Injection

To minimize exposure of oocytes to 4-(2-hydroxyethyl) piperazine-1-ethanesulfonic acid (HEPES)-containing medium, inject oocytes in batches of 10 to 20 at a time. Before and after injection incubate the oocytes in mCZB in drops under oil, and ensure that the first oocytes to be injected are also the first

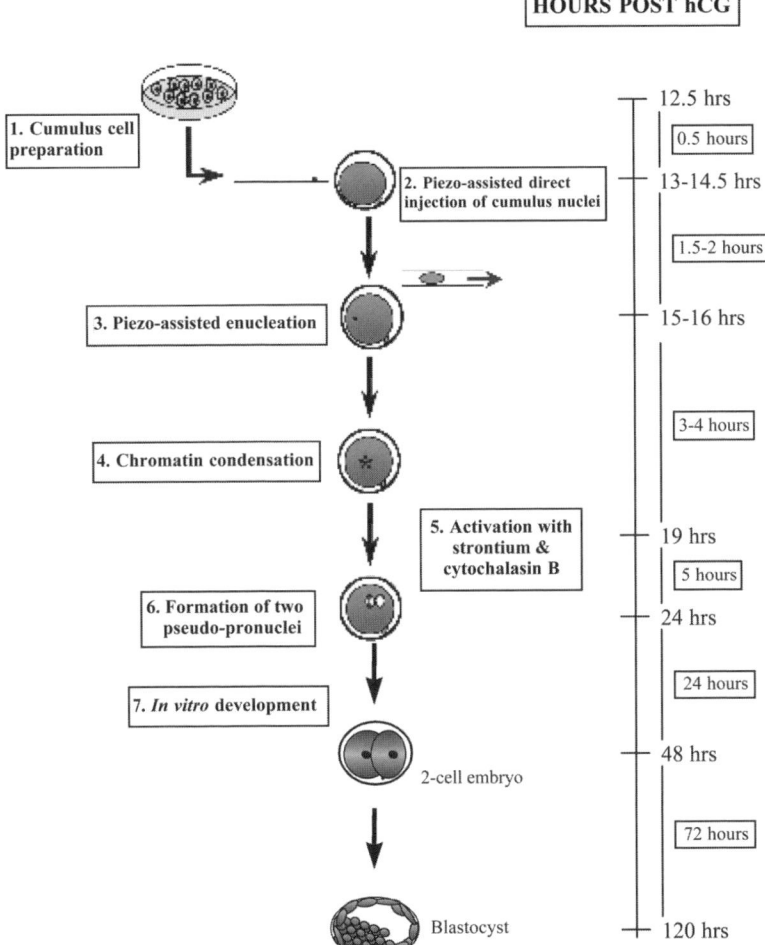

Fig. 2. Steps involved in piezo-assisted direct nuclear injection of cumulus cell nuclei. The timing of each step is indicated in terms of age of the recipient oocytes and is expressed as the number of hours after the administration of hCG. *1.* Cumulus/ oocyte masses are collected 12.5 h after hCG, treated with hyaluronidase and cumulus cells suspended in PVP until required (maximum 2 h). *2.* MII oocytes are injected with mechanically isolated cumulus cell nuclei using piezo-assisted driver 13 to 14.5 h after hCG. *3.* Injected oocytes are enucleated 15 to 16 h after hCG in the presence of cytochalasin B. Enucleation is confirmed by observing the metaphase plate in the removed cytoplasm following Hoechst staining and UV illumination. *4.* After enucleation, reconstituted oocytes are returned to culture for 3 to 4 h where the donor chromosomes condense. *5.* At 19 h after hCG, the reconstituted oocytes are activated by exposure to strontium chloride in the presence of cytochalasin B. *6.* Two pseudo-pronuclei form 5 to 6 h after activation. *7.* Reconstituted oocytes are returned to culture for up to 4 d with embryonic development observed daily.

to be enucleated. When ready to commence injection, position the holding and injection pipets in the large rinse drop of H-mCZB in the injection chamber and perform final alignment of pipets (*see* **Note 9**).

1. Force the mercury column to the tip of the injection pipet and aspirate a small volume of H-mCZB into both pipets.
2. Ensure that the mercury/medium interface in the injection pipet is moving smoothly and is just out of the field of view (magnification ×200).
3. Raise the holding pipet clear of the drop and reposition the injection chamber so that only the tip of the injection pipet is in the PVP drop containing the donor cumulus cells.
4. Aspirate cells either individually or in succession, up to six at a time, into the injection pipet.
5. Move the loaded injection pipet into the drop of H-mCZB containing the oocytes and lower the holding pipet.
6. Hold each oocyte in turn by suction on the holding pipet and orientate it so that the metaphase plate (visible as a slightly transparent region) is neither close to the holding pipet nor to the point of entry of the injection pipet. Orientating the oocyte with the metaphase plate at the 6- or 12-o'clock position is ideal.
7. Apply a slight pulse using the piezo to allow the injection pipet to penetrate the zona pellucida without touching the oolema. Ensure that each cumulus cell membrane is ruptured before insertion into the oocyte; use additional pulsing if necessary.
8. Bring a ruptured cell to the tip of the injection pipet.
9. Gently push the tip of the injection pipet into the oocyte toward the holding pipet (*see* **Fig. 3**).

Fig. 3. *(opposite page)* Illustration of piezo-assisted direct injection of cumulus cell nuclei and enucleation of MII oocytes. The position of the cumulus cell nucleus and the metaphase plate is indicated by an arrow and arrowhead, respectively. Cumulus cells are held in microdrops of PVP (**A**) and individual cumulus cells are aspirated gently and expelled from a blunt-ended 7- to 8-µm injection pipet until the cell membrane ruptures leaving an intact nucleus. For injection each oocyte is positioned so that the metaphase plate is in the 6-o'clock position (**B,C,D**). A series of piezo pulses bore a small hole in each zona allowing the penetration of a blunt-ended injection pipet (**B**) containing the cumulus cell nuclei. The nucleus is forced to the tip of the pipet and the pipet pushed to the opposite side of the oocyte (**C**). A single low-frequency piezo pulse is then applied causing a localised rupture of the oolema. The nucleus is expelled at this site and the pipet rapidly withdrawn leaving an indentation of the oolema (**D**). For piezo-assisted enucleation of injected MII oocytes, the metaphase plate of each oocyte is positioned in the 3-o'clock position (**E,F**). A series of high frequency piezo pulses are applied to the 9- to 10-µm enucleation pipet to core a hole through the zona (**F**). Once inside the perivitelline space, cytoplasm containing the metaphase plate is drawn into the pipet and removed (**G**). The removal of only the maternal chromosomes is confirmed by Hoechst staining and ultraviolet illumination

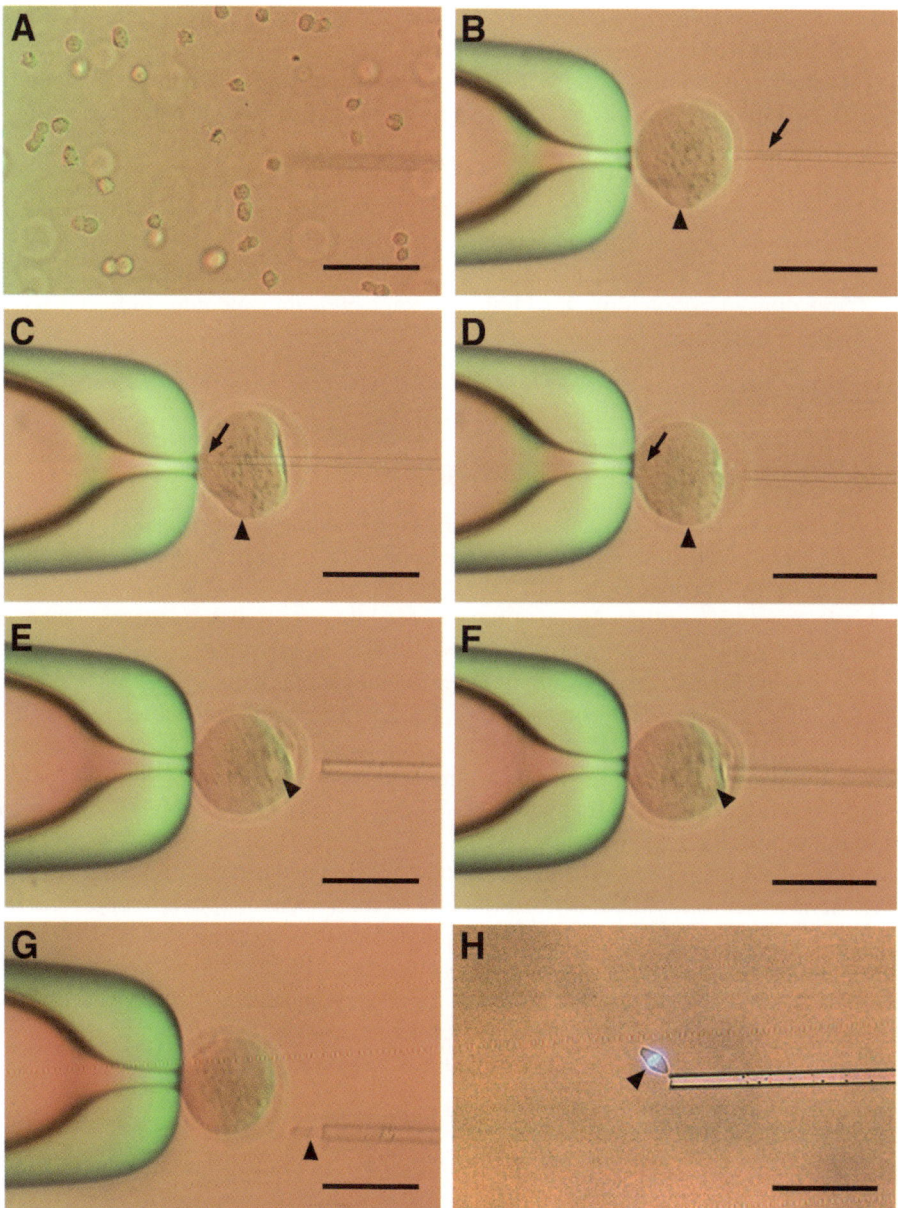

Fig. 3. *(continued)* of the removed portion of cytoplasm (**H**). For purposes of illustration, the fluorescently labeled metaphase plate is shown within the enucleation pipet. Normally the oocyte must not be exposed to ultraviolet irradiation, so enucleation is confirmed within the drop in the chamber from which the oocyte has been removed. All enucleations are performed in the presence of cytochalasin B. Scale bar represents 100 μm.

10. Administer a low-frequency pulse to cause the oolema to rupture neatly; this can be recognized as a retraction of the oolema to resume its original spherical shape.
11. Deposit the cumulus nucleus in the cytoplasm and gently withdraw the pipet.
12. Repeat the process for each oocyte in the drop; refill or replace the injection pipet when necessary.
13. Return the batch of oocytes to a fresh drop of mCZB in the incubator.
14. Bring the next batch of 10 to 20 oocytes into the injection chamber and repeat the process. No longer than 2 h should be spent on the injection phase of the procedure (*see* **Fig. 2**). Nonmanipulated oocytes are not wasted, but can be activated and allowed to develop to the blastocyst stage of development to confirm that the culture conditions are optimal.

3.3.3. Enucleation

Perform all enucleation under ×200 magnification. Enucleate in batches of six to eight oocytes in 10-min intervals.

1. Before enucleation, incubate each batch of injected oocytes in H-mCZB containing 2.5 µg/mL cytochalasin B for 5 min in individual drops of the enucleation dish.
2. During the incubation period, exchange the injection pipet for an enucleation pipet and lower it into the large rinse drop of the prepared enucleation chamber.
3. Aspirate H-mCZB containing cytochalasin B into each pipet.
4. Align the pipets and reposition them in an enucleation drop.
5. Anchor each oocyte on the holding pipet by suction and orientate it so that the metaphase plate is next to the enucleation pipet (*see* **Fig. 3**).
6. Drive the tip of the enucleation pipet through the zona with the aid of the piezo pulse.
7. Aspirate the region of the oocyte containing the chromosomes into the tip with minimal extra cytoplasm.
8. Withdraw the pipet so that the metaphase plate is pinched off from the oocyte.
9. Return reconstructed oocytes to mCZB in the incubator for recovery. Enucleation may be confirmed by labeling each of the drops with Hoechst or DAPI after removal of the oocytes and observing with a fluorescence microscope. Examination of murine oocytes with fluorescence is not compatible with subsequent normal development, but **Fig. 3** shows, for demonstration purposes only, successful removal of a metaphase plate labelled *in situ* with Hoechst.

3.4. Development of Reconstructed Eggs

After a recovery period of 1 to 2 h in mCZB, transfer the reconstructed and non-manipulated oocytes to Sr-mCZB for 5 h at 37°C in a humidified atmosphere containing 5% carbon dioxide for activation. Subsequent incubation is in fresh mCZB until the most advanced embryos have attained the late two-cell or early four-cell stage of development. Transfer them to G2 medium at 37°C in a humidified atmosphere containing 5% carbon dioxide and allow them to develop into blastocysts. The status of reprogramming as indicated by the expression of transgenes may be monitored at any preimplantation stage. This may be achieved by staining with X-gal (*see* **Subheading 3.5.**, **Fig. 4** or by observation with fluorescence. The obvious advantage of fluorescently labeled transgenes is that gene expression can be monitored without termination of development. However, murine preimplantation embryos are very sensitive to fluorescence, and subsequent development may be compromised. Nonmanipulated oocytes serve to confirm the quality of the media and conditions. Oocytes from D2 mice can be activated with very high efficiency. However, most transgenic donor mice are not bred on such a permissive background; consequently, the oocytes will develop very poorly after activation. Fertilized embryos from the transgenic donor strain must therefore be used to show the expression pattern of the transgene. These embryos may develop at a different rate from the reconstructed embryos.

3.5. Staining With X-Gal

Incubate reconstructed embryos in X-gal fix for 5 min at room temperature, wash three times in X-gal wash, then transfer them into staining solution until the color develops. The cumulus cells from ZIN40 mice stain with X-gal, whereas the oocytes within cumulus masses are negative (*see* **Fig. 4A**). Oocytes from D2 mice do not stain with X-gal (*see* **Fig. 4B**). **Figure 4C** shows a blastocyst developing from a D2 oocyte in which the nucleus was replaced with one from a ZIN40 cumulus cell. The staining in all nuclei of the blastocyst indicates reactivation of the ZIN40 genome. Blastocysts recovered from normally fertilized ZIN40 females also stain in this way (not shown). The blastocyst shown in **Fig. 4C** would have been a suitable candidate for derivation of ES cells expressing the transgenic marker from the donor nucleus.

4. Notes

1. It is very important that the osmolarity of all manipulation and culture media is within the recommended range. Too high a value may result in suboptimal development, whereas too low a value will cause lysis of oocytes during injection. It may be helpful to use a high osmolarity (~280 mosmoles) for CZB during the manipulation stages and transfer to a lower osmolarity (~265 mosmoles) after activation. The osmolarity can be increased by cautious addition of sucrose.
2. The mice from which the oocytes are harvested ideally are 4 to 12 wk old. Oocytes from older females are more likely to be low in number and have a tendency to abnormality after superovulation. It is less critical that the cumulus donor mice fall within this age range.
3. The quality of hormones used for superovulation may affect the quality of oocytes. Do not store frozen aliquots for more than a few months and thaw immediately before injection.

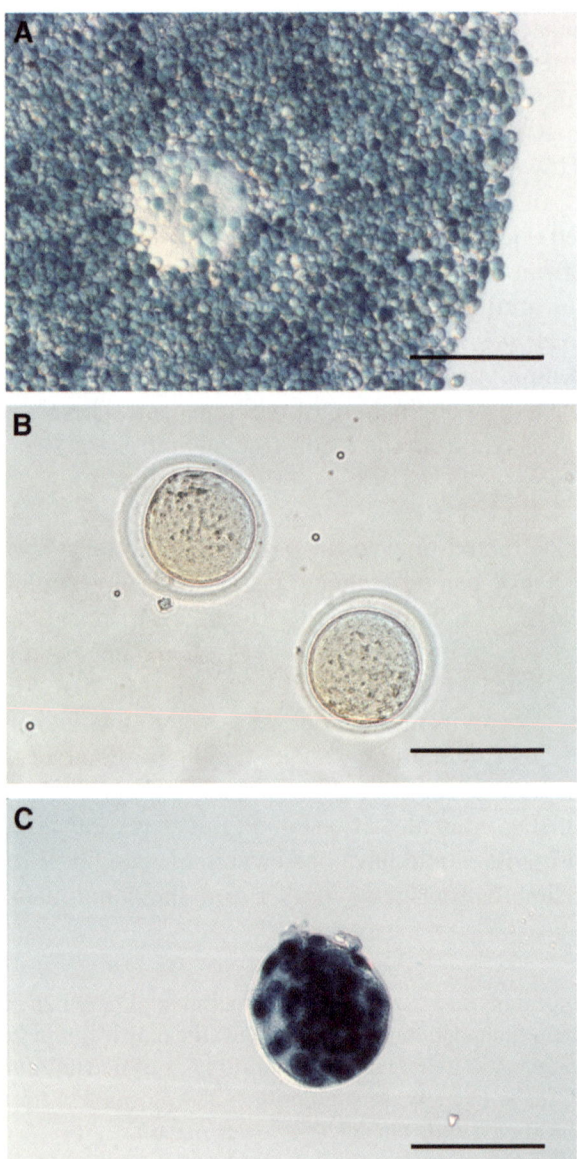

Fig. 4. Expression of *lacZ* reporter gene confirms nuclear contribution in nuclear transfer embryos derived from transgenic somatic cell nuclei. Cumulus cells derived from ZIN40 transgenic mice display characteristic blue X-gal stained nuclei (**A**), whereas D2 oocytes do not stain (**B**). Nuclear transfer blastocyst, derived from direct injection of ZIN40 cumulus nuclei into D2 oocytes that were subsequently enucleated, displayed blue X-gal stained nuclei characteristic of the donor cumulus cell (**C**). Scale bar represents 100 µm.

4. Hydrofluoric acid is extremely toxic and will destroy any glass container. Store safely in a plastic container and wear protective clothing during handling. Ensure that acid, water and isopropanol solutions contain no impurities.
5. Mercury is toxic and must be handled carefully. Wear gloves at all times and dispense in an area where spills can be easily contained. Mercury spillage kits are available from Scientific Laboratory Supplies.
6. The shape of the injection and enucleation pipets is critical. They must be long enough to allow control of flow within the fine part of the barrel but not so long that their flexibility prevents controlled penetration of the zona. The correct shape of pipets can be decided only by trial and error.
7. Creating the ideal pipet for injection or enucleation requires some skill and practice. Fusing the filament to the appropriate point near the tip of the needle without causing thickening or distortion of its walls may present some difficulty at first.
8. Acid etching of injection and enucleation pipets will help guard against blockages during manipulation. The thinness of the walls at the tips of the pipets that is achieved by acid etching allows the penetration of the zona and oolema with minimal intensity and velocity of the piezo pulse, which reduces lysis of the oocyte and damage to the donor nucleus.
9. It is important to align holding and manipulation pipets correctly before performing either enucleation or injection. Failure to do so may increase the risk of lysis.

Acknowledgment

This work was supported by Stem Cell Sciences plc and the Biotechnology and Biological Sciences Research Council of the UK.

References

1. Mountford, P. and Smith, A. G. (1995) Internal ribosome entry sites and dicistronic RNAs in mammalian transgenesis. *Trends Genet.* **11,** 179–184.
2. Munsie, M. J., Michalska, A. E., O'Brien, C. M., Trounson, A. O., Pera, M. F., and Mountford, P. S. (2000) Isolation of pluripotent embryonic stem cells from reprogrammed adult mouse somatic cell nuclei. *Curr. Biol.* **10,** 989–992.
3. Munsie, M., O'Brien, C., and Mountford, P. (2002) Transgenic strategy for demonstrating nuclear reprogramming in the mouse. *Cloning Stem Cells* **4,** 121–130.
4. Tada, M., Takahama, Y., Abe, K., Nakatsuji, N., and Tada, T. (2001) Nuclear reprogramming of somatic cells by in vitro hybridization with ES cells. *Curr. Biol.* **11,** 1553–1558.
5. Terada, N., Hamazaki, T., Oka, M., Hoki, M., Mastalerz, D. M., Nakano, Y., et al. (2002) Bone marrow cells adopt the phenotype of other cells by spontaneous cell fusion. *Nature* **416,** 542–545.
6. Ying, Q. L., Nichols, J., Evans, E. P., and Smith, A. G. (2002) Changing potency by spontaneous fusion. *Nature* **416,** 545–548.
7. Eggan, K., Akutsu, H., Hochedlinger, K., Rideout, W., 3rd, Yanagimachi, R., and Jaenisch, R. (2000) X-Chromosome inactivation in cloned mouse embryos. *Science* **290,** 1578–1581.

8. Boiani, M., Eckardt, S., Scholer, H. R., and McLaughlin, K. J. (2002) Oct4 distribution and level in mouse clones: consequences for pluripotency. *Genes Dev.* **16,** 1209–1219.

9. Chatot, C. L., Lewis, J. L., Torres, I., and Ziomek, C. A. (1990) Development of 1-cell embryos from different strains of mice in CZB medium. *Biol. Reprod.* **42,** 432–440.

10. Kimura, Y., and Yanagimachi, R. (1995) Intracytoplasmic sperm injection in the mouse. *Biol. Reprod.* **52,** 709–720.

11. Barnes, F. L., Crombie, A., Gardner, D. K., Kausche, A., Lacham-Kaplan, O., Suikkari, A. M., et al. (1995) Blastocyst development and birth after in-vitro maturation of human primary oocytes, intracytoplasmic sperm injection and assisted hatching. *Hum. Reprod.* **10,** 3243–3247.

11

Using Immunofluorescence to Observe Methylation Changes in Mammalian Preimplantation Embryos

Fátima Santos and Wendy Dean

Summary

DNA methylation is an important epigenetic mark involved in gene silencing, X chromosome and transposon inactivation, genomic imprinting, and chromosome stability. Recently, it has been increasingly recognized that DNA methylation plays an essential regulatory function in mammalian development and cancer biology, serving to repress nontranscribed genes stably in differentiated adult somatic cells. DNA methylation is subject to reprogramming during development, involving both demethylation (active and passive) and *de novo* methylation phases. Recent data show deregulation of genome-wide DNA methylation levels in embryos generated by somatic nuclear transfer, suggesting this epigenetic mark and its regulation are essential to ensure viable embryos. Immunofluorescence using antibodies against 5-methylcytosine is an invaluable method for the large-scale screening of genome-wide methylation. This method has a high degree of reproducibility, essential for the analysis of small numbers of valuable samples, providing information on methylation profiles of individual cells and embryos.

Key Words: DNA methylation; immunofluorescence; preimplantation; mammalian; embryos; reprogramming.

1. Introduction

DNA methylation at the C-5 position of cytosine (5-MeC) in CpG dinucleotides is the major form of DNA modification in vertebrate animals and its function has been implicated in a diverse range of biological processes. Molecular and genetic studies have demonstrated that DNA methylation plays critical roles in gene silencing, X chromosome and transposon inactivation, genomic imprinting, and chromosome stability (reviewed in **refs. *1–3***). DNA methylation has been shown to be essential for mammalian development as inactivation of Dnmt1, the major maintenance DNA cytosine methyltransferase, results in genome-wide demethylation, and embryonic lethality (reviewed in

From: *Methods in Molecular Biology, vol. 325: Nuclear Reprogramming: Methods and Protocols*
Edited by: S. Pells © Humana Press Inc., Totowa, NJ

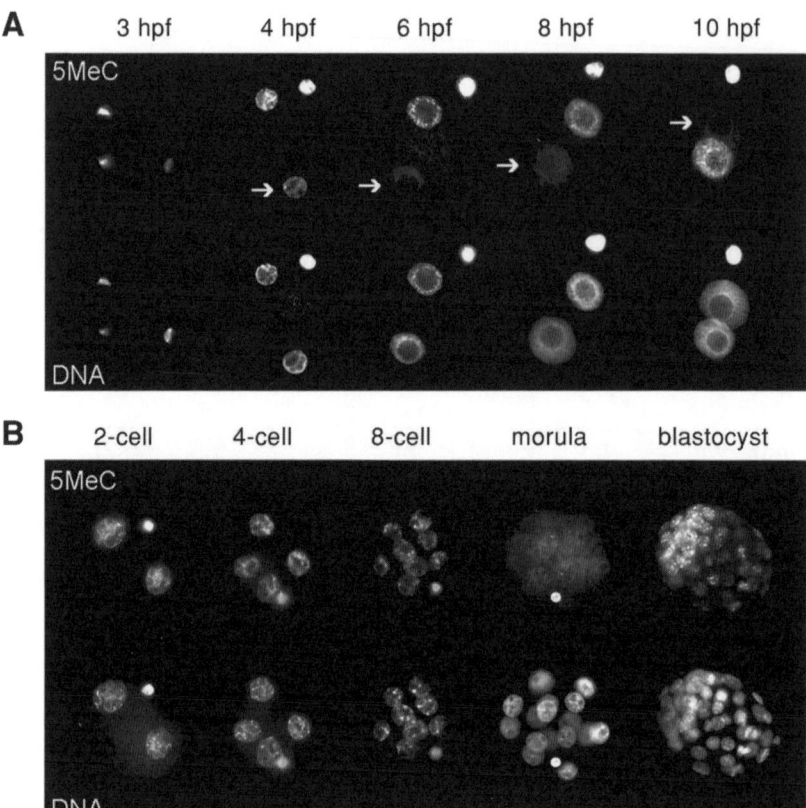

Fig. 1. DNA methylation reprogramming in mouse preimplantation development. **(A)** Upon fertilization, the paternal genome undergoes a rapid and specific active demethylation (upper row, 5-MeC staining; lower row DNA staining [YOYO1 iodide] of the same embryo; →, male pronucleus; hpf, hours post fertilization). **(B)** From the two-cell stage onwards, there is passive loss of DNA methylation (5-MeC) resulting from the exclusion of the maintenance DNA methyltransferase from the nucleus, with *de novo* methylation evident in the inner cell mass at the blastocyst stage (upper row, 5-MeC staining; lower row DNA staining [YOYO1 iodide] of the same embryo). The 5-MeC antibody reveals patterns of repeat sequences and densely methylated regions but not single copy genes and their potential regulatory sequences. Large, intensely stained heterochromatic foci are a characteristic pattern observed by the blastocyst stage.

ref. *4*). In early mouse embryos, DNA methylation undergoes genome-wide reprogramming *(5)*. As soon as the fertilizing sperm starts decondensing, specific active demethylation of its genome is initiated, being completed within 6 to 8 h *(6)*, whereas the maternal genome remains highly methylated throughout **(Fig. 1A)**.

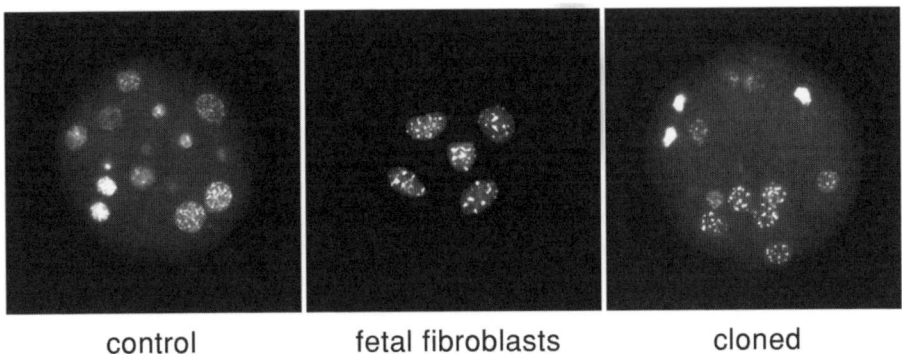

control   fetal fibroblasts   cloned

Fig. 2. Genome-wide DNA methylation patterns in cloned bovine embryos. Organization of DNA methylation in cloned and control bovine embryos between the 8 and 16-cell stages. The heterogeneous population of nuclei observed in the control embryo (**left**) was replaced in the cloned (**right**) by a highly homogeneous population with a methylation pattern strongly reminiscent of the donor nuclei (**center**), suggesting a failure to erase cellular memory. Interestingly, characterization of DNA methylation organization in embryonic nuclei showed species-specific distinctive patterns *(11)*.

Thereafter, a second phase of DNA demethylation occurs through a passive, replication-dependent process, this time affecting equally the maternal and paternal genomes (**Fig. 1B** *[6–9]*). Finally, a third phase of DNA methylation reprogramming occurs with the restoration of DNA methylation levels by *de novo* methylation by the blastocyst stage (**Fig. 1B** *[6]*). It is still unclear what the significance of this reprogramming process may be, but it seems certain that it is essential for the success of mammalian development as it has been observed that DNA methylation patterns fail to be reprogrammed in the majority of cloned preimplantation embryos *(10–12)*, presumably a major factor contributing to the low efficiency of cloning (**Fig. 2**).

Immunofluorescence using antibodies against 5-methylcytosine, described in this chapter, is a method for the large-scale screening of genome-wide methylation that has several advantages over other global methylation analysis techniques using extracted genomic DNA. Not only is it amenable to the analysis of very small and heterogeneous samples, but also methylation profiles can be evaluated in individual cells or embryos and even in parental sets of chromosomes uniquely visible at syngamy. The advantage of topological assessment of DNA methylation in the context of chromatin architecture, such as heterochromatin and euchromatin, in interphase nuclei, makes it possible to identify relationships between functional and structural parameters on the largest scale of genomic organization and investigate major epigenetic changes related to such crucial events as development and cellular memory.

2. Materials

2.1. Production of Preimplantation Embryos

2.1.1. Superovulation

1. Female mice, 3- to 4-wk-old, C57B1/6J × CBA/Ca-F_1 (Harlan; Charles River, UK).
2. Phosphate-buffered saline (PBS; Sigma).
3. Pregnant mare serum gonadotropin (Folligon), 1500 U per vial (Intervet).
4. Human chorionic gonadotropin (hCG-Chorulon) 500 U per vial (Intervet).

2.1.2. Collection of Preimplantation Embryos

1. Dissecting tools; No. 5 watchmaker's forceps; fine-pointed scissors (iris; Weiss). Sterile plastic Petri dishes (50 mm).
2. Sterile 27-gage × 0.5-in. needles.
3. Flushing and handling medium, M2 (Sigma M-7167) supplemented with 100 U/ mL penicillin and 50 µg/mL streptomycin.
4. Hyaluronidase (Sigma H-6254), 300 µg/mL in M2 (from stock solution).
5. Drawn 9-in (0.2 mm diameter) Pasteur pipets, plugged and flame polished.
6. Mouth pipets (aspirator mouthpiece).
7. Dissection microscope with understage illumination.

2.2. Fixation of Preimplantation Embryos

1. Glass embryo dishes (distributed by VWR International in UK, cat. no. 406/0204/01).
2. Aspirator mouthpiece and drawn Pasteur pipets (*see* **Subheading 2.1.2.**).
3. PBS (cat. no. D-7030; Sigma) pH 7.5, filtered before use.
4. Fixative: 4% paraformaldehyde (PFA; cat. no. P-6148, Sigma) in PBS, pH adjusted to 7.5, made fresh (*see* **Note 1**).
5. PBT: 0.05% Tween-20 (cat. no. P-7949; Sigma) in PBS, filtered before use.
6. Permeabilization solution: 0.2% Triton X-100 (cat. no. T-9284; Sigma) in PBS.

2.3. 5-MeC Antibody Staining of Preimplantation Embryos

1. Glass embryo dishes (*see* **Subheading 2.2.**).
2. Aspirator mouthpiece and drawn Pasteur pipets (*see* **Subheading 2.1.2.**).
3. Depurination solution: 4 *N* HCl, 0.1% Triton X-100.
4. PBT solution (*see* **Subheading 2.2.**).
5. Blocking solution: 0.2% bovine serum albumin (BSA; cat. no. A-9085, Sigma) in PBT.
6. Antibodies: mouse monoclonal antibody against 5-methylcytidine (MMS-900P-B; Eurogentec *[13]*), Alexa Fluor 594 goat anti-mouse IgG (cat. no. A-11005, Molecular Probes; *see* **Note 2**).
7. DNA dye: YOYO1 iodide 1 m*M* solution in dimethyl sulfoxide (cat. no. Y-3601, Molecular Probes).
8. Polylysine (trade name: "Polysine")-coated microscope slides (cat. no. 406/0178/ 00, BDH).
9. Glass coverslips 22 × 32 mm (cat. no. 406/0188/24; BDH).

10. Mounting medium (cat. no. 1000-4, Sigma).
11. Colorless nail polish.

2.4. Microscopy

1. Conventional wide-field microscope equipped for epifluorescence and transmitted light illumination (e.g., Olympus BX41).
2. Filter sets specific for the fluorochromes (fluorescein and Texas Red; *see* **Note 2**).
3. Digital CCD monochrome camera F-ViewII (Soft Imaging System GmbH), AnalySIS 3.2 (Soft Imaging System GmbH), and Adobe Photoshop software.

3. Methods
3.1. Production of Embryos
3.1.1. Superovulation

Immature females B6CB-F$_1$, 3- to 4-wk old, are best suited for high response to hormonal regimes. Because a restricted number of strains respond to this procedure, there may be some limitations imposed where genetic rigor is paramount (*see* **Note 3**). Consistent high yields can be obtained from hybrid F$_1$ animals derived from a number of widely available mouse strains *(14)*.

1. Pregnant mare serum 5.0 to 7.5 U (Folligon, Intervet) is injected intraperitoneally (ip) followed by 5.0 to 7.5 U of human chorionic gonadotrophin (hCG; Chorulon, Intervet) 44 to 48 h later. Timing of initiation is dictated by light/dark cycles.
2. After hCG injection, individual females are incarcerated with mature, that is, greater than 8-wk-old, B6CB-F$_1$ males.
3. Presence of a copulation plug is checked for the following morning. This is day 0.5 of gestation.

3.1.2. Collection of Preimplantation Embryos

One-cell embryos are collected no earlier than 18 h post-hCG to ensure high rates of fertilization.

1. Sacrifice the animals by cervical dislocation, excise the swollen ampullae of the oviduct intact, and place into a plastic Petri dish.
2. Place oviduct in 50-µL drops of hyaluronidase solution (300 µg/mL) in M2 and tear open the swollen portion of the ampullae with a sterile 27-gage × 0.5-inch needle while holding the oviduct by No. 5 watchmaker's forceps. Incubate at room temperature until the fertilized oocytes are denuded of cumulus cells. (This should take no more than 5 min.) Wash in a series of 10 drops of M2 to clean away cumulus cells, spermatozoa, and other cellular debris. Discard obviously abnormal embryos and unfertilized oocytes (*see* **Note 4**).

Later-staged preimplantation embryos can be flushed out of the oviducts (two-cell to morulae) or uteri (blastocysts) in M2 medium. Detailed information on this procedure can be found in **ref. *14***.

3.2. Fixation of Preimplantation Embryos

A good fixation protocol (1) must prevent antigen leakage, (2) permeabilize the cells to allow access of the antibody, (3) keep the antigen in a form that will allow it to be recognised by the antibody as efficiently as possible, and (4) maintain the cell structure intact *(15)*. Fixatives can be broadly categorized as either organic solvents or crosslinking reagents. Organic solvents such as alcohols and acetone remove lipids and dehydrate the cells distorting them and destroying the authentic subcellular structure *(15)* and, although not significantly reducing permeability, the loss of nucleic acids can be a problem *(16)*. Crosslinking reagents, on the other hand, form intermolecular bridges, usually through free amino groups, and thereby create a network of linked antigens. This preserves fairly well the tissue morphology and retains nucleic acids but leads to a reduction of the tissue permeability, therefore calling for a permeabilization step. The conditions of fixation are a compromise: stronger fixation yields a better preservation of cellular morphology, but the increased crosslinking lowers the accessibility of the target. Fixation times should therefore be restricted to the minimum required for good cellular preservation as prolonged fixation may increase masking of the epitopes. We have devised a method that allows a 3D study of the relationships between the components of preimplantation embryos while insuring high-quality antigen detection. All embryo manipulations are performed by mouth pipeting, under a dissection microscope with understage illumination, and all incubations are performed in glass embryo dishes.

1. Mouth pipet the embryos from the M2 medium and wash them briefly in PBS.
2. Fix the embryos in ice-cold 4% PFA (*see* **Note 1**) for 15 min at room temperature (*see* **Note 5**).
3. Wash the embryos briefly in PBS and leave them in PBT for 15 min at room temperature.
4. Permeabilize the embryos in 0.2% Triton X-100 solution for 30 min at room temperature (*see* **Note 6**).
5. Wash the embryos in PBT (*see* **Note 7**).

3.3. 5-MeC Antibody Staining of Preimplantation Embryos

For the antibody to get access to the epitope it is necessary to create an appropriate space as the 5-MeC is situated within the major groove of the DNA. Several procedures for increasing the accessibility of methylated chromatin regions to 5-MeC antibody have been proposed, from alkaline, pepsin/HCl, and other enzyme treatments *(17)* to ultraviolet *(9)* and formamide *(18)* DNA denaturation. Most of these methods are constrained by the fixation protocol of choice. In our experience optimal and highly reproducible results were obtained using a simple HCl DNA-depurinating step *(6,11,19)*.

1. Wash the embryos briefly in PBS.
2. Depurinate the embryos in 4 *N* HCl, 0.1% Triton X-100 for 10 min at room temperature.
3. Wash the embryos briefly in PBS and leave them in PBT for 15 min at room temperature.
4. Incubate the embryos in blocking solution (2% BSA in PBT) overnight at 4°C.
5. Incubate the embryos in a 1:500 dilution of mouse monoclonal antibody against 5-MeC in blocking solution for 1 h at room temperature.
6. Wash the embryos in blocking solution three times for a minimum of 30 min in total.
7. Incubate the embryos in a 1:500 dilution of Alexa Fluor 594 goat anti-mouse IgG in blocking solution for 1 h at room temperature in the dark.
8. Wash the embryos in blocking solution three times for a minimum of 30 min.
9. Incubate the embryos in 100 n*M* YOYO1 iodide (in PBT) for 10 min at room temperature in the dark.
10. Transfer the embryos onto a clean poly-lysine microscope slide. With the help of the pipet make sure the embryos are sufficiently separated from one another and remove the excess liquid (*see* **Note 8**).
11. Cover with 10 µL of Mounting Medium. Carefully place a coverslip on top (*see* **Note 9**). Seal the coverslip with nail polish. Store the slides at 4°C for a few hours before microscope observation (*see* **Note 10**).

3.4. Microscopy

Using the appropriate bandpass filters for YOYO1 iodide (fluorescein) and Alexa Fluor 594 (Texas Red) sequential images can be digitally captured and if necessary later pseudocoloured and merged using Adobe Photoshop software.

4. Notes

1. Alternatively, the 4% PFA solution can be stored frozen (–20°C) in conveniently-sized aliquots for at least 6 mo. When using animal models other than mouse, some modification of the fixation protocol may be necessary. In our experience, this 4% PFA concentration will work equally well with bovine, pig, and rat preimplantation embryos. Sheep embryos gave better results using 2.5% PFA, 0.1% Triton-X100 in PBS, pH 7.5.
2. Special consideration should be taken when selecting both the fluorochromes and the filter sets. When analyzing signals of likely unequal intensity, there is the possibility of bleeding between fluorescent channels during observation, making it impossible to assess both colocalization and purity of signal. To minimize this problem, it is important not only to choose fluorochromes that are sufficiently apart in both excitation and emission spectra, but also to titrate the concentration of antibodies and fluorescent probes. Bleed-through can also be minimized with the appropriate choice of bandpass excitation and emission filters. A filter that blocks the first color also can be inserted into the light path when viewing the second color (cut-off).

3. We have made extensive use of superovulated females to generate considerable numbers of embryos using the least number of animals. However, we have compared the staining patterns of superovulated vs naturally produced preimplantation embryos and have found no difference between them.

4. This step is conducted under relatively low power using a dissection microscope. Thus, abnormal embryos are defined as those that are either fragmented or have grossly overrepresented perivitelline space.

5, If using 2.5% PFA, 0.1% Triton-X100 in PBS, pH 7.5, as fixative, increase the fixation time to 30 min at room temperature.

6. The embryos can become fairly sticky both toward each other and to the glass pipet itself. Be *extra* careful and handle the embryos in groups no bigger than 10 to minimize losses from this step onward.

7. At this point, embryos can be stored in PBT (to which 0.05% sodium azide has been added as preservative) at 4°C for at least 6 mo.

8. Although it is important to remove as much liquid as possible during this step (so the embryos will attach to the slide), **it is essential not to allow the embryos to dry**.

9. Once the cover slip is placed over the embryos, it should never be repositioned as that will cause serious distortions. Also, no attempt should be made to remove any trapped air bubbles as that will disrupt the embryos.

10. Cooling down the fluorochromes will enhance their quantum emission allowing for shorter exposure times and thus less fading.

Acknowledgments

The work conducted by the authors tookplace in the Laboratory of Developmental Genetics and Imprinting, Babraham Institute. The authors are indebted to A. Niveleau for generously supplying the 5-methyl-cytosine antibody. This work is funded by the BBSRC and MRC.

References

1. Robertson, K. D. and Wolffe, A. P. (2000) DNA methylation in health and disease. *Nat. Rev. Genet.* **1,** 11–19.

2. Bird, A. (2002) DNA methylation patterns and epigenetic memory. *Genes Dev.* **16,** 6–21.

3. Meehan, R. R. (2003) DNA methylation in animal development. *Semin. Cell Dev. Biol.* **14,** 53–65.

4. Bestor, T. H. (2000) The DNA methyltransferases of mammals. *Hum. Mol. Genet.* **9,** 2395–2402.

5. Reik, W., and Walter, J. (2001) Genomic imprinting: parental influence on the genome. *Nat. Rev. Genet.* **2,** 21–32.

6. Santos, F., Hendrich, B., Reik, W., and Dean, W. (2002) Dynamic reprogramming of DNA methylation in the early mouse embryo. *Dev. Biol.* **241,** 172–182.

7. Monk, M., Boubelik, M., and Lehnert, S. (1987) Temporal and regional changes in DNA methylation in the embryonic, extraembryonic and germ cell lineages during mouse embryo development. *Development* **99,** 371–382.

8. Oswald, J., Engemann, S., Lane, N., Mayer, W., Olek, A., Fundele, R., et al. (2000) Active demethylation of the paternal genome in the mouse zygote. *Curr. Biol.* **10**, 475–478.

9. Rougier, N., Bourc'his, D., Gomes, D. M., Niveleau, A., Plachot, M., Paldi, A., et al. (1998) Chromosome methylation patterns during mammalian preimplantation development. *Genes Dev.* **12**, 2108–2113.

10. Bourc'his, D., Le Bourhis, D., Patin, D., Niveleau, A., Comizzoli, P., Renard, J. P., et al. (2001) Delayed and incomplete reprogramming of chromosome methylation patterns in bovine cloned embryos. *Curr. Biol.* **11**, 1542–1546.

11. Dean, W., Santos, F., Stojkovic, M., Zakhartchenko, V., Walter, J., Wolf, E., et al. (2001) Conservation of methylation reprogramming in mammalian development: aberrant reprogramming in cloned embryos. *Proc. Natl. Acad. Sci. USA* **98**, 13734–13738.

12. Kang, Y. K., Koo, D. B., Park, J. S., Choi, Y. H., Lee, K. K., and Han, Y. M. (2001) Influence of oocyte nuclei on demethylation of donor genome in cloned bovine embryos. *FEBS Lett.* **499**, 55–58.

13. Reynaud, C., Bruno, C., Boullanger, P., Grange, J., Barbesti, S., and Niveleau, A. (1992) Monitoring of urinary excretion of modified nucleosides in cancer patients using a set of six monoclonal antibodies. *Cancer Lett.* **61**, 255–262.

14. Hogan, B., Beddington, R., Constantini, F., and Lacy, E. (1994) *Manipulating the Mouse Embryo: A Laboratory Manual*, 2nd Ed. Cold Spring Harbor Laboratory Press, Cold Spring Harbor, New York.

15. Harlow, E., and Lane, D. (1999) *Using Antibodies: A Laboratory Manual*. Cold Spring Harbor Laboratory Press, Cold Spring Harbor, New York.

16. Leitch, A. R., Schwarzacher, T., Jackson, D., and Leitch, I. J. (1994) *In Situ Hybridization: A Practical Guide*. Royal Microcopical Society Microscopy Handbooks., BIOS Scientific Publishers Limited, Oxford, UK.

17. Barbin, A., Montpellier, C., Kokalj-Vokac, N., Gibaud, A., Niveleau, A., Malfoy, B., et al. (1994) New sites of methylcytosine-rich DNA detected on metaphase chromosomes. *Hum. Genet.* **94**, 684–692.

18. Mayer, W., Niveleau, A., Walter, J., Fundele, R., and Haaf, T. (2000) Demethylation of the zygotic paternal genome. *Nature* **403**, 501–502.

19. Santos, F., Zakhartchenko, V., Stojkovic, M., Peters, A., Jenuwein, T., Wolf, E., et al. (2003) Epigenetic marking correlates with developmental potential in cloned bovine preimplantation embryos. *Curr. Biol.* **13**, 1116–1121.

12

Observing S-Phase Dynamics of Histone Modifications With Fluorescently Labeled Antibodies

Rong Wu, Anna V. Terry, and David M. Gilbert

Summary

Histone modifications are central to epigenetic regulation and must be reestablished with each round of DNA replication. Here we describe methods to localize these modifications within mammalian nuclei and to relate them to specific spatiotemporal patterns of DNA replication.

Key Words: Nuclear structure; histone modification; histone methylation; immunofluorescence microscopy.

1. Introduction

The postgenomic era has brought with it an appreciation of the limitations of DNA sequence information and the importance of epigenetic regulation through nuclear structure/chromatin dynamics. Histone proteins, the major components of chromatin, are subject to dynamic covalent modifications, mostly at their N-terminal tails, and different combinations of these modifications are thought to represent different "histone codes" that regulate processes such as transcription, replication, recombination, and repair by forming myriad interactive surfaces for regulatory proteins (*1,2*). Layered upon this histone modification code are various isoforms of the core histones, known as histone variants, as well as many linker histones, all of which appear to have specialized roles in modulating chromosomal functions (*3*). One tool to probe the functional roles of these various epigenetic regulators is to determine their subcellular localization and how this localization changes during the cell cycle and development. An important opportunity for plasticity in epigenomic structure occurs during DNA replication, when histones are disrupted and then reassembled (*4*). Here, we describe methods to examine histone modifications and their relationship with DNA replication during S-phase.

From: *Methods in Molecular Biology, vol. 325: Nuclear Reprogramming: Methods and Protocols*
Edited by: S. Pells © Humana Press Inc., Totowa, NJ

A typical immunofluorescence protocol includes the following steps: (1) cellular fixation to immobilize the proteins of interest, (2) cellular permeabilization to allow antibodies to access antigens, (3) blocking nonspecific immunoglobulin binding, (4) incubation with antibody to recognize proteins of interest, and (5) visualization by microscopy. Visualization is achieved either by conjugating the specific antibody itself with a fluorochrome (direct immunofluorescence) or by including a subsequent incubation with a fluorescently tagged anti-immunoglobin secondary antibody whose specificity matches the primary antibody isotype (indirect immunofluorescence). Indirect immunofluorescence is used more commonly because the same fluorescently conjugated secondary antibody can be used for many different primary antibodies, avoiding the need to fluorescently conjugate each new antibody and providing continuity in experiments through the use of the same reagent for many different visualizations. In addition, because the secondary antibody recognizes multiple epitopes on the primary antibody, an amplification of signal is achieved over direct immunofluorescence. The primary drawback of indirect immunofluorescence is the added risk of crossreactivity and increased background resulting from the use of two antibodies rather than one.

As with all methods, certain limitations should be considered when interpreting the results of immunofluorescence experiments. These limitations and potential approaches to overcome them have been discussed extensively elsewhere *(5,6)*. However, two of the most important issues related to immunofluorescence microscopy of histone modification are stressed here. First, antibody specificity is the single most important determinant for successful immunofluorescence microscopy, yet this determinant is often ignored as investigators all too frequently assume that their reagents are highly specific. Importantly for our discussion here, some sites of histone modification (for example, H3-K9 and H3-K27) have similar neighboring amino acid sequences that can easily cause crossreactivity between respective antibodies. Thus, quality control for antibody specificity, for example, by using histone modification-peptide arrays *(7)* to spot potential crossreactivity, is an invaluable asset for interpretation of results. Second, it is important to remember that histone modifications are embedded inside nuclei and that epitopes may not be presented in the same way as they are in Western blots or even peptide arrays. Hence, the fixation and permeabilization steps should be controlled carefully to preserve in vivo distribution of histone modification and allow for antibody access as well. If the same pattern is obtained with multiple fixation and permeabilization methods, one can feel more comfortable with the results. If not, more complicated methods, such as mass spectrometry and mutational analyses *(8,9,16)* may be needed to aid in the interpretation of results.

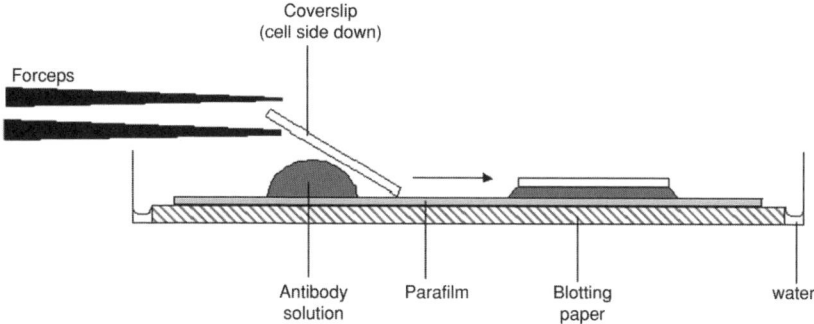

Fig. 1. Summary of antibody reaction in humidified chamber. A piece of filter paper slightly smaller than the box size is placed in the box and moistened with ddH$_2$O. After draining off excess ddH$_2$O, a piece of parafilm is placed on top of the filter paper. For antibody incubation, a drop of antibody solution is pipetted onto the parafilm and the cover slip is placed onto it, cell side facing down, by gently lowering one side of the cover slip.

2. Materials

2.1. Supplies and Equipment

1. Microscope cover glasses (12CIR; cat. no. 12-545-80; Fisher Scientific): wrapped in aluminum foil, autoclaved and transferred to sterile dishes.
2. Sterile culture dishes and 24-well plates.
3. Microscope slides.
4. 24-well cell culture plates (e.g., Costar brand, cat no. 07-200-84; Fisher Scientific).
5. Forceps with fine tips (e.g., Jewelers Microforceps from Fisher Scientific, cat. no. 08-953E straight end or 08-953F curved end).
6. Humidified chamber (see **Fig. 1**): any small box with a lid can be made into a humidified chamber. For example, we use the shipping box (~6.5 × 9.0 cm) for Vectashield mounting medium from Vector laboratories, Inc. Alternatively, plastic microscope slide storage boxes make another suitable chamber. A piece of blotting paper slightly smaller than the box size is placed in the box and moistened with ddH$_2$O. After draining off excess ddH$_2$O, a piece of parafilm is placed on top of the blotting paper and the chamber is ready for use.
7. Conventional or confocal-equipped epifluorescence microscope with appropriate filter set and high power, high numerical aperture lenses.

2.2. Reagents

1. 5-Bromo-2'-deoxyuridine (BrdU; Sigma Chemical, St. Louis, MO): make a 10 mg/mL stock solution in ddH$_2$O, filter sterilize (0.2 μm; cat. no. 21062-25; Corning), aliquot, and store at –20°C. After thawing the stock solution, the white precipitate should be redissolved in solution by vortexing and/or heating to 37°C.

2. 2% and 4% Paraformaldehyde (PFA) in phosphate-buffered saline (PBS), pH 7.4. It is optimal to prepare the fixative freshly. However, this is very time-consuming. For most applications, we make an 8% paraformaldehyde in PBS stock and keep it frozen at –20°C. When needed, the stock is thawed at 60°C and diluted with PBS to the desired concentration.

3. 8% (w/v) PFA in PBS, pH 7.4. Add 8 g of solid paraformaldehyde to approx 80 mL of ddH$_2$O. Add a few drops of 1 N NaOH to the solution and heat to 60°C while stirring. Cool the solution to room temperature, add 10 mL of 10X PBS, adjust pH to 7.4, and bring the total volume to 100 mL. Aliquot and freeze at –20°C in the dark for up to several months.

4. 1X PBS.

5. 70% EtOH (ice cold).

6. PBS/T: 0.5% Tween-20 in PBS.

7. 0.2% Triton X-100 in PBS.

8. Blocking buffer (*see* **Note 1**): 3% bovine serum albumin\PBS\0.5% Tween-20.

9. Primary antibody solution: dilute antibody with blocking buffer. For different batches of antibodies, titration is needed to determine the optimal dilution. In this protocol, we describe the use of a rabbit polyclonal antibody raised against histone H3 tri-methylated at lysine 9 (rabbit αMe$_3$K9H3 *[10]*).

10. Secondary antibody solution: choose fluorescence as required by experiments (*see* **Note 2**). For example, here we use Alexa 488-conjugated goat anti-rabbit IgG (Invitrogen, cat. no. A-11034) that is diluted 1/400 in blocking buffer.

11. 0.3 M Glycine in PBS.

12. 0.5% NP40 (a nonionic detergent) in PBS.

13. 1.5 N Hydrochloric acid.

14. Mouse anti-BrdU antibody (Becton Dickinson, cat. no. 347580): diluted 1/10 to 1/20 with blocking buffer.

15. Secondary antibody for BrdU: Alexa 592-conjugated goat anti-mouse IgG diluted 1/400 with blocking buffer.

16. 4',6-Diamidino-2-phenylindole dihydrochloride (DAPI; cat. no. D9540; Sigma Co.): make a 10 mg/mL stock by dissolving the powder in ddH$_2$O, aliquot, and store at –20°C in the dark. (It is stable for more than 1 yr.) Also, make a 0.1 mg/mL solution and keep at 4°C for routine use. (This solution also is very stable.)

17. Vectashield (cat. no. H-1000; Vector Laboratory) or other antifading reagent.

3. Methods

3.1. Preparation of Cells Attached to Cover Slips (see Note 3)

1. Place several sterile cover slips in the culture dish either with sterile tweezers or by glass pipet aspiration (the vacuum is sufficient to hold the cover slip).

2. Seed the cells on the dish and let the cells grow to the desired cell density. Usually 50 to 70% cell confluency is desirable. High cell densities will facilitate microscopic analysis, but too high a cell density will inhibit BrdU incorporation and introduce staining background.

3. Label the cells by adding BrdU (from stock solution) to a final concentration 10 to 30 μg/mL. Mix well and return to the incubator for 10 min. The minimum labeling time is 2 min at higher concentrations *(11)*, but do not exceed 30 min because each replication site takes 45 to 60 min to complete its replication *(12)*.

3.2. Fixation and Permeabilization of Cells (see Notes 4 and 5)

Cells on cover slips can either be fixed directly in dishes or the cover slips can be removed one at a time and fixed in 24-well plates.

3.2.1. Paraformaldehyde Fixation and Triton Permeabiliztion

1. Remove the medium from the dish and quickly rinse the cells with PBS twice to remove cell debris.
2. Incubate with 2% PFA in PBS at room temperature for 10 min (*see* **Note 4**).
3. Wash the cells twice for 5 min each in PBS. Cover slips can be stored at 4°C for up to a week.
4. Incubate for 5 min with 0.2% Triton X-100 in PBS to permeabilize the cell membrane (*see* **Note 6**).
5. Wash the cells twice for 5 min each in PBS.

3.2.2. Ethanol Fixation

1. Remove the medium from the dish and quickly rinse the cells with PBS twice to remove cell debris.
2. Incubate with ice-cold 70% ethanol at 4°C for 10 min.
3. Wash the cells twice for 5 min each in PBS. Cover slips can be stored at 4°C for up to a week.

3.3. Immunostaining of Cells for Histone Modification (see Notes 7 and 8)

Do not allow cover slips to dry at any step to avoid high background staining.

1. Incubate the cells with blocking buffer for 20 min (*see* **Note 1**).
2. Rinse the cells with PBS/T.
3. Incubate the cells with primary antibody for 1 h.

All antibody incubations should be conducted in a humidified chamber, usually at room temperature (*see* **Note 9**). As shown in **Fig. 1**, pipet drops of antibody solution (as little as 4 μL for each drop) onto the parafilm, being careful to avoid air bubbles. One cover slip should be placed on top of each drop with the cell side facing down.

4. Add PBS/T to a 24-well cell culture plate and transfer the cover slips to the 24-well plate with the cell side facing up. Wash with PBS/T three times for 5 min each.
5. Incubate with Alexa 488-conjugated secondary antibody solution for 1 h (*see* **Note 2**).

OPTIONAL: if staining is to be done only for protein and not for BrdU:

6. Transfer the cover slips back to a 24-well plate and incubate with 0.01 μg/mL DAPI in PBS/T for 5 min.
7. Wash with PBS/T three times for 5 min each.
8. Pipet approx 2 to 3 μL of Vectashield onto slides and place the cover slips on top of the Vectashield with the cell side facing down. Seal the cover slips with nail varnish and store the specimens in a sealed dark box at 4°C or –20°C. Significant loss of fluorescence signal may be experienced after 1 wk of storage at 4°C or longer than 1 mo at –20°C.

3.4. Postfixation and Stain for BrdU Staining

If colocalization with sites of DNA replication is desired, the initial antibodies must be covalently fixed with formaldehyde to preserve their localization throughout the denaturation step that is necessary to visualize sites of BrdU incorporation. In this case, cover slips after **step 5** (in **Subheading 3.3.**) should be treated as follows:

1. Wash with PBS/T twice for 5 min each.
2. Wash with PBS once for 5 min.
3. Incubate with 4% PFA in PBS at room temperature for 15 min.
4. Wash in PBS once for 5 min.
5. Incubate with 0.3 M glycine in PBS for 5 min to quench unreacted formaldehyde.
6. Wash in PBS once for 5 min.
7. Incubate with 0.5% NP40 in PBS for 15 min.
8. Wash in PBS twice for 5 min each.
9. Denature DNA in 1.5 N hydrochloric acid at room temperature for 30 min.
10. Wash in PBS twice for 5 min each.
11. Wash in PBS/T once for 5 min.
12. Incubate with mouse anti-BrdU diluted 1/10 to 1/20 in blocking buffer for 1 h.
13. Wash in PBS/T three times for 5 min each.
14. Incubate with Alexa 592-conjugated goat anti-mouse IgG diluted 1/400 in blocking buffer for 1 h.
15. Proceed to **step 6** in **Subheading 3.3. Figure 2** shows an example of the kinds of results obtained with this protocol.

4. Notes

1. Bovine serum albumin is a practical and economical blocking agent that is usually sufficient. Ideally, however, to eliminate background staining, commercially available nonimmune serum from the species of the secondary antibody (e.g., rabbit, sheep, horse, donkey) should be used to block nonspecific antibody binding. We use 10% serum in PBS\Tween-20.
2. For the secondary antibody, you can choose the specific fluorochrome according to your experimental requirement. Fluorescein and Texas Red are used commonly. However, they are prone to photobleaching and are not the best choice for

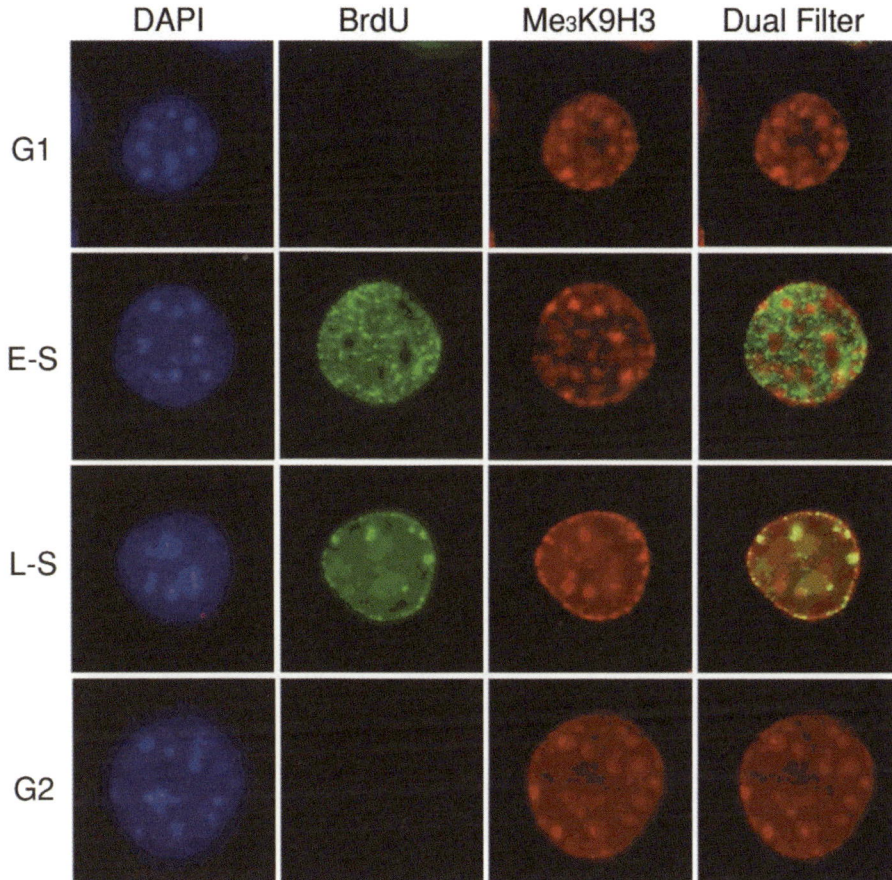

| | DAPI | BrdU | Me₃K9H3 | Dual Filter |

Fig. 2. Localization of Me₃K9H3 during cell cycle. Exponentially growing mouse C127 cells were pulse-labeled with BrdU and then costained for anti-BrdU and anti-methyl K9 H3 *(10,16)*, as described herein. Although we recommend cell synchronization to verify the spatiotemporal order of BrdU staining patterns, once established, one can easily distinguish cells in different phases of the cell-cycle based upon these patterns. Shown are examples of G1-phase (small BrdU-negative nuclei), early and late-S-phase (distinguished by the spatial pattern of BrdU labeling), and G2-phase (large BrdU-negative nuclei).

image acquisition requiring long exposure times, such as Z- stack acquisition by confocal (or deconvolution) microscopy. Companies are now introducing more photostable fluorochromes. Among them, our favorite is the Alexa dye series (Invitrogen, Inc).

3. Adherent cells can be grown on slides or cover slips before fixation. In our hands, growing cells directly on 12-mm round glass cover slips is easier and more convenient for subsequent immunostaining manipulation. For those cells that adhere

less strongly, cover slips should be pretreated with poly-(L-lysine) or with calgon metasilicate to facilitate cell attachment (*6*). Suspension cells can be stained in suspension, as is done for flow cytometry. However, for convenience, suspension cells can be adhered to clean microscope slides using a cytocentrifuge (e.g., cytospin2 or 3; Shandon) and the same procedures described here for adherent cells can be followed. Cytocentrifuges are typically designed for slides. However, a round cover slip can be placed atop the slide to spin cells onto a cover slip. One drawback of adhering cells with a cytocentrifuge is the apparent change in cell morphology (flattening) that occurs as a result of the cytospin method, which needs to be taken into account when interpreting results.

4. Sometimes, triton extraction can be performed before cell fixation to remove any soluble pool of proteins that are present in excess within nuclei and mask the fluorescence pattern of the chromatin-bound fraction (*13*). This is not necessary for most histones because there is very little soluble histone pool.

5. Choosing a suitable fixative may be the most important and the most challenging task during the development of a protocol for a new antigen because the choice of fixative will directly affect the accuracy and interpretation of the final result. Very different localization patterns can be obtained with different fixatives and confirming the true in vivo localization of a protein can be difficult, for example, by comparison with the localization of a GFP-tagged version of the same protein (e.g., *see* **Fig. 2** in **ref.** *14*). For the localization of histone modifications, such confirmation is not available, and many of the existing results will need to be confirmed by microdissection or other purification methods coupled with modern mass spectrometry. There are two classes of fixative: crosslinking reagents and organic solvents. Crosslinking reagents, including formaldehyde and glutaraldehyde, covalently crosslink molecules and generally are believed to more accurately preserve cell morphology. Glutaraldehyde fixation produces poor accessibility of antibodies to nuclear antigen and is rarely used to study nuclear antigen. Formaldehyde (~1–4%) fixation works relatively well to preserve cell morphology and antigen accessibility and can be chosen as a starting fixation protocol for new antibodies. Longer fixation times can introduce artifacts and therefore it is recommended to use the shortest fixation time that maintains antigen localization. Obviously, if the epitope of the antigen is crosslinked by formaldehyde or if it is masked within the structure of the protein or within protein complexes, the antigenicity of that protein will be reduced. Hence, in cases of poor antigenicity, organic solvent fixation can be tried. The most commonly used organic solvents are cold methanol (100%) and ethanol (70%), which maintain insoluble components and large soluble components (>100 kDa) in place while washing away small soluble components (*15*). Although organic solvents cause cell shrinkage and disrupt cytoplasmic structure dramatically, they can produce very good results with insoluble nuclear proteins. Unfortunately, there is no perfect fixation method and the best results must be determined empirically for each new antibody.

6. After formaldehyde fixation, the cell and nuclear membrane should be permeabilized to allow antibodies to penetrate the nuclei. The most commonly used permeabilization agents for nuclear antigens are the nonionic detergents Triton X-100, NP40, Bij35, and Tween-20. In our hands, Triton X-100 and NP40 are more efficient whereas Tween-20 is less efficient for permeabilizing nuclear membrane. Conditions presented here are very robust for most nuclear antigens but permeabilization conditions should be optimized for each new antibody.

7. Several experiments should be included to control for the specificity of the antibody reaction, e.g., (1) compete the primary antibody with the antigen peptide to check the specificity of the primary antibody; (2) omit the secondary antibody to examine the autofluorescence from the specimen; (3) omit the primary antibody to check the specificity of the secondary antibody.

8. Here, we describe the protocol for one protein antigen, co-localized with BrdU. Quite often, two protein antigens need to be stained simultaneously to evaluate co-localization. Often, one can mix two antibodies together and follow the same protocol. However, sometimes the presence of one antibody can inhibit another antigen's antibody reaction, particularly when they co-localize very closely. In this case (i.e., if one antigen is not observed when simultaneously stained with another), sequential staining with each antibody added in both orders should be tried *(11)*. When co-localizing with BrdU, the harsh denaturation steps required to visualize BrdU necessitate staining for protein first.

9. The length and temperature of incubation can be varied. In many cases, reactions can be accelerated (15–30 min) at 37°C, but we routinely incubate for 1 h at room temperature. However, if poor signals are obtained it is recommended that antibody incubations be carried out overnight at 4°C. With either of these deviations, the risk of sample dehydration is increased; hence, it is advised to add more antibody solution.

Acknowledgments

We thank D. Cameron for critical evaluation of the manuscript. Research in the Gilbert lab is supported by NIH grant GM-57233-01, NSF grant MCB-0077507, and American Cancer Society grant RPG-97-098-04-CCG.

References

1. Jenuwein, T. and Allis, C. D. (2001) Translating the histone code. *Science* **293,** 1074–1080.
2. Strahl, B. D. and Allis, C. D. (2000) The language of covalent histone modifications. *Nature* **403,** 41–45
3. Malik, H. S. and Henikoff, S. (2003) Phylogenomics of the nucleosome. *Nat. Struct. Biol.* **10,** 882–891
4. McNairn, A. J. and Gilbert, D. M. (2003) Epigenomic replication: linking epigenetics to DNA replication. *Bioessays* **25,** 647–656
5. Harlow, E. and Lane, D. (1999) *Using Antibodies: A Laboratory Manual.* Cold Spring Harbor Laboratory Press, Cold Spring Harbor, New York.

6. Pines, J. (1997) Localization of cell cycle regulators by immunofluorescence. *Methods Enzymol.* **283,** 99–113
7. Perez-Burgos, L., Peters, A., Opravil, S., Kauer, M., Mechtler, K., and Jenuwein, T. (2004) Generation and characterization of methyl l-lysine histone antibodies. *Methods Enzymol.* **376,** 234–254.
8. Peters, A. H., Kubicek, S., Mechtler, K., O'Sullivan, R. J., Derijck, A. A., Perez-Burgos, L., Kohlmaier, A., Opravil, S., Tachibana, M., Shinkai, Y., Martens, J. H., and Jenuwein, T. (2003) Partitioning and plasticity of repressive histone methylation states in mammalian chromatin. *Mol. Cell* **12,** 1577–1589.
9. Rice, J. C., Briggs, S. D., Ueberheide, B., Barber, C. M., Shabanowitz, J., Hunt, D. F., et al. (2003) Histone methyltransferases direct different degrees of methylation to define distinct chromatin domains. *Mol. Cell* **12,** 1591–1598.
10. Cowell, I. G., Aucott, R., Mahadevaiah, S. K., Burgoyne, P. S., Huskisson, N., Bongiorni, S., et al. (2002) Heterochromatin, HP1 and methylation at lysine 9 of histone H3 in animals. *Chromosoma* **111,** 22–36.
11. Dimitrova, D. S., Todorov, I. T., Melendy, T., and Gilbert, D. M. (1999) Mcm2, but not RPA, is a component of the mammalian early G1-phase prereplication complex. *J. Cell Biol.* **146,** 709–722.
12. Dimitrova, D. S., and Gilbert, D. M. (1999) The spatial position and replication timing of chromosomal domains are both established in early G1 phase. *Mol. Cell* **4,** 983–993.
13. Dimitrova, D. S., and Gilbert, D. M. (2000) Stability and nuclear distribution of mammalian replication protein A heterotrimeric complex. *Exp. Cell Res.* **254,** 321–327.
14. Okuno, Y., McNairn, A. J., den Elzen, N., Pines, J., and Gilbert, D. M. (2001) Stability, chromatin association and functional activity of mammalian pre-replication complex proteins during the cell cycle. *EMBO J.* **20,** 4263–4277.
15. Melan, M. (1999) Overview of Cell Fixatives and Cell Membrane Permeants, in: *Immunocytochemical Methods and Protocols* (Javois, L., ed.), Methods in Molecular Biology, Vol. 115, Humana Press Inc., Totowa, NJ, pp. 45–55.
16. Wu, R., Terry, A. V., Singh, P. B., and Gilbert, D. M. (2005) Localization and replication timing of histone H3 lysine 9 methylation states. *Mol. Biol. Cell* **16,** 2872–2881.

13

Quantitative Analysis of Telomerase Activity and Telomere Length in Domestic Animal Clones

Dean H. Betts, Steven Perrault, Lea Harrington, and W. Allan King

Summary

It has been speculated that incomplete epigenetic reprogramming of the somatic cell genome is the primary reason behind the developmental inefficiencies and postnatal abnormalities observed after nuclear transplantation in domestic animal clones. One chromosome structure that is altered in dividing somatic cells is telomere length—the terminal ends of linear chromosomes capped by repetitive sequences of G-rich noncoding DNA, $(TTAGGG)^n$, and specific binding proteins. Telomeres are critical structures that function in maintaining chromosome stability and ensure the full replication of coding DNA by acting as a buffer to terminal DNA attrition due to the end replication problem. Telomere shortening limits cellular proliferation through a DNA damage signal activating permanent cell cycle arrest at a critical telomere length or through structural telomere alterations that prevents effective chromosome capping. Telomere-mediated signaling of cellular senescence has been established for many somatic cell types in vitro, except for germ cells, cancer lines, and regenerative tissues in which telomere length is maintained primarily by the ribonucleoprotein telomerase, a reverse transcriptase that synthesizes TTAGGG repeats de novo onto the chromosome ends. Telomere length discrepancies have been reported in animal clones as being shorter, no different, and even longer than in age-matched control animals, but the etiology is not yet understood. Possible explanations include differences in donor cell type and the efficiency of telomerase reprogramming. This chapter summarizes the conventional protocols and recent advances in telomere length and telomerase activity measurement that will help elucidate the mechanism(s) behind telomere length deregulation in somatic cell clones and its role in chromosomal instability, cellular senescence, and organismal aging in vivo.

Key Words: Telomere; TRF; Q-FISH; telomerase; TRAP; RQ-TRAP; animal cloning; SCNT; aging.

1. Introduction

Somatic cell nuclear transfer (SCNT) technology has resulted successfully in the production of live, fertile offspring from a variety of vertebrate species *(1–4)*. However, with high embryonic, fetal, and neonatal death rates and de-

From: *Methods in Molecular Biology, vol. 325: Nuclear Reprogramming: Methods and Protocols*
Edited by: S. Pells © Humana Press Inc., Totowa, NJ

fects, including abnormalities affecting the placenta, the respiratory and circulatory systems and growth of the fetus and neonate, its efficiency is poor *(5)*. Variations in DNA methylation *(6)*, patterns of X-inactivation *(7)*, imprinted gene expression *(8)*, telomerase activity *(9)*, and telomere lengths *(9,10)* have been observed in the somatic cells of some SCNT clones, which has led to the speculation that incomplete genetic and epigenetic reprogramming of the donor cells is the main cause of these developmental irregularities and postnatal abnormalities *(5)*. Sufficient reversal of chromatin modifications to an undifferentiated, totipotent state must take place in the transplanted somatic cell nucleus to permit the temporal and spatial re-expression of genes required for proper embryonic and fetal development of a clone.

1.1. Biology of Telomeres

One example of a chromatin structure that is altered in differentiated somatic cells is telomere length. In all vertebrate cells, the terminal ends of linear chromosomes are capped by repetitive sequences of G-rich noncoding DNA, $(TTAGGG)^n$, and by specific binding proteins that are repeated up to many kilobases in length *(11)*. The length of the telomere varies widely among species and also among cells within a single population (reviewed in **ref. *12***). The G-rich telomeric strand runs 5' to 3' towards the terminus and protrudes 12 to 16 nucleotides beyond the complementary C-rich strand in ciliates *(13)* and as many as 150 to 200 nucleotides in some human cells *(14,15)*. The presence of the G-rich overhangs is both conserved and believed to be essential for the maintenance of chromosome end structure and function *(16,17)*.

Telomeres are critical structures that function in chromosome stability maintenance, function in the proper segregation of chromosomes during meiotic and mitotic divisions, and ensure the full replication of coding DNA during cell division (reviewed in **ref. *18***). However, conventional DNA polymerases cannot replicate the extreme 5' ends of chromosomes because removal of the most terminal RNA primer from the lagging strand leaves a small region of uncopied DNA *(19)*. Telomeric DNA shortening has been calculated to be, on average, between 50 and 200 base pairs (bp) per cell division *(19,20)* and has been proposed to act as a "mitotic clock" that limits proliferative capacity through a DNA damage signal activating permanent cell cycle arrest (cellular senescence) at a critical telomere length *(21)*, or through structural telomere alterations *(22–24)* that prevent effective chromosome "capping."

1.2. Telomerase

Cells can overcome this "end-replication problem" by the activation/expression of the ribonuceloprotein enzyme called telomerase, which synthesizes TTAGGG repeats *de novo* onto the ends of chromosomes *(25)*. The RNA subunit (TR) of the enzyme contains a region of homology with telomeric DNA

(25,26). This domain allows the enzyme to align with its substrate and to provide a template for the addition of deoxynucleotides to the terminal telomeric DNA *(27)*. The catalytic subunit (TERT) of telomerase was identified in yeast, cilates, and the human as Est2p, p123, and hEST2 respectively *(28–31)*. All three proteins contain several sequence motifs that are characteristic of the catalytic regions of reverse transcriptases. The catalytic subunit is upregulated upon activation of telomerase activity and downregulated when telomerase activity is decreased *(31,32)*. These results suggest that induction of hEST2/ TERT messenger ribonucleic acid (mRNA) expression is required for telomerase activity. However, the regulation of telomerase activity now appears more complex. Recently, a number of RNA variants were identified from the single copy telomerase catalytic subunit gene *(31)*. Alternative splicing of the transcript may be important for the regulation of telomerase activity and may give rise to proteins with different biochemical properties *(31)*. Indeed, some of these isoforms are catalytically inactive and may titrate telomerase RNA away from interactions with other TERT isoforms, causing cells to lack telomerase activity *(33)*.

Telomerase is downregulated upon differentiation *(34)*; hence, most adult somatic cell types do not express telomerase activity and therefore have a limited replicative lifespan in vitro *(20,21)*. In humans, telomerase activity, with telomere length maintenance, has been detected in germline tissues, cells of renewal tissues, cancer cells, and immortalized cell lines but not in most somatic tissues *(35–38)*. The introduction of the telomerase catalytic subunit (hTERT) into normal human diploid cells succeeded in extending replicative lifespan but the cells displayed normal growth controls and normal karyotypes *(39,40)*. Late-generation mice lacking the telomerase RNA (mTR$^{-/-}$) component displayed shortened telomeres, chromosomal abnormalities, increased apoptosis, and decreased proliferation in the testis, bone marrow, and spleen, and exhibited defective spermatogenesis *(41)*. Thus telomerase activity is thought to maintain sufficiently long telomeres from generation to generation through the germline and helps counteract telomere loss while sustaining chromosome integrity for highly proliferative cell types *(41)*.

1.3. Cellular Aging In Vivo?

Although telomeres are involved in the regulation of cellular replicative lifespan and aging in vitro, their role(s) in these processes in vivo are not as clear. Telomere length does shorten with age in many tissues *(42)*. There is an apparent loss of division potential in certain tissues over time *(43,44)*, and senescent fibroblast cells have been shown to accumulate with age in skin biopsies *(45)*. However, it is not known whether shortened/disrupted telomere lengths trigger permanent cell growth arrest in vivo and contributes to organismal aging.

The telomere hypothesis of cellular aging can be indirectly assessed in vivo by using somatic cell cloning since it involves the production of animals from aged adult and cultured somatic cells without the involvement of the germline. To date, telomere lengths in animal clones have been reported as being shorter *(10,46–49)*, no different *(9,10,50–52)*, and even longer *(10,53)* than those from age-matched control animals. Sheep cloned by nuclear transfer of cultured cells from embryonic or fetal origin displayed shortened telomeres compared with those from age-matched controls (*[46–48]*; Betts and King, unpublished data). Dolly, cloned from a cultured mammary cell from a 6-yr-dd ewe, displayed significantly shortened telomere lengths of approx 20% at 1 yr of age *(46,47)* and 13% at the time of her death *(49)* compared with her age-matched counterparts. Therefore, both the age of the donor nucleus and the proliferation in culture contributed to the telomeric loss observed in this small sample of sheep clones. In contrast, Lanza et al. *(53)* have demonstrated the reversal of cellular aging using near-senescent donor somatic cells as the nuclear donor. Fibroblasts from bovine fetuses cloned from these cells displayed an extended replicative lifespan and rebuilding of telomere length compared with control fetal fibroblasts and senescent fibroblasts, respectively. Nucleated blood cells from 5- to 10-mo-old cattle clones appeared to have longer telomere lengths compared with newborn and age-matched control animals *(53)*. We and others have measured telomere lengths in cattle *(9,10,50,52)* and pig *(51)* clones that are no different from those measured from age-matched controls; however, we have also observed shortened telomere lengths in cattle and sheep clones derived from adult and fetal fibroblasts, respectively (Betts and King, unpublished). Miyashita et al. *(10)* revealed variable telomere length differences in clones, both longer and shorter than age-matched controls, that were dependent on the type of donor somatic cell used for nuclear transfer.

Evidently, telomere length is abnormal in clones but the etiology is not clear. Possible explanations include differences in donor cell type *(10)*, efficiency of telomerase reprogramming *(9)*, individual-to-individual variations *(9,10)*, effects of the nuclear transfer procedure itself *(9,10)*, the tissue/cell type analyzed *(9,10)*, and the disparity among species *(9,10,46–53)*. Although clones harbor altered telomere lengths and display early onset of age-related phenotypes (reviewed in **ref.** *54*), further research in telomerase and telomere length regulation must be carried out in clones to determine the mechanism(s) behind telomere length deregulation and its role in chromosomal instability, cellular senescence in vivo and organismal aging.

2. Materials

2.1. Telomeric Repeat Amplification Protocol

The reagents for telomeric repeat amplification protocol (TRAP) are as follows:

Material/Reagent	Supplier	Cat. no.
2-Propanol (Isopropanol)	Sigma	I-9516
40% Polyacrylamide/bisacrylamide stock (19/1)	Bio-Rad	161-0144EDU
Ammonium persulfate	Bio-Rad	161-0700EDU
Bio-Rad Protein Assay Kit	Bio-Rad	500-0002
Boric acid	Sigma	B-0252
Bromophenol blue	Sigma	B-6131
EDTA	Sigma	E-5134
EGTA	Sigma	E-3839
Ethanol	—	—
Filter paper	—	—
Fuji Medical X-ray Film (Super RX)	—	03G050
Glacial acetic acid	—	—
Glycerol	Sigma	G-2025
Heat block	—	—
Magnesium chloride ($MgCl_2 \cdot 6H_2O$)	Sigma	M-2670
Nonidet P-40 (NP40)	Sigma	N-6507
PAGE vertical gel apparatus	—	—
PhosphorImager™	—	—
Phenylmethyl sulfonyl fluoride	Sigma	P-7626
Potassium chloride (KCl)	Sigma	P-4504
Potassium phosphate (KH_2PO_4)	Sigma	P-5655
Power supply	—	—
Pronase	Calbiochem	537088
Ribonuclease (RNase) inhibitor	Gibco-BRL	15518-012
Sodium acetate	Sigma	S-8625
Sodium chloride (NaCl)	Sigma	S-3014
Sodium deoxycholate	Sigma	D-6750
Sodium phosphate ($Na_2HPO_4 \cdot 7H_2O$)	Sigma	S-9390
Sterile H_2O	—	—
SYBR® Green	Molecular Probes	S-7563
T4 polynucleotide kinase	Gibco-BRL	18004-010
Taq polymerase	Gibco-BRL	10342020
TEMED	Bio-Rad	161-0800EDU
Thermocycler (PCR)	—	—
TRAPEZE® Telomerase Detection Kit	Intergen Company	S7700
Tris base	Sigma	T-1503
Xylene cyanol	Sigma	X-4126
β-mercaptoethanol	Sigma	M-7522
^{32}P-ATP	PerkinElmer	BLU502H

EDTA, ethylenediaminetetraacetic acid; EGTA, ethylenebis(oxyethylenenitrilo)tetraacetic acid.

2.1.1. TRAP Solutions

1. NP40 Lysis buffer:

   a. 100 mL Stock buffer

10 mM Tris-HCl, pH 7.5	1 mL of 1.0 M stock
1 mM MgCl$_2$	1 mL of 0.1 M stock
1 mM EGTA	1 mL of 0.1 M stock
150 mM NaCl	10 mL of 1.5 M stock
10% glycerol	10 mL
Sterile H$_2$O	77 mL
	100 mL

   b. Before use, make up the working solution as follows: 4 µL of 0.1 M phenylmethyl sulfonyl fluoride (17.4 mg/mL in 2-propanol), 1.4 µL β-mer-captoethanol, 0.84 µL of 0.12 M sodium deoxycholate, 10 µL NP-40, 40 µL RNase inhibitor (10 U/µL). Bring to 4 mL with NP40 stock buffer.

2. 10X Phosphate-buffered saline (PBS): 80 g NaCl, 2 g KCl, 11.5 g Na$_2$HPO$_4$·7H$_2$O, 2 g KH$_2$PO$_4$, pH 7.3, to 1 L with sterile H$_2$O.

3. 10% Polyacrylamide stock in 400 mL of 0.5X TBE: 100 mL of 40% polyacrlamide solution (19:1), 40 mL of 5X TBE buffer, 260 mL deionized H$_2$O. Final vol 400 mL.

4. 50 mL of 10% Polyacrylamide gel: 49.5 mL of 10% polyacrylamide stock (19:1), 0.5 mL of 10% ammonium persulfate, 0.05 mL TEMED. Final vol 50 mL.

5. 50X TAE buffer: 242 g Tris base, 57.1 mL glacial acetic acid, 100 mL of 0.5 M EDTA, pH 8.0. Add sterile H$_2$O to 1 L and autoclave.

6. 5X TBE buffer: 54 g Tris base, 27.5 g boric acid, 20 mL of 0.5 M EDTA. Add deionized water to 1 L.

7. TRAP 1 L Gel fixative: 29.2 g NaCl, 3.28 g Na·acetate, 800 mL 50% ethanol; pH 4.2, to 1 L with 50% ethanol.

8. 5 mL Gel loading dye solution: 2.5 mL of glycerol, 1.0 mL of 1.25% bromophenol blue, 1.0 mL of 1.25% Xylene cyanol, 0.5 mL of 0.5 M EDTA, pH 8.0. Final vol 5.0 mL.

2.2. Quantification of Telomerase Activity by Real-Time SYBR Green Telomeric Repeat Amplification Protocol

The reagents for Real-Time SYBR Green Telomeric Repeat Amplification Protocol (RQ-TRAP) are as follows:

Material/Reagent	Supplier	Cat. no.
1X CHAPS lysis buffer	Chemicon	S7705
RNase inhibitor	Roche Diagnostics	3335399
Oligonucleotide TS primer (5'-AATCCGTCGAGCAGAGTT-3')	—	—
Oligonucleotide ACX primer (5'-GCGCGG[CTTACC]$_3$CTAACC-3')	—	—
SYBR Green PCR Master Mix	Roche Diagnostics	2158817
Thermal Cycler	Roche Diagnostics	03531414201
Positive control 293T cells	Chemicon	S7701

2.2.1. RQ-TRAP Solutions

1. 1X CHAPS lysis buffer: 10 mM Tris-HCl, pH 7.5, 1 mM MgCl$_2$, 1 mM EGTA, 0.1 mM benzamidine, 5 mM β-mercaptoethanol, 0.5% CHAPS, 10% glycerol. Store at –20°C.

2.3. Terminal Repeat Fragment Analysis

The materials used for terminal repeat fragment (TRF) analysis are as follows:

Material/Reagent	Supplier	Cat. no.
Telomere Length Assay Kit	BD PharMingen	559838
TeloTAGGG Telomere Length Assay Kit	Roche Molecular Biochemicals	2 209 136
Hybond-N+ Nylon Transfer Membrane	Amersham Pharmacia Biotech	RPN 303B
*Hinf*I	Gibco-BRL	15223-019
*Rsa*I	Gibco-BRL	15424-013
Fuji Medical X-ray Film (Super RX)	—	03G050
Agarose	Gibco-BRL	15510027
NaOH pellets	Sigma	S-5881
Hydrochloric Acid, 50% v/v (1+1)	Fisher Scientific	LC15130-1
Lauryl sulfate (SDS)	Sigma	L-4509

2.3.1. TRF Solutions

1. 20X SSC: 175.3 g NaCl, 88.2 g Na·citrate·2H$_2$O, 800 mL sterile H$_2$O, pH 7.0 (few drops of 10 N NaOH). Volume to 1 L with sterile H$_2$O and autoclave.
2. 50X TAE Buffer: 242 g Tris base, 57.1 mL glacial acetic acid, 100 mL 0.5 M EDTA, pH 8.0. Add sterile H$_2$O to 1 L and autoclave.

2.4. Quantitative Fluorescent In Situ Hybridization Measurement of Telomere Length

Materials for Q-FISH

Material/Reagent	Supplier	Cat. no.
Acetic acid	—	—
Blocking reagent	Roche Molecular Biochemicals	1096176
Colcemid	Boehringer-Mannheim	295892
Cover slips (24 × 55 mm)	—	—
Ethanol	—	—
Formaldehyde	—	—
Formamide (Ultrapure, pH 7.0–7.5)	—	—

(continued)

Materials for Q-FISH *(Continued)*

Material/Reagent	Supplier	Cat. no.
KCl	—	—
Methanol	—	—
MgCl$_2$ buffer	—	—
PBS	—	—
Pepsin	Boehringer-Manheim	1693387
Peptide nucleic acid (PNA) Telomere-FITC/Cy3 probe	PE Biosystems	8501
PermaFluor Aqueous Mountant	Immunon Shandon	434990
0.2-µm Orange fluorescent beads	Molecular Probes	F8809
Sodium citrate	—	—
TRIS buffer	—	—
Trypsin	—	—
Vectashield Mounting Medium	Vector Laboratories	H-1000

2.4.1. Solutions or Reagents for Use in Q-FISH

1. Hybridization buffer, 100 µL (10 slides): 70 µL of formamide (70% formamide), 20 µL of 100 mM Tris, pH 7.2 plus 10 µL of water in 10 mM Tris, pH 7.2, 0.001 g of blocking reagent (Roche Molecular Biochemicals), 1% (w/v) blocking reagent, 100 mM Tris pH 7.2, 1.211 g of Tris base, pH to 7.2 and QS to 100 mL. This solution gives one a final concentration of Tris at 20 mM because an equal amount of probe, which cuts the concentration in half, is added.
2. 100 mL Formamide buffer: 70 mL formamide (70% formamide), 10 mL 100 mM Tris, pH 7.2 (10 mM Tris, pH 7.2), 20 mL of water. Store at 4°C.
3. 500 mL Tris buffer: 3.0275 g of Tris base (50 mM Tris), 4.383 g of NaCl (0.15 M NaCl), pH to between 7.0 and 7.5, 250 µL of Tween-20 (0.05% Tween-20).
4. 10X PBS: 80 g NaCl, 2 g KCl, 11.5 g Na$_2$HPO$_4$·7H$_2$O, 2 g KH$_2$PO$_4$; pH 7.3, to 1 L with sterile H$_2$O.
5. 4% of Formaldehyde in PBS:

Formaldehyde	35.4 mL	47.2 mL	70.8 mL
PBS	265.6 mL	354 mL	531.2 mL
	300 mL	450 mL	600 mL

6. Pepsin in HCl:

Pepsin	125 mg	250 mL	300 mg	400 mg
Distilled water	125 mL	250 mL	300 mL	400 mL
HCl	104 µL	208 µL	250 µL	333.3 µL
	125 mL	250 mL	300 mL	400 mL

Store at 4°C. Pepsin from Boehringer Manheim, cat. no. 1693387, 5 g. Keep it at 4°C.

7. $MgCl_2$ buffer: 25 mM $MgCl_2$, 9 mM citric acid, 82 mM Na_2HPO_4; pH 7.4.
8. 50 g of 2.5% Blocking reagent in water (cat. no. 1096176; Roche Applied Science. Store at 4°C. It is not necessary to dissolve the blocking protein in the water; just make a suspension, and then mix it well just before taking some to add to the formamide. It will dissolve rapidly in this solution.
9. 1 M Tris-HCl: dissolve 121.1 g of Tris in 800 mL of water. Add 72 mL of HCl to reach pH 7.1. Make up to 1000 mL with water.
10. TE buffer: 10 mM Tris-HCl, pH 8.0, 1 mM EDTA, pH 8.0.
11. Wash solution I:

Formamide	175 mL	315 mL
10% Bovine serum albumin	2.5 mL	4.8 mL
1 M Tris	2.5 mL	4.8 mL
Distilled water	70 mL	126 mL
	250 mL	450 mL

12. Wash solution II:

10X TBS	35 mL	70 mL	105 mL
10% Tween-20	2.5 mL	5 mL	7.5 mL
Distilled water	312.5 mL	625 mL	937.5 mL
	350 mL	700 mL	1050 mL

13. 10X TBS (Tris-sodium chloride buffer): 60.54 g of 1 M Tris, 43.82 g of 1.5 M NaCl. Dissolve in 350 mL of distilled water. Add 35 to 36 mL of HCl to adjust the pH to 7.2. Make the final volume up to 500 mL by adding distilled water.
14. DAPI in Vectashield: 3 to 5 µL of DAPI stock (0.2 µg/mL distilled water [ddH_2O]) in 1 mL of Vectashield®.
15. 1 M $MgCl_2$: 203.3 g of $MgCl_2$·6H_2O in 800 mL of distilled water and make up to 1 L.
16. 30 mL PermaFluor Aqueous Mountant (cat. no. 434990; Immunon Shandon). Keep at 4°C.
17. PNA Probe: stock solution is 100X, 950 µg/mL in 1/1 N,N-dimethylformamide : water (or just water), in many aliquots, frozen in liquid nitrogen. After thawing, keep at 4°C. (Ordering information for the PNA probe is as follows: PE Biosystems, 500 Old Connecticut Path, Framingham, MA 01701, Tel: 508-383-7700 Fax; 508-383-7880. P.O. No: y08907 Sequence Name: Tel/cy3, CRD No: 8501. Sequence from N-terminus-to C-terminus): Cy3-OO-CCC-TAA-CCC-TAA-CCC-TAA. Amount of PNA: 4 OD.
18. 20 mL Colcemid (cat. no. 295892; Boehringer-Mannheim). Store at 4°C.

3. Methods

3.1. Detection of Telomerase Activity Using TRAP

The development of TRAP has allowed the large-scale screening of telomerase activity in samples of small cell numbers and limited tissue (*36*). TRAP is a polymerase chain reaction (PCR)-based assay that involves the addition of telomeric repeats $(TTAGGG)^n$ to the 3' end of a radiolabeled-oligonucleotide substrate (^{32}P-TS primer) that telomerase recognizes and binds to. The extended TRAP products are amplified by the PCR using the substrate oligonucleotide and reverse primers, generating a ladder of products with six-base increments starting at 50 nucleotides (i.e., 50, 56, 62, 68, etc.). TRAP reaction products are resolved by gel electrophoresis and then analyzed by densitometry.

3.1.1. Measurement of Telomerase Activity in Oocytes and Embryos

Telomerase activity has been detected in bovine oocytes and in vitro-produced embryos *(55)* and nuclear transfer embryos *(9)* using the TRAPeze® telomerase detection kit (Intergen Company, Purchase, NY) with minor modifications *(56)* of the original TRAP assay *(36)*. This protocol encompasses three primary steps: (1) Preparation of telomerase extracts from oocytes/embryos (2) telomerase assay, and (3) polyacrylamide gel electrophoresis of TRAP products.

3.1.2. Preparation of Telomerase Extracts From Oocytes/Embryos

1. Rinse pools of 10 to 25 oocytes/embryos three times in PBS and transfer to a 0.5-mL microcentrifuge tube. Pulse centrifuge to spin down cells and remove remaining PBS with a narrow-bore mouth pipet. Proceed to **step 2** or snap-freeze sample in liquid nitrogen. Store at –80°C or in liquid nitrogen.
2. Add NP-40 lysis buffer (working solution made fresh) to a concentration of one oocyte/embryo per microliter of lysis buffer (*see* **Note 1**). Aspirate the lysis buffer repeatedly with a narrow-bore pipet to lyse the zona-intact oocytes/embryos. Freeze-thaw the sample three times using liquid nitrogen and let sit on ice for a minimum of 30 min.

Alternatively, remove the zona pellucida surrounding live (not frozen) oocytes/embryos by a brief treatment (2–4 min) in 0.1% pronase (dissolved in culture medium) pre-equilibrated at 38.5°C in a 5% CO_2 in air atmosphere. Wash three times in PBS, transfer to a centrifuge tube, and remove remaining PBS by a quick spin and mouth pipet removal. Add NP-40 lysis buffer (1 oocyte/embryo per microliter) and vortex for 30 s. Freeze-thaw the sample three times using liquid nitrogen and let sit on ice for a minimum of 30 min.

3. Centrifuge the samples at 12,000*g* for 20 min at 4°C. Transfer the supernatant, leaving behind the cellular debris, into a new tube, snap-freeze in liquid nitrogen and store at –80°C.

3.1.3. Telomerase Assay

The assay is performed in two steps: Telomerase-mediated extension of an oligonucleotide primer (TS), which serves as a substrate for telomerase, and a PCR amplification of the resultant product using the TS and reverse primers.

Telomerase is a heat sensitive enzyme. Therefore, as a negative control, sample extracts are tested for heat sensitivity by incubating 5 µL at 85°C for 10 min. A negative lysis control consisting of 2.0 µL of NP40 lysis buffer and a positive telomerase extract control of an immortal telomerase/hTERT-positive 293T cell extract (0.5 µg) are used in each experiment. The optimal embryo/oocyte equivalents and protein content utilized for comparing relative telomerase activities between samples is determined by way of a standard curve (*see* **Fig. 1**). Two microliters of a serial dilution of embryo/oocyte extracts (0.625–5.0 embryo equivalents) and control samples are each added to 48 µL of the "Master Mix."

1. Thaw reagents, then on ice:
2. Prepare master mix:

	Radioactive per reaction	Nonradioactive per reaction
10X TRAP buffer	5 mL	5 mL
50X dNTP mix	1 mL	1 mL
³²P-TS primer[a]	2 mL	1 mL
TRAP primer mix	1 mL	1 mL
Taq Poly. (5U/mL)	0.4 mL	0.4 mL
dH₂O	38.6 mL	9.6 mL
	48 mL	48 mL

[a]32P-TS Primer (per reaction).

³²P-ATP	0.25 µL
TS primer	1 µL
5X exchange buffer	0.4 µL
T4 kinase (10 U/µL)	0.05 µL
dH₂O	0.3 µL
	2 µL

Place TS primer to be labeled at: (1) 37°C for 20 min; (2) 85°C for 5 min; and (3) ice (4°C). (*See* **Note 2**.)

3. Make up heat-inactivated control samples: for each sample extract, incubate 4 to 5 µL at 85°C for 10 min.
4. Into each tube pipet 48 µL of "Master Mix," plus any one of the following:
   a. 2 µL of sample extract
   b. 2 µL of heat-inactivated sample extract
   c. 2 µL of positive control (telomerase-positive cell line extract)
   d. 2 µL of negative control (NP-40 lysis buffer)
5. TS-Primer extension and PCR amplification (*see* **Note 3**):
   a. Telomerase extension of TS primer: place tubes at 30°C for 30 min and then transfer immediately to 94°C to inactivate the telomerase enzyme.

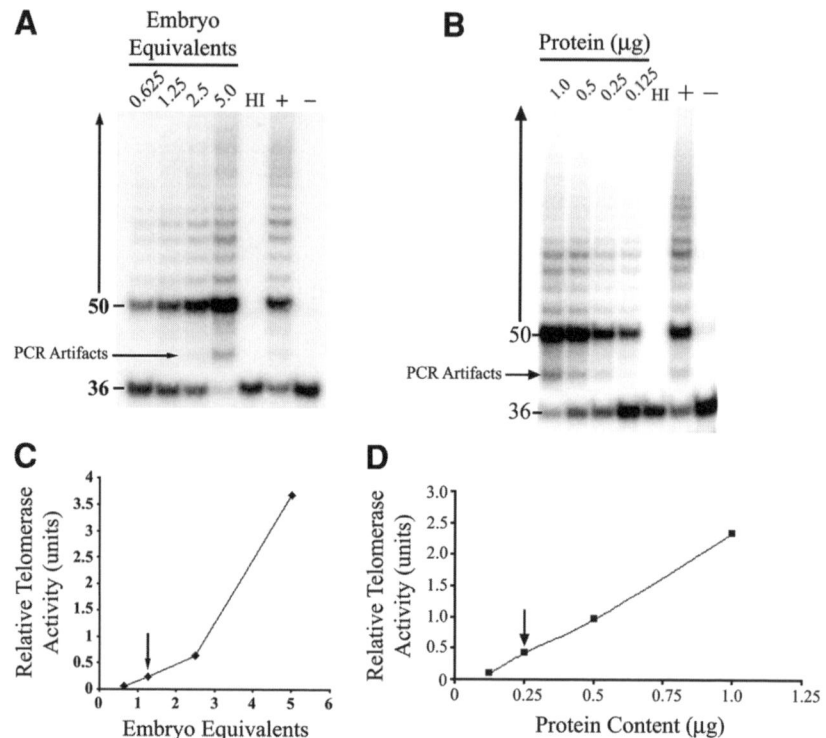

Fig 1. Serial embryo and protein dilution series for telomerase detection by the TRAP assay. To determine the optimal embryo equivalents and protein content to use in the TRAP assay, serial dilutions of cell extracts are required. (**A,B**) TRAP products as shown by a ladder of DNA bands with six base increments starting from 50 bp, with a 36-bp internal standard to control for equal sample loading and aberrant PCR amplifications. PCR artifacts are apparent at high embryo/protein concentrations due to primer dimers and *Taq* inhibitors that also reduced the 36 bp internal control signal. Heat inactivated (HI) control; positive 293T-cell extract control (+); and negative lysis buffer control (–) were used in each TRAP assay. (**C,D**) Densitometric analyses of TRAP reaction products reveal relative telomerase activities between samples of different embryo and protein content. A linear distribution was evident with the chosen optimal TRAP embryo (1.25 embryo equivalents) and protein (0.25 µg) content for these samples indicated by the arrows.

b. PCR amplification: two-step PCR: 94°C, 4 min; then 27 PCR cycles of 94°C, 30 s, followed by 60°C, 30 s; 4°C, soak. For nonradioactive TRAP: 30 PCR cycles.

As a PCR amplification control, the TRAPᴇᴢᴇ® Primer Mix contains internal control oligonucleotides K1 and TSK1 that together with radiolabeled TS oligonucleotide, produces a 36-bp band in every lane.

3.1.4. Polyacrylamide Gel Electrophoresis of TRAP Products

To visualize and quantify telomerase reaction products, samples are loaded and electrophoresed on a 10% nondenaturing polyacrylamide gel electrophoresis (PAGE). The gel is then fixed and placed on a PhosphorImager™ (Molecular Dynamics) screen or X-ray film overnight. The PhosphorImage™ screen or digitally scanned X-ray film is analyzed using the Molecular Analyst Software (Bio-Rad) or any image analysis system.

1. Add 5 μL of loading dye to each reaction.
2. Load 25 μl onto 10% ND-PAGE (no urea) in 0.6X TBE buffer.
3. Run 30 min at 100 V.
4. Run 3.0 h at 300 V, until the second dye is near the bottom.
5. Fix the gel in TRAP gel fixative, 25 min
6. Expose the gel to a PhosphorImager™ plate (or X-ray film) for 24 h.
7. Scan PhosphorImage or develop film.

3.1.5. Quantification of Relative Telomerase Activity

The relative telomerase activity is determined by the densitometric measurement of the ladder of TRAP reaction products for each sample(s), positive control extract (pc), background (b) signal and the 36-bp internal PCR control products for each sample (ic_s), and for the positive control (ic_{pc}). The internal PCR control product adjusts for sample-to-sample variation and equal gel loading whereas the positive control accounts for any gel-to-gel differences in densitometric readings and provides the reference point for the relative telomerase activity of each test sample. Densitometry readings should cover the entire TRAP product profile (50, 56, 62, 72 bp, etc.) for each lane/sample and the densitometry area should remain the same from lane to lane for each analysis. Therefore, the densitometry area should be as large as the lane with the strongest TRAP ladder intensities and processivity (i.e., the lane with the longest TRAP product). A smaller densitometry box will suffice for each 36-bp internal control signal. Quantification of the relative amount of telomerase activity is calculated using the following formula:

$$\text{Relative Telomerase Assay (units)} = [(s - b)/ic_s]/[(pc - b)/ic_{pc}]$$

3.1.6. Optimization of the TRAP Assay and Its Applications

The method described above has been optimized for extracts prepared from pools of 10 to 25 bovine oocytes and embryos (*9,55*). Quantifiable TRAP products have been obtained from sample extracts derived from as few as five pooled oocytes (data not published). The assay can indisputably compare relative telomerase activities from extracts equivalent to a concentration of less than one oocyte/embryo per reaction. However, there is opportunity within the protocol for improving the sensitivity of the TRAP assay. Increasing the incubation time for telomerase extension of the TS primer from 30 min to an hour or greater could increase the ability to detect and quantify telomerase activity from a single-cell sample. Additionally, increasing the number of PCR cycles could increase the TRAP amplification products (*also see* **Subheading 3.2.**). With any modification to the assay, a new standard curve must be derived from a serial dilution of sample extracts to determine the new linear range of telomerase detection (*see* **Fig. 1**).

3.2. Quantification of Telomerase Activity by Real-Time Quantitative TRAP

The TRAP assay has been used successfully to survey telomerase activity in cells and tissues from many species and specifically to demonstrate the dynamics of telomerase activity during early development within in vitro fertilized, parthenogenetically activated, and nuclear transfer-produced embryos *(9,10,55,57)*. A typical two-step TRAP methodology involves the telomerase-dependent addition of telomere repeats onto a substrate oligonucleotide followed by a PCR-based amplification of the extended products (*see* **Subheading 3.1.3.**). Previously, the assay also required post-PCR analysis steps involving polyacrylamide gel electrophoresis, autoradiography, and quantification by densitometry *(36)*. The optimization of this assay for use in a SYBR Green real-time protocol allows for a faster, more sensitive and more objective quantification of telomerase activity, making it ideal for the analysis of cells used in nuclear transfer and of the embryos so produced *(58)*.

The analysis of cultured cells and embryos will be used to demonstrate the methodology used in quantifying telomerase activity by real-time quantitative (RQ)-TRAP. Although the SYBR Green PCR Master Mix and Thermal Cycler used is from Roche Diagnostics, similar reagents and systems can be used with optimization.

The following method details: (1) the preparation of positive control cell extracts; (2) the preparation of cell extracts from culture; (3) the preparation of embryo extracts from culture; (4) the RQ-TRAP assay; and (5) quantification of telomerase activity.

3.2.1. Preparation of Positive Control Cell Extracts

Normally, transformed human kidney cells (293T) are used as a high telomerase-positive control cell sample. Cryopreserved 293T cells can be purchased from American Type Cell Culture (ATCC). Additionally, HeLa cells can be obtained either cryopreserved (ATCC) or as a frozen cell pellet ready for lysis (Chemicon). After the preparation of 10^6 cells in 200 µL of 1X CHAPS lysis buffer (5000 cell/µL) as described in **Subheading 3.2.2.**, aliquot and store at –80°C.

3.2.2. Preparation of Cell Extracts From Culture

An aliquot of 1X CHAPS lysis buffer is prepared immediately before beginning cell work by adding RNase inhibitor to a final concentration of 200 U/mL. A total of 200 µL of lysis buffer is needed for 10^6 cells of each sample. More or fewer cells can be prepared by adjusting the lysis buffer volume.

1. Cells are collected by centrifugation ($250g$, 5 min), washed with PBS, and recentrifuged into a pellet. The pellet can be stored at $-80°C$ or lower and is stable for one year. If frozen, cell pellets should be resuspended immediately upon thaw on ice in 1X CHAPS lysis buffer containing 200 U/mL RNase inhibitor.
2. Resuspend the sample in lysis buffer at 200 µL per 10^6 cells.
3. Incubate the lysate on ice for 30 min.
4. After incubation, spin the sample in a microcentrifuge at $12,000g$ for 20 min at 4°C.
5. Transfer 160 µL of the supernatant into a fresh tube, being careful to avoid disturbing the pellet.
6. Quickly aliquot the extract and freeze at $-80°C$. When ready to use, dilute aliquots 1:10 with lysis buffer and use 2 µL per reaction (2 µL = 1000 cells).

Note: the volume of 2 µL is critical as 1X CHAPS lysis buffer contains 1 mM MgCl$_2$, adjustment of which will alter the kinetics of the PCR.

3.2.3. Preparation of Embryo Extracts From Culture

An aliquot of 1X CHAPS lysis buffer is prepared immediately before beginning embryo work by adding RNase inhibitor to a final concentration of 200 U/mL. A total of 2.5 µL of of lysis buffer is needed for each embryo. For proper lysis, the zona pellucida must be removed before being incubated in lysis buffer.

1. Assess the embryos for quality and thoroughly wash in warm PBS. Removal of the zona pellucida is achieved either chemically by a short incubation in 0.1% pronase at 37°C, mechanically by use of a narrow-bore pipet, or a combination of the two. If a pronase solution is used, it is desirable to quickly wash the embryos in PBS before the addition of lysis buffer.
2. Transfer the embryos into a fresh tube containing a minimum of 2.5 µL of lysis buffer. Telomerase activity can be assessed on individual embryos, or on pools of embryos within one sample.
3. Incubate the lysate on ice for 30 min.
4. After incubation, spin the sample in a microcentrifuge at $12,000g$ for 20 min at 4°C.
5. Transfer 80% of the supernatant (2 µL) into a fresh tube, being careful to avoid disturbing the pellet. The lysate is now ready for the RQ-TRAP assay.

3.2.4. The RQ-TRAP Assay

RQ-TRAP is performed through two steps in the light-cycler *(58)*. In the first step, telomerase mediates extension of an oligonucleotide primer (TS), which serves as a substrate. After this, the sample is incubated at 95°C to inactivate telomerase and heat-activate Taq polymerase present in the master mix. Amplification of the extended product using PCR is then conducted with the TS and ACX primers.

Telomerase is a heat-sensitive enzyme. Each sample to be run requires a heat-inactivated negative control. Because embryo samples are valuable, a negative control for embryos is prepared by heat-inactivation of positive control cell lysate only (293T or HeLa cells). Fresh lysis buffer should be prepared by adding 200 U/mL RNase inhibitor to 1X CHAPS lysis buffer for 1/10 dilutions of cell lysates. Embryo lysates should *not* be diluted.

1. Thaw and dilute positive control and any other cell samples 1/10 with lysis buffer on ice. Lysates will now be at a concentration of 500 cells/µL.
2. Prepare lysate samples for a standard curve of the positive control by making serial dilutions in lysis buffer. Final concentrations of 500, 50, 5, and 0.5 cells/ µL will be required.
3. For a negative control, heat-inactivate by incubation one-half of each sample's lysate at 85°C for 10 min.
4. Assemble the SYBR Green Master Mix (per sample):
   a. 11.6 µL of H_2O.
   b. 1.2 µL of $MgCl_2$ (25 mM).
   c. 1.6 µL of primer TS (0.5 µg/µL).
   d. 1.6 µL of primer ACX (0.25 µg/µL).
   e. 2 µL of LC DNA Master SYBR Green 1.
5. Prepare light cycler capillaries as normal. To each, add 18 µL of master mix, plus any one of the following:
   a. 2 µL of standard curve serial dilutions (2 µL = 1000, 100, 10, 1 cell equivalents).
   b. 2 µL of positive control.
   c. 2 µL of sample.
   d. 2 µL of sample negative heat-inactivated control.
6. Run the following reaction in the light cycler:
   a. 20 min at 25°C for primer TS extension.
   b. 10 min at 95°C for telomerase inactivation and Taq Polymerase activation.
   c. 35 two-step PCR cycles consisting of 30 s at 95°C and 90 s at 60°C.

3.2.5. Quantification of Relative Telomerase Activity

Relative telomerase activity of samples is quantified by determining threshold cycle values (C_t) from semi-log amplification plots (log increase in fluorescence versus cycle number) and comparison to a standard curve generated from serial dilutions of the positive control sample. The Roche Light Cycler software was used for analysis, using both an arithmetic baseline adjustment and second derivative maximum algorithm (*see* **Fig. 2**). Cell samples are typically expressed as the relative number of cells to the positive control.

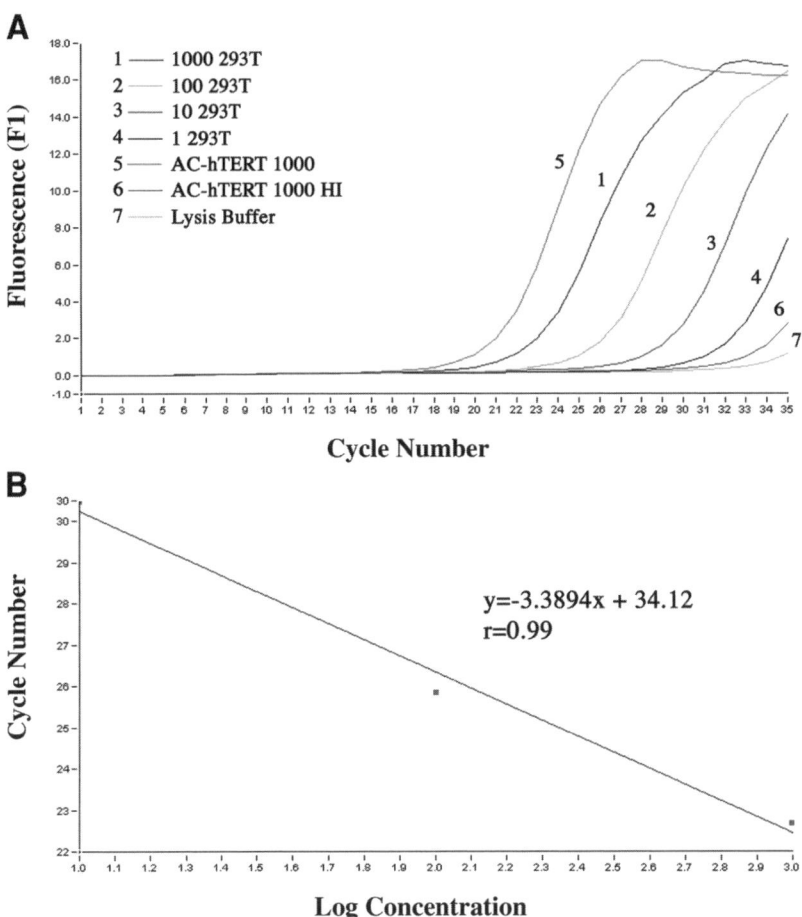

Fig. 2. Analysis of data obtained from the RQ-TRAP assay. **(A)** Charting cycle number versus fluorescence units shows a representative cell number and sample dependent amplification of 293T cell serial dilutions; a transgenic cell line expressing telomerase under a CMV promoter (AC-hTERT); a heat-inactivated (HI) positive control; and a lysis buffer negative control. **(B)** Amplification of a serial dilution of 293T cells is charted as log concentration versus cycle number with a linear trend line for regression analysis (analysis of variance).

3.3. Measurement of Telomere Length by TRF Analysis

Three main approaches to measure telomere length have been described; Southern blot analysis, quantitative fluorescent *in situ* hybridization (Q-FISH), and a flow cytometry method using fluorescent *in situ* hybridization (flow FISH). Each procedure has a broad range of applications that are limited by their inherent advantages and disadvantages.

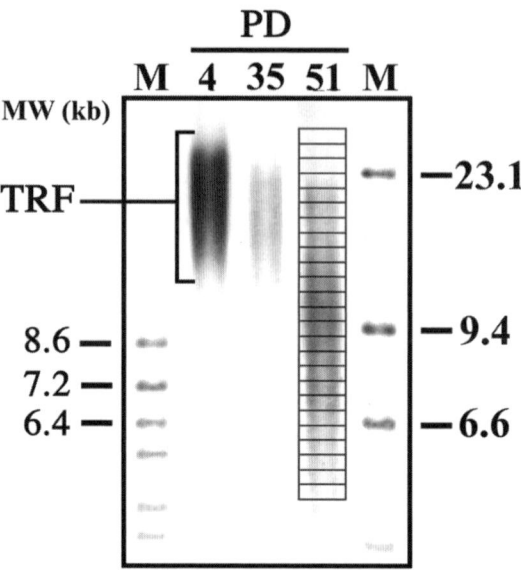

Fig. 3. Measurement of telomere length by TRF analysis. The absence of restriction enzyme recognition sequences within the telomere allows the determination of telomere length. Double digestion of genomic DNA with *Rsa*I and *Hin*fI restriction enzymes cut the chromosomal DNA into small fragments. The distance from the last restriction site to the end of the telomere comprises the TRF. The digested DNA (e.g. from bovine fibroblasts at 4, 35, and 51 population doublings [PD]) is separated on a 0.6% agarose gel and transferred to positively charged membrane for Southern blot analysis. The blot is probed with a biotinylated telomere-specific probe and the terminal restriction fragments are detected by a chemiluminescent detection system. The mean TRF length is calculated using a densitometry grid from the position of the detected signal relative to the position of known size standards (M) (Telomere Length Assay Kit; BD PharMingen™).

The traditional technique for measuring telomere length uses Southern blot analysis of TRFs obtained by restriction endonuclease digestion of genomic DNA (*see* **Fig. 3**). The absence of restriction recognition sequences within the telomere allows the isolation of intact telomeres and enables the estimation of telomere length. The chromosomal DNA is cut into small fragments except for the telomeres and subtelomeric regions (DNA adjacent to the telomere generated from the last restriction cut site), which together comprise the TRF. After digestion, the DNA fragments are separated by gel electrophoresis, blotted onto a positively charged membrane, and TRF profiles are visualized directly or indirectly by hybridization with a labeled-telomeric DNA oligonucleotide that is complementary to the telomeric repeat sequence. Finally, the size distribution of the TRFs can be compared to a DNA length standard *(20)*. A few commercial kits have been developed that use a nonradioactive chemiluminescent probe to determine telomere length. TRF analysis can be conducted using either the TeloQuant™ Telomere Length Assay Kit (BD PharMingen) or *Telo TAGGG* Telomere Length Assay (Roche Molecular Biochemicals).

The Southern blotting TRF method is the simplest approach to measure telomere length within cell and tissue samples. However, because of the requirement of a minimal quantity of genomic DNA (~2 μg), there are limitations with respect to the types of samples that can be analyzed. Therefore, early developmental stages of embryos cannot be conveniently analyzed for TRF length, but data have been obtained from 2-wk-old in vivo bovine embryos that have been flushed out of reproductive tracts (data not published). Reliable data can be obtained from later-staged embryos, embryo-derived cell lines (e.g., embryonic stem cells), fetal and newborn tissues. The protocol described below is a summary of the procedure described in the Telomere Length Assay Kit (BD PharMingen), emphasizing the key steps and listing helpful tips to generate consistent results.

3.3.1. TRF Protocol (Adapted From the Telomere Length Assay Kit; BD Pharmingen)

1. Isolate genomic DNA using standard protocols. There are numerous commercial kits (i.e., Roche Molecular Biochemicals; Qiagen) now available to isolate genomic DNA from cells, blood and tissues (*see* **Note 4**).
2. Digest genomic DNA using a *Rsa*I/*Hinf*I enzyme mixture in a 1X final concentration of enzyme reaction buffer. Digest for 12 to 18 h at 37°C using 4 U of the enzyme mix per microgram of DNA. At least 2 μg per sample should be used for cells with large telomere lengths.
3. Resolve the TRFs on 0.6% agarose gel using 1X TAE buffer (*see* **Notes 5** and **6**).
4. Load at least 2 μg/lane of digested genomic DNA for long telomeres and 7.5 μg/lane for very short telomeres (e.g., from tumor cells).
5. Load 12 μL/marker lane of biotinylated marker (*Bst*EII and *Hin*dII). Heat at 60°C for 3 min before loading.
6. Run gel at 1 V/cm overnight (24–28 h) to resolve the TRF fragments.
7. Soak the gel in 0.25 *M* HCl for 15 min with gentle agitation. Repeat (*see* Note 7).
8. Rinse in dH₂O.
9. Soak the gel in 0.4 *N* NaOH for 15 min with gentle agitation. Repeat.
10. Using 0.4 *N* NaOH as transfer buffer, prepare a Southern transfer (as depicted in Maniatis *[59]*) to a positively charged nylon membrane (Amersham Hybond-N⁺). Transfer at room temperature for 1 h with 2 in of paper towels, then change the paper towels and continue the transfer for 1 to 2 h (*see* **Note 8**).
11. Rinse the membrane in 2X SSC. Air-dry the filter for storage if desired. Store in a desiccator at room temperature.

12. Denature the biotinylated Telomere Probe by boiling (100°C) for 10 min, then chill on ice for 5 min before use.
13. Warm the hybridization buffer (part no. 4533KC) to bring all components back into solution. Soak the filter using 0.1 mL per cm² of membrane. Rotate in a hybridization tube at 55°C for 30 min
14. Add the biotinylated Telomere Probe (part no. 4527KC) to the hybridization buffer and rotate overnight at 55°C. The final concentration of the probe can range from 1 to 5 ng/mL (although 5 ng/mL is the most common concentration).
15. Prewarm the hybridization stringency wash buffer (2X) to bring all components back into solution. Dilute the buffer 1/1 with sterile H₂O. The resulting buffer contains 2X SSC/0.1% SDS.
16. Wash the membrane three times for 5 min each in 1X stringency wash buffer (0.2 mL per cm² of membrane) with gentle agitation at 55°C.
17. Pour off hybridization stringency wash buffer and add membrane-blocking buffer (part no. 4562KC) to cover the membrane (0.25 mL per cm² of membrane). Incubate for 30 min at room temperature.
18. Pour off some block buffer into a separate tube and add stabilized Streptavidin-HRP (part no. 4528KC) to make a 1/300 final dilution when added to the membrane. Add diluted Streptavidin–HRP to the membrane and incubate for 30 min at room temperature with gentle agitation.
19. Dilute wash buffer (4X) (part no. 4563KC) to 1X with sterile H₂O. Wash the membrane four times for 5 min per wash with gentle agitation.
20. In a clean wash tray, incubate the membrane in the North2South™ Substrate Equilibrium Buffer (0.25 mL per cm² of membrane) for 5 min at room temperature with gentle agitation.
21. Prepare North2South™ Substrate working solution (0.1 mL per cm² of membrane) by mixing equal volumes of the North2South™ Stable Peroxide Solution and North2South™ Luminol/Enhance Solution.
22. Transfer the membrane to the substrate working solution and incubate for 5 to 10 min at room temperature.
23. Pour off excess substrate from the membrane and place the filter between two pieces of plastic wrap. Remove any trapped air bubbles by rolling a pipet over the membrane with paper towels underneath and on top of it. Dry the outside of the plastic wrap with paper towel.
24. Place the covered membrane in an X-ray film cassette and expose the blot to film (Fuji Medical X-ray Film) for 10 s, 30 s, 2 min or longer until a clear signal with low background is obtained on development.
25. Calculate the mean TRF length. Using Molecular Analyst Software (Bio-Rad), divide the scanned image into a digital grid consisting of approx 30 boxes per column. Position the grid over the entire vertical length of a lane so that many boxes overlay a signal telomere smear. For each sample, optical density (OD) and length (L) are computed for each grid box, where OD is the total signal intensity within a grid box and L is the molecular weight at the mid-point of the grid box. A standard curve of molecular weight versus migration is determined from a

densitometric analysis of the molecular weight markers on the scanned image. The mean TRF length for each sample is calculated using the formula:

$$L = \Sigma (ODi \cdot Li) / \Sigma (ODi)$$

where ODi and Li are the signal intensity and TRF length, respectively, at position i on the gel image (*see* **Fig. 3**).

3.4. Measurement of Telomere Length Using Q-FISH

Conducting FISH using a telomere DNA probe (CCCTAA)[3] localizes telomeric DNA to bovine metaphase chromosomes *(55)*. The use of Q-FISH, alongside standard curves of fluorescent intensity of telomeric DNA of known length, can determine telomere lengths of individual chromosomes from metaphase nuclei *(60)*. To measure the average length of telomere repeats in individual cells a flow cytometry method using fluorescence *in situ* hybridization (flow FISH) has also been developed *(61)*.

The underlying principle for the FISH methods of telomere length determination is the quantification of a fluorescent signal obtained from a telomere-specific probe labeled with a flourochrome. To ensure specificity of binding to the telomeric repeat (TTAGGG)[n] region of the chromosome an 18-bp probe consisting of (AATCCC)[3] is used. To increase hybridization efficiency and specificity, the probes for Q-FISH are constructed using a synthetic DNA/RNA analog, PNA. In PNA probes, the sugar phosphate backbone characteristic of nucleic acid probes is replaced by a neutral peptide/polyamine backbone that keeps an even distance between the bases and makes the probe highly resistant to degradation by DNases, RNases, proteinases, and peptidases. PNA probes labeled with FITC or Cy3 for telomere detection can be obtained from various commercial sources (e.g., PE Biosystems, Framingham MA), or telomere PNA FISH kits are also available (e.g., DakoCytomation).

The use of Q-FISH to study telomeres allows the examination of individual chromosomes in single cells (*see* **Fig. 4**). Although labor-intensive, it permits the detection of chromosome and/or cell-specific variation. This technique is of particular use when applied to small samples, such as pre-placental stage embryos, which yield too little DNA for telomere length determination by southern blot analysis (TRF method). This protocol comprises three essential steps: chromosome preparation, *in situ* hybridization, and densitometry.

3.4.1. Chromosome Preparations

3.4.1.1. Air-Dried Chromosome Spreads From Lymphocytes and Solid Tissues

1. Establish lymphocyte or whole blood culture according to standard protocols (*see* **Note 9**).
2. 1 to 2 h before terminating the culture, introduce a mitotic inhibitor (e.g., colcemide at a final concentration of 0.05 μg/mL).

3. Transfer cultures to centrifuge tubes and centrifuge at 100*g* for 2 to 5 min to concentrate the cells.
4. Remove all but 1 mL of medium and resuspend cells in 10 to 15 mL of 37°C hypotonic (0.075 *M*) KCl. Let stand for 10 to 15 min. Note that the volume of KCl and the duration should be optimized for each cell type and species.
5. Centrifuge tubes at 100*g* for 10 min or until a pellet of cells appears on the bottom.
6. Remove all but 0.5 mL of KCl and resuspend the cell pellet in freshly prepared methanol/acetic acid fixative (3/1 v/v) and let stand for 30 min.
7. Centrifuge at 100*g* for 10 min, discard fixative and replace with fresh fixative. Cell suspensions can be stored overnight at 4°C or for long periods at –20°C.
8. The day before hybridization, centrifuge the cells to form a pellet, remove the old fixative and resuspend the cells in a small volume of freshly prepared fixative (0.5–2 mL).
9. Drop cell suspension on precleaned slides to allow metaphase chromosomes to spread. Air dry the slides overnight. Note the volume of fixative in **step 8** should be adjusted to the size of the pellet to ensure those chromosomes are well spread out with few overlapping chromatids.

3.4.1.2. Chromosome Spreads for Tissue Cultures

1. Establish and maintain tissue cultures according to protocol of choice.
2. When cells are in rapid growth phase introduce a mitotic inhibitor (e.g., colcemid at a final concentration of 0.05 µg/mL) to the culture vessels.
3. Continue incubation for 1 to 4 h.
4. Detach dividing cells by gentle trypsinization and agitation.
5. Centrifuge and proceed as for lymphocytes above beginning at **step 3**.

3.4.1.3. Air-Dried Chromosome Spreads From Zona Pellucida Enclosed Oocytes and Embryos

1. Introduce embryos to equilibrated prewarmed culture medium containing a microtubule inhibitor such as colcemid at a final concentration of 0.05 µg/mL medium (*see* **Note 10**).
2. Incubate at 39°C in 5% CO_2 for up to 5 h (minimum of 30 min).
3. Transfer the embryo into a hypotonic solution (0.88% sodium citrate) and keep at room temperature for 3 to 5 min (requires optimization).
4. Place the embryo with a small drop of hypotonic solution onto a pre-cleaned slide. The hypotonic solution should not be allowed to evaporate as this will dry the embryo or (oocyte) and prevent spreading of the chromosomes.
5. Drop one to two drops (~50 µL) of freshly-prepared fixative onto the embryo (1/1 methanol/acetic acid for bovine embryos and 3/1 for mouse embryos)
6. Mark a circle on the back of the slide to indicate the location of the embryo.
7. Air-dry by gently blowing on the slide.
8. Place the slide into fresh fixative (3/1 methanol/acetic acid) for 30 min (if bovine embryos).
9. Remove the slide from the fixative and air-dry at room temperature in a fume hood or well-ventilated area.

3.4.2. In Situ *Hybridization and Image Analysis*

1. Prepare the slides the day before FISH and let them dry overnight at room temperature.
2. Rehydrate the cells on the slides in 1X PBS (Ca^{2+} and Mg^{2+} free, pH 7.0–7.5,) for 15 min.
3. Fix in 4% formaldehyde in 1X PBS for 2 min.
4. Wash in 1X PBS 3 times for 5 min.
5. Treat with 1 mg/mL pepsin at 37°C for 10 min (note: pepsin should be prepared freshly in acidified water pH 2.0).
6. Wash twice in 1X PBS for 2 min.
7. Fix in 4% formaldehyde in 1X PBS for 2 min.
8. Wash three times for 5 min in 1X PBS.
9. Serially dehydrate the cells in ethanol: 5 min 70%, 5 min 90%, 5 min 100%.
10. Air dry the slides.
11. Prepare the hybridization mixture:

Stock	Final concentration	For 250 mL
Formamide (ultrapure, pH 7.0–5.0)	70%	175 μL
Blocking protein in water 2.5%	0.25%	25 μL
0.2 *M* Tris	10 m*M*	2.5 μL
PNA Telomere—FITC/Cy3 probe	0.5 μg/mL	2.5 μL (100X stock)
MgCl₂ buffer, pH 7.4	5%	21.4 μL
dd water		till 250 μL

12. Pipet two drops of 10 μL of hybridization mix onto a cover slip (24 × 55 mm), cover carefully with the slide (upside down), avoiding air bubbles, and turn the slide over.
13. Denature in a preheated oven (80°C) for 3 min
14. Hybridize in a humidified chamber at room temperature for 2 h.
15. Remove the coverslip carefully in wash solution I (70% formamide, 10 m*M* Tris, 0.1% bovine serum albumin, pH 7.0–7.5).
16. Wash the slides twice for 15 min in wash solution I.
17. Wash three times for 5 min in wash solution II (0.1 *M* Tris, 0.15 *M* NaCl, 0.08% Tween-20, pH 7.0–7.5).
18. Serially dehydrate the cells in ethanol: 5 min 70%, 5 min 90%, and 5 min 100%.
19. Air dry the slides.
20. Place 20-μL drops of Vectashield containing 200 ng DAPI on the coverslip.
21. Keep the slide in a light-protected storage box at 4°C.
22. View under an epifluorescent microscope at a wavelength of 405 and 546 nm, which are used to excite the DAPI (chromosomes) and Cy3 probes (telomeres) respectively, and capture images separately with a CCD monochromatic camera.
23. Digitally captured images of telomeres are then analyzed to estimate telomere length by a variety of image analysis programs (e.g., TFL-TELO *[60]*; Optimas image analysis software (Media Cybernetics, Silver Spring, MD; *[62]*), that are designed to compare selected pixel densities identified (*see* **Fig. 4** and **Note 11**).

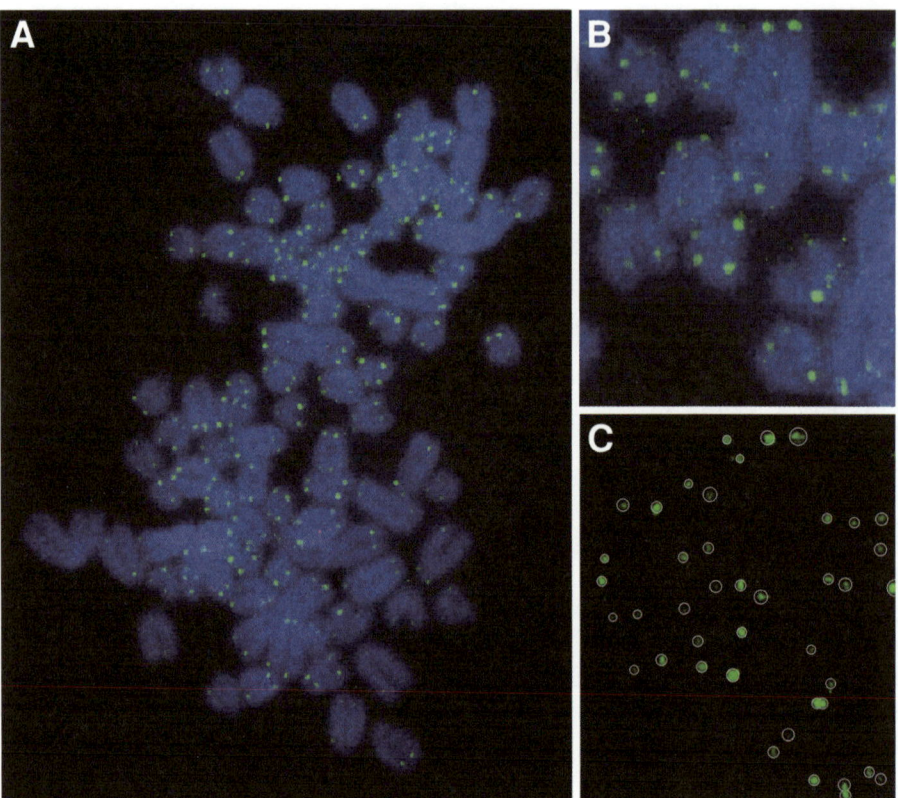

Fig. 4. Telomere length measurements using Q-FISH of ovine metaphase chromosomes. (**A**) Punctate fluorescent signals localized to the end regions of both the long and short arms of both chromatids of each ovine chromosome. (**B**) Enlarged section of (A) showing a typical fluorescence image of the telomeres and representative segmentation boundaries (**C**) that are generated for telomere fluorescent intensity quantification. For more information of telomere length measurement using digital fluorescence microscopy, *see* **ref. 60**.

3.5. A Quantitative PCR-Based Assay for Telomere Length Measurements

Both the TRF and Q-FISH methods of telomere length determination entail multiple techniques and are arduous, requiring multiple days of work. Additionally, biological variability is introduced in TRFs, as polymorphisms exist between individuals in subtelomeric restriction sites and in subtelomeric lengths causing variability in attempts to measure the true length of telomeric DNA within samples *(63)*. Telomere length measurement by quantitative PCR (Q-PCR) measures, for each DNA sample, the factor by which the sample differed from a dilution curve of a reference DNA sample in its ratio of telomere

Table 1
Primer Sequences for Telomere Measurement by Q-PCR

Primer	Sequence (5' → 3')
Tel 1	GGTTTTTGAGGGTGAGGGTGAGGGTGAGGGTGAGGGT
Tel 2	TCCCGACTATCCCTATCCCTATCCCTATCCCTATCCCTA

From **ref. 64**.

repeat copy number to single-copy gene copy number (e.g., 36B4 gene encoding acidic ribosomal phosphoprotein PO; *see* **ref. 64**). With the clever design of an oligonucleotide Tel 1 and Tel 2 primer set (**Table 1**) the Q-PCR assay specifically amplifies telomeric DNA without forming primer dimer products. The PCR crossing point (C_p) of the amplified telomere signal (T) is normalized against the C_p of a single copy gene (S) amplified from the same sample. The T/S ratio is defined as the relative telomere to single copy signal amplified, and therefore an unknown sample with a T/S ratio equal to 1 would have an identical telomere length to the known reference control at a specific DNA concentration.

Therefore, the Q-PCR assay allows for a faster, more objective and reproducible measurement of telomere lengths using relatively small amounts of DNA (ng quantities). This level of sensitivity would allow for accurate measurements of average telomere lengths in early embryos and other samples of low DNA yield. However, the assay requires expensive real-time PCR equipment and reagents. In addition, the assay calls for a control sample of known telomere length to which unknown experimental samples are expressed relative to. The Q-PCR technique for telomere measurement is described in detail in *(64)*.

3.6. Future Perspectives and Conclusions

Although telomerase activity appears to be reprogrammed, albeit delayed, in various embryo, cell, and tissue products analyzed from SCNT clones thus far *(9)*, a thorough analysis of telomerase activity and telomere length reprogramming at various developmental time points using different donor cell types has not been conducted. Using the RQ-TRAP and Q-PCR methods for telomerase activity and telomere length determination respectively should expedite this survey.

However, the differential telomere length restoration observed in various animal clones suggests that besides telomerase reactivation, other epigenetic regulatory mechanisms may be involved. Each vertebrate species has a characteristic maximum telomere length that is maintained during germline transmission by unidentified factors *(65)*. Telomere-specific binding proteins such as TRF1, TRF2, and TIN2 are mediators of telomere length that directly inhibit/facilitate the binding of telomerase to telomeric DNA or provide structural changes within the telomere that prevent/promote telomerase binding. During spermatogenesis, the telomere regions undergo structural reorganization that includes the apparent loss of somatic TRF1 and the appearance of a sperm-specific telomere binding protein *(66)*. Recent findings indicate that substantial epigenetic changes in the state of telomeric heterochromatin are associated with abnormal telomere lengths *(67)*. Therefore chromosome immunoprecipitation analyses using various antibodies to telomere binding proteins/modifications and probing the immunoprecipitates with a labeled telomeric DNA probe could reveal quantitative epigenetic alterations within the telomeric chromatin of animal clones.

Although abnormal telomere lengths have been detected in somatic cell clones, the proliferative lifespan of somatic cells appears to be conserved after nuclear transfer *(48)*. Primary sheep fibroblasts derived from fetal clones displayed the same proliferative capacity and rate of telomere shortening as the donor cell lines even though initial telomere lengths were significantly shorter *(48)*. These results suggest that cellular senescence is not triggered by a predetermined short telomere but is genetically determined. Recent studies have shown that senescence is triggered through abrogation of the telomere structure rather than its overall length *(22–24)*. However, the ramifications of shorter telomere lengths before entry into cellular senescence would be an increased likelihood of chromosomal instabilities in animal clones. Indeed, higher chromosomal abnormalities have been observed after somatic cell cloning in both primates *(68)* and cattle *([69]*; Betts and King, unpublished data) suggesting that clones do harbor genetic instabilities. Although possessing altered telomere lengths, most clones and their offspring appear physically normal and healthy for their age. It is still an unresolved issue whether cellular aging has relevance to organismal aging. Nevertheless, the premature deaths of other animal clones *(70)*, including "Dolly" *(49)*, are intriguing and therefore warrant the long-term evaluation of the cellular and animal products of animal cloning to determine how/if cellular senescence, and aging, are affected by SCNT.

4. Notes

1. The NP-40 lysis-buffer is used as the cell extraction solution rather than 1X CHAPS, supplied in the TRAPEZE® kit (Intergen) because it is a more potent extraction buffer. The sensitivity of the assay is therefore increased, allowing for the detection of telomerase activity from fewer oocytes and/or embryos.
2. If a nonradioactive method of detection (i.e., SYBR® Green stain) is used, this end-labeling step is omitted.
3. These TRAP reaction conditions have been optimized for use with extracts of pooled in vitro derived bovine oocytes or embryos. Nevertheless, before each series of experiments a standard curve of relative telomerase activities from a serial dilution of each sample extract should be conducted to determine the opti-

mal extract concentration for telomerase activity comparisons. The chosen concentration should be within the linear range of detection, excluding samples with excessive PCR artifacts and diminished 36-bp internal control products (*see* **Fig. 1**).

4. To accurately quantify the DNA concentration, RNase (Roche Molecular Biochemicals) treatment will remove any contaminating RNA from your samples. Precise DNA quantification will produce more consistent results since TRF length determination is not only dependent on the length of the fragments, but on the intensities of the TRF profiles. DNA overloading may hinder the comparison of relative TRF lengths between samples.

5. Use a horizontal gel apparatus that uses gel casts of a least 15 cm in length.

6. Pulsed-field electrophoresis for TRF separation: Typically, a standard electrophoresis system is used to separate TRF after the restriction digest of samples. Depending on the source of the sample, TRF lengths can be as short as a few kilobases or as long as 50. Standard electrophoresis systems, in which DNA travels at a constant rate in a fixed field, only migrates fragments as a function of size up to approx 20 kb. Fragments greater than 20 kb migrate through agarose gels with difficulty, and so do not follow this relationship. To counter this, a pulsed-field electrophoresis system may be used. While relatively more expensive, these systems are capable of properly separating DNA of any size up to 10 mb. Using pulsed field electrophoresis also reduces observed artifacts in TRF separations, allowing for a more accurate quantification by densitometry.

   If a pulsed-field electrophoresis system is available, the entire range of TRF sizes can be separated under the following conditions: 14°C, 6 V/cm, 10–12 h, 120° angle, 0.5 TBE, constant switch time.

7. It is important to make fresh 0.25 *M* HCl from a concentrated stock solution just before use.

8. The efficiency of DNA transfer is optimal at transfer times of 3 h or greater.

9. The Q-FISH procedure is dependent on the presence of high quality metaphase chromosomes. The frequency of cells in metaphase can be increased by accumulation and synchronization of the cell and blood cultures at G_1/S of the cell cycle by the addition of excess amounts of thymidine (0.8 mg/mL). After 16 h of incubation, the block is released by washing the cultures twice with fresh culture medium. The cells are again allowed to grow for 7 h. At this time, the cells are exposed to colcemid (0.05 µg/mL) and processed as above.

10. Oocytes do not require incubation with spindle inhibitors. Begin at **step 3**. If the chromosomes do not spread or are overspread, adjust the duration of the hypotonic treatment (**step 3**).

11. Image acquisition and fluorescent intensity measurements: Although relatively laborious compared with the TRF measurements, the telomere Q-FISH technique allows telomere length determination from a limited number of cells and provides results for each chromatid arm of every chromosome (*60*). The telomere Q-FISH algorithm requires two spectral fluorescence images to generate a length estimate for each telomere in a metaphase chromosome spread: one of DAPI-stained chromosomes, and one of the associated Cy3-stained telomeres. The al-

gorithm performs telomere segmentation, telomere fluorescence measurement, and chromosome segmentation (*see* **Fig. 4**; *[60]*). The integrated fluorescence intensity for each telomere is calculated after correction for background, based on the values of the surrounding pixels, and after correction for image acquisition exposure time. For proper image acquisition it is suggested that one use a wide-field microscope (e.g., Zeiss Axioplan) over confocal microscopy to avoid photobleaching. A fluorescence 63x magnification objective lens with a numerical aperture of 1.4 (Plan Apochromat 63×/1.4, Zeiss or equivalent) is recommended along with a DAPI/Cy3 filter set to overcome the pixel shift problem of using multiple filter blocks. Replacing the standard fluorescence illumination source with a hybrid mercury/xenon lamp (200 W) will produce more consistent results. To compare telomere fluorescent intensity data from one experiment to another performed at a different time, the data collected must be calibrated to account for day-to-day variations in the lamp and alignment of the optics, etc. The fluorescent intensity value of fluorescent beads (orange beads, size 0.2 μm; Molecular Probes) serves as the normalization constant in each separate experiment. In addition, to relate fluorescence intensity to number of TAAGGG repeats, plasmids with a defined $(TTAGGG)^n$ length of 0.15, 0.40, 0.80, and 1.60 kb can be hybridized and analyzed *(63)*. Please see Poon et al. *(60)* and O'Sullivan et al. *(62)* for detailed descriptions of telomere length measurements using fluorescent in situ hybridization.

Acknowledgments

The financial support of the Natural Sciences and Engineering Research Council of Canada (NSERC), the Canadian Institutes of Health Research (CIHR), the Food Systems Biotechnology Centre (FSBC), the Ontario Ministry of Agriculture, Food and Rural Affairs (OMAFRA), and the Ontario Research and Development Challenge Fund is gratefully acknowledged. We thank Gianfranco Coppola for assistance with the preparation of this chapter and Dr. Lena King for reviewing the chapter and for helpful discussions.

References

1. Wilmut, I., Schnieke, A. E., McWhir, J., Kind, A. J., and Campbell, K. H. S. (1997) Viable offspring derived from fetal and adult mammalian cells. *Nature* **385,** 810–813.
2. Wakayama, T., Perry, A. C.F., Zuccotti, M., Johnson, K. R., and Yanagimachi, R. (1998) Full-term development of mice from enucleated oocytes injected with cumulus cell nuclei. *Nature* **394,** 369–374.
3. Kato, Y., Tani, T., Sotomaru, Y., Kurokawa. K,, Kato. J,, Doguchi. H,, Yasue. H,, et al. (1998) Eight calves cloned from somatic cells of a single adult. *Science* **282,** 2095–2098.
4. Woods, G. L., White, K. L., Vanderwall, D. K., Li, G. P., Aston, K. I., Bunch, T. D., et al. (2003) A mule cloned from fetal cells by nuclear transfer. *Science* **301,** 1063.

5. Rideout, W. M. III, Eggan, K., and Jaenisch, R. (2001) Nuclear cloning and epigenetic reprogramming of the genome. *Science* **293**, 1093–1098.

6. Dean. W., Santos, F., Stojkovic, M., Zakhartchenko, V., Walter, J., Wolf, E., et al. (2001) Conservation of methylation reprogramming in mammalian development: aberrant reprogramming in cloned embryos. *Proc. Natl. Acad. Sci. USA* **98**, 13734–13738.

7. Xue, F., Tian, X. C., Du, F., Kubota, C., Taneja, M., Dinnyes, A., Dai, Y., et al. (2002) Aberrant patterns of X chromosome inactivation in bovine clones. *Nat. Genet.* **31**, 216–220.

8. Humpherys, D., Eggan, K., Akutsu, H., Hochedlinger, K., Rideout, W. M. III, Biniszkiewicz, D., et al. (2001) Epigenetic instability in ES cells and cloned mice. *Science* **293**, 95–97.

9. Betts, D. H., Bordignon, V., Hill, J., Winger, Q., Westhusin, M., Smith, L., King, W. (2001) Reprogramming of telomerase activity and rebuilding of telomere length in cloned cattle. *Proc. Natl. Acad. Sci. USA* **98**, 1077–1082.

10. Miyashita, N., et al. (2002) Remarkable differences in telomere lengths among cloned cattle derived from different cell types. *Biol. Reprod.* **66**, 1649–1655.

11. Allshire, R. C., Dempster, M., and Hastie, N. D. (1989) Human telomeres contain at least three types of G-rich repeats distributed non-randomLy. *Nucleic Acids Res.* **17**, 4611–4627.

12. Preston, R. J. (1997) Telomeres, telomerase and chromosome stability. *Radiat. Res.* **147**, 529–534.

13. Klobutcher, L. A., Swanton, M. T., Donini, P., and Prescott, D. M. (1981) All gene-sized DNA molecules in four species of hypotrichs have the same terminal sequence and an unusual 3' terminus. *Proc. Natl. Acad. Sci. USA* **78**, 3015–3019.

14. Wright, W. E., Tesmer, V. M., Huffman, K. E., Levene, S. D., and Shay, J. W. (1997) Normal human chromosomes have long G-rich telomeric overhangs at one end. *Genes Dev.* **11**, 2801–2809.

15. McElligott, R., and Wellinger, R. J. (1997) The terminal DNA structure of mammalian chromosomes. *EMBO J.* **16**, 3705–3714.

16. van Steensel, B., Smogorzewska, A., and de Lange, T. (1998) TRF2 protects human telomeres from end-to-end fusions. *Cell* **92**, 401–413.

17. Huffman, K. E., Levene, S. D., Tesmer, V. M., Shay, J. W., and Wright, W. E. (2000) Telomere shortening is proportional to the size of the 3' G-rich telomeric overhang. *J. Biol. Chem.* **275**, 19719–19722.

18. McKee, B. D. (2004) Homologous pairing and chromosome dynamics in meiosis and mitosis. *Biochim. Biophys. Acta* **1677**, 165–180.

19. Allsopp, R. C., and Harley, C. B. (1995) Evidence for a critical telomere length in senescent human fibroblasts. *Exp. Cell Res.* **219**, 130–136.

20. Harley, C. B., Flutcher, A. B., and Greider, C. W. (1990) Telomeres shorten during ageing of human fibroblasts. *Nature* **345**, 458–460.

21. Harley, C. B., Vaziri, H., Counter, C. M., and Allsopp, R. C. (1992) The telomere hypothesis of cellular aging. *Exp. Gerontol.* **27**, 375–382.

22. Stewart, S. A., Ben-Porath, I., Carey, V. J., O'Connor, B. F., Hahn, W. C., and Weinberg, R. A. (2003) Erosion of the telomeric single-strand overhang at replicative senescence. *Nat. Genet.* **33**, 492–496.

23. Karlseder, J., Smogorzewska, A., de Lange, T. (2002) Senescence induced by altered telomere state, not telomere loss. *Science* **295**, 2446–2449.

24. Li, G. Z., Eller, M. S., Firoozabadi, R., and Gilchrest, B. A. (2003) Evidence that exposure of the telomere 3' overhang sequence induces senescence. *Proc. Natl. Acad. Sci. USA* **100**, 527–531.

25. Greider, C. W., and Blackburn, E. H. (1985) Identification of a specific telomere terminal transferase activity in Tetrahymena extracts. *Cell* **43**, 405–413.

26. Greider, C. W., and Blackburn, E. H. (1989) A telomeric sequence in the RNA of Tetrahymena telomerase required for telomere repeat synthesis. *Nature* **337**, 331–337.

27. Collins, K., Kobayashy, R., and Greider, C. W. (1995) Purification of Tetrahymena telomerase and cloning of genes encoding the two protein components of the enzyme. *Cell* **81**, 677–686.

28. Counter, C. M., Meyerson, M., Eaton, E. N., and Weinberg, R. A. (1997) The catalytic subunit of yeast telomerase. *Proc. Natl. Acad. Sci. USA* **94**, 9202–9207.

29. Lingner, J., Hughes, T. R., Shevchenko, A., Mann, M., Lundblad, V., and Cech, T. R. (1997) Reverse transcriptase motifs in the catalytic subunit of telomerase. *Science* **276**, 561–567.

30. Nakayama, J., Saito, M., Nakamura, H., Matsuura, A., and Ishikawa, F. (1997) TLP1: a gene encoding a protein component of mammalian telomerase is a novel member of WD repeats family. *Cell* **88**, 875–884.

31. Kilian, A., Bowtell. D. D. L., Abud, H. E., Hime, G. R., Venter, D. J., Keese, P. K., et al. (1997) Isolation of a candidate human telomerase catalytic subunit gene, which reveals complex splicing patterns in different cell types. *Hum. Mol. Genet.* **6**, 2011–2019.

32. Meyerson, M., Counter, C. M., Eaton, E. N., Ellisen, L. W., Steiner, P., Caddle, S. D., et al. (1997) hEST2, the putative human telomerase catalytic subunit gene is up-regulated in tumor cells and during immortalization. *Cell* **90**, 785–795.

33. Ulaner, G. A., and Giudice, L. C. (1997) Developmental regulation of telomerase activity in human fetal tissues during gestation. *Mol. Hum. Reprod.* **3**, 769–773.

34. Sharma, H. W., Sokoloski, J. A., Perez, J. R., Maltese, J. Y., Sartorelli, A. C., Stein, C. A., et al. (1995) Differentiation of immortal cells inhibits telomerase activity. *Proc. Natl. Acad. Sci. USA* **92**, 12343–12346.

35. Counter, C. M., Avillion, A. A., Lefeuvre, C. E., Stewart, N. G., Greider, C. W., Harley, C. B., et al. (1992) Telomere shortening associated with chromosome instability is arrested in immortal cells which express telomerase activity. *EMBO J.* **11**, 1921–1929.

36. Kim, N. W., Piatyszek, M. A., Prowse, K. R., Harley, C. B., West, M. D., Ho, P. L., et al. (1994) Specific association of human telomerase activity with immortal cells and cancer. *Science* **266**, 2011–2015.

37. Harle-Bachor, C., and Boukamp, P. (1996) Telomerase activity in the regenerative basal layer of the epidermis in human skin and in immortal and carcinoma-derived skin keratinocytes. *Proc. Natl. Acad. Sci. USA* **93,** 6476–6481.

38. Wright, W. E., Brasiskyte, D., Piatyszek, M. A., and Shay, J. W. (1996) Experimental elongation of telomeres extends the lifespan of immortal x normal cell hybrids. *EMBO J.* **15,** 1734–1741.

39. Bodnar, A. G., Ouellette, M., Frolkis, M., Holt, S. E., Chiu, C.-P., Morin, G. B., et al. (1998) Extension of life-span by introduction of telomerase into normal human cells. *Science* **279,** 349–352.

40. Nakayama, J.-I., Tahara, H., Tahara, E., Saito, M., Ito, K., Nakamura, H., et al. (1998) Telomerase activation by hTRT in human normal fibroblasts and hepatocellular carcinomas. *Nat. Genet.* **18,** 65–68.

41. Lee, H.-W., Blasco, M. A., Gottlieb, G. J., Horner, J. W. II, Greider, C. W., and DePinho, R. A. (1998) Essential role of mouse telomerase in highly proliferative organs. *Nature* **392,** 569–574.

42. Cherif, H., Tarry, J. L., Ozanne, S. E., and Hales, C. N. (2003) Ageing and telomeres: a study into organ- and gender-specific telomere shortening. *Nucleic Acids Res.* **31,** 1576–1583.

43. Miller, R. A. (1996) The aging immune system: primer and prospectus. *Science* **273,** 70–74.

44. D'Ippolito, G., Schiller, P. C., Ricordi, C., Roos, B. A., and Howard, G. A. (1999) Age-related osteogenic potential of mesenchymal stromal stem cells from human vertebral bone marrow. *J. Bone Miner. Res.* **14,** 1115–1122.

45. Dimri, G. P., Lee, X., Basile, G., Acosta, M., Scott, G., Roskelley, C., et al. (1995) A biomarker that identifies senescent human cells in culture and in aging skin *in vivo. Proc. Natl. Acad. Sci. USA* **92,** 9363–9367.

46. Shiels, P. G., Kind, A. J., Campbell, K. H. S., Waddington, D., Wilmut, I., Colman, A., et al (1999a) Analysis of telomere lengths in cloned sheep. *Nature* **399,** 316–317.

47. Shiels, P. G., Kind, A. J., Campbell, K. H. S., Wilmut, I., Waddington, D., Colman, A., et al. (1999b) Analysis of telomere length in dolly, a sheep derived by nuclear transfer. *Cloning* **1,** 119–125.

48. Clark, A. J., Ferrier P., Aslam S., Burl, S., Denning, C., Wylie, D., et al. (2003) Proliferative lifespan is conserved after nuclear transfer. *Nat. Cell Biol.* **5,** 535 538.

49. Rhind, S., Cui, W., King, T., Ritchie, W., Wylie, D., and Wilmut, I. (2004) Dolly: A final Report. *Reprod. Fertil. Devel.* **16,** 156.

50. Kubota, C., Tian, X. C., and Yang, X. (2004) Serial bull cloning by somatic cell nuclear transfer. *Nat. Biotechnol.* **22,** 693–694.

51. Jiang, L., Carter, D. B., Xu, J., Yang, X., Prather, R. S., and Tian, X. C. (2004) Telomere lengths in cloned transgenic pigs. *Biol. Reprod.* **70,** 1589–1593.

52. Tian, X. C., Xu, J., and Yang, X. (2000) Normal telomere lengths found in cloned cattle. *Nat. Genet.* **26,** 272–273.

53. Lanza, R. P., Cibelli, J. B., Blackwell, C., Cristofalo, V. J., Francis, M. K., Baerlocher, G. M., et al. (2000) Extension of cell life-span and telomere length in animals cloned from senescent somatic cells. *Science* **288,** 665–669.

54. Kuhholzer-Cabot, B., and Brem, G. (2002) Aging of animals produced by somatic cell nuclear transfer. *Exp. Gerontol.* **37,** 1317–1323.
55. Betts, D. H., and King, W. A. (1999) Telomerase activity and telomere detection during early bovine development. *Dev. Genet.* **25,** 397–403.
56. Holt, S. E., Aisner, D. L., Shay, J. W., and Wright, W. E. (1997) Lack of cell cycle regulation of telomerase activity in human cells. *Proc. Natl. Acad. Sci. USA* **944,** 10687–10692.
57. Xu, J., and Yang, X. (2001) Telomerase activity in early bovine embryos derived from parthenogenetic activation and nuclear transfer. *Biol. Reprod.* **64,** 770–774.
58. Wege, H., Chui, M. S., Le, H. T., Tran, J. M., and Zern, M. A. (2003) SYBR Green real-time telomeric repeat amplification protocol for the rapid quantification of telomerase activity. *Nucleic Acids Res.* **31,** e3.
59. Maniatis, T., Fritsch, E. F., Sambrook, J. (1982) *Molecular Cloning: A Laboratory Manual.* Cold Spring Harbor Laboratory Press, Cold Spring Harbor, NY.
60. Poon, S. S., Martens, U. M., Ward, R. K., and Lansdorp, P. M. (1999) Telomere length measurements using digital fluorescence microscopy. *Cytometry* **36,** 267–278.
61. Rufer, N., Dragowska, W., Thornbury, G., Roosnek, E., and Lansdorp, P. M. (1998) Telomere length dynamics in human lymphocyte subpopulations measured by flow cytometry. *Nat. Biotechnol.* **16,** 743–747.
62. O'Sullivan, J. N., Finley, J. C., Risques, R. A., Shen, W. T., Gollahon, K. A., Moskovitz, A. H., et al. (2004) Telomere length assessment in tissue sections by quantitative FISH: image analysis algorithms. *Cytometry* **58A,** 120–131.
63. Zijlmans, J. M., Marten, U. M. Poon, S. S. S., Raap, A. K., Tanke, H. J., Ward, R. K., et al. (1997) Telomeres in the mouse have large inter-chromosomal variations in the number of T2AG3 repeats. *Proc. Natl. Acad. Sci. USA* **94,** 7423–7428.
64. Cawthon, R. M. (2002) Telomere measurment by quantitative PCR. *Nucleic Acids Res.* **30,** e47.
65. Zhu, L., Hathcock, K. S., Hande, P., Lansdorp, P. M., Seldin, M. F., and Hodes, R. J. (1998) Telomere length regulation in mice is linked to a novel chromosome locus. *Proc. Natl. Acad. Sci. USA* **95,** 8648–8653.
66. Kozik, A., Bradbury, E. M., and Zalensky, A. O. (2000) Identification and characterization of a bovine sperm protein that binds specifically to single-stranded telomeric deoxyribonucleic acid. *Biol. Reprod.* **62,** 340–346.
67. Garcia-Cao, M., O'Sullivan, R., Peters, A. H., Jenuwein, T., and Blasco, M. A. (2004) Epigenetic regulation of telomere length in mammalian cells by the Suv39h1 and Suv39h2 histone methyltransferases. *Nat. Genet.* **36,** 94–99.
68. Simerly, C., Dominko, T., Navara, C., Payne, C., Capuano, S., Gosman, G., et al. (2003) Molecular correlates of primate nuclear transfer failures. *Science* **300,** 297.
69. Bureau, W. S., Bordignon, V., Leveille, C., Smith, L. C., and King, W. A. (2003) Assessment of chromosomal abnormalities in bovine nuclear transfer embryos and in their donor cells. *Cloning Stem Cells* **5,** 123–132.
70. Ogonuki, N., Inoue, K., Yamamoto, Y., Noguchi, Y., Tanemura, K., Suzuki, O., et al. (2002) Early death of mice cloned from somatic cells. *Nat. Genet.* **30,** 253–254.

14

Pluripotency

Capacity for In Vitro Differentiation of Undifferentiated Embryonic Stem Cells

Cornelia Wiese, Gabriela Kania, Alexandra Rolletschek, Przemyslaw Blyszczuk, and Anna M. Wobus

Summary

Embryonic stem (ES) cells, the pluripotent cells of early embryos have been successfully cultured as undifferentiated cells. The cells are characterized by two unique properties, unlimited self-renewal capacity and the ability to differentiate into all cells of the body. Because of the high in vitro differentiation potential, ES cells have been used as model system in cell and developmental biology. Here we present methods that use mouse embryonic stem cells for the in vitro differentiation and characterization of neuronal, cardiac, pancreatic and hepatic cells, derivatives of the ectoderm, mesoderm and endoderm, respectively. In the future, differentiated cells may be also generated from human ES cells by cultivation of early embryos or from reprogrammed cells derived by nuclear transfer. Such cells could represent potential sources for tissue repair of serious human diseases.

Key Words: Embryonic stem cells; in vitro differentiation; neurons; cardiomyocytes; hepatic cells; pancreatic cells.

1. Introduction

Pluripotent embryonic stem (ES) cells have the potential to self-renew and to differentiate into virtually any cell type of somatic and germ cell lineages (for review, *see* **ref. *1***). In the early 1980s, undifferentiated embryonic cells from the inner cell mass (ICM) of mouse blastocysts were cultured as pluripotent ES cell lines *(2–5)*. Pluripotent embryonic cell lines also were established from mouse primordial germ cells (EG cells, i.e., **ref. *6***), livestock, and laboratory animals *(reviewed in* **rcf. *7***), and human (h) blastocysts generated by in vitro fertilization *(*hES cells; **ref. *8***).

From: *Methods in Molecular Biology, vol. 325: Nuclear Reprogramming: Methods and Protocols*
Edited by: S. Pells © Humana Press Inc., Totowa, NJ

Self-renewal capacity and multilineage differentiation ability are the unique properties of ES cells. Therefore, ES cells are suitable tools for studying pluripotency of reprogrammed cells. Mouse ES cells are the most efficient experimental system with respect to ES cell derivation, long-term cultivation, genetic manipulation, and multilineage differentiation. In the following, we describe methods of in vitro cultivation and differentiation of mouse ES cells.

To maintain the undifferentiated state, mouse ES cells are cultured routinely on a feeder layer (FL) of mouse embryonic fibroblasts (MEF) or in the presence of leukemia inhibitory factor (LIF; ref. *9*). LIF is a member of the interleukin-6 (IL-6) family of cytokines that act via gp130 receptor-mediated pathways *(10)*. Undifferentiated ES cells are characterized by a normal (euploid) karyotype and a short generation time of 12 to 15 h with a short G1 cell cycle phase *(11)*. Undifferentiated mouse ES cells express specific cell surface antigens (SSEA-1; **ref.** *12*), and enzyme activities such as the tissue nonspecific alkaline phosphatase *(4)* and telomerase (reviewed in **refs.** *7* and *13*).

Undifferentiated mouse ES cells are further characterized by the expression of molecular markers that define "stemness," such as the transcription factor Oct 3/4 *(14)* and the homeodomain protein Nanog *([15,16], see also* **ref.** *17)*. Both Nanog and Oct 3/4 *(18)* are essential to maintain ES cell identity, but STAT3, after LIF activation, and the activation of BMP (bone morphogenetic protein) and MEK/ERK-dependent signaling pathways play an accessory role *(10,19)*.

To test for pluripotency or multilineage differentiation capacity, three experimental models are accepted for mouse ES cells: (1) contribution of a given ES cell line to all cell lineages, including the germ line after retransfer into mouse blastocysts *(20)*; (2) the induction of benign teratomas and/or malignant teratocarcinomas containing various somatic cell types by extrauterine transplantation of ES cells into appropriate mouse strains *(4)*; and (3) the differentiation of ES cells in vitro *(reviewed in **ref.** 1)*.

Experimental protocols for the in vitro differentiation of ES cells have been established based on the "hanging drop" method *(21–24)*, "mass culture" *(5)*, cultivation in methylcellulose *(25)*, or by co-culture with stromal cell line activity *(26)* and direct differentiation induction in adherent monolayer cultures in the absence of LIF *(27)*.

Here, we present differentiation protocols based on "hanging drop" cultures, where aggregates called embryoid bodies (EBs) of a defined number of mouse ES cells are formed. After plating to adhesive substrates, cells within EBs undergo specific morphological changes. First, an outer layer of endoderm-like cells is formed, followed by the development of an ectodermal rim and the differentiation of mesodermal cells. The differentiation is promoted into a variety of specialized cell types, including cardiac muscle *(5,23,28)*, smooth muscle *(29,30)*, skeletal muscle *(31)*, hematopoietic *(25,32)*, adipogenic *(33)*, chondrogenic *(34)*, endothelial *(30,35)*, neuronal *(26,36–40)*, epithelial *(41)*, hepatic *(42–46)*, and pancreatic *(46–49)* cells. The temporal expression of tissue-specific genes and proteins in ES-derived cells during in vitro differentiation indicates that the early processes of in vivo development into ectoderm, mesoderm and endoderm lineages are recapitulated in vitro (reviewed in **refs. 17, 50, 51**). It also was found that the differentiated cells develop functional properties like ion channels, receptors and cell–cell junctions *(29,52,53)*.

The culture conditions and specific parameters, such as ES cell density, media components, growth factors and additives and the quality of fetal calf serum specifically influence the differentiation capacity of ES cells. In addition, different cell lines are characterized by different developmental properties in vitro *(24)*. Differentiation factors have to be added or, alternatively, genetic modifications have to be introduced into ES cells to increase the capacity to develop into the desired phenotype (e.g., **refs.** *24,49,54,55*).

This chapter describes the cultivation of undifferentiated mouse ES cells and the differentiation into derivatives of the three primary germ layers, ectoderm (neuronal cells; **ref.** *56)*, mesoderm (cardiomyocytes; **refs.** *23,24,28)*, and endoderm (hepatic and pancreatic cells; **refs.** *46,49)*.

2. Materials

2.1. Media, Reagents, and Stock Solutions for Cell Culture

2.1.1. ES Cell Culture

1. Phosphate-buffered saline (PBS): containing 10 g of NaCl, 0.25 g of KCl, 1.44 g of Na_2HPO_4, and 0.25 g of $KH_2PO_4 \cdot 2H_2O$/L, filter-sterilized through a 0.22-μm filter.
2. Trypsin solution: 0.2% trypsin 1/250 (Serva, Heidelberg, Germany) in PBS for routine passaging, 0.1% trypsin 1/500 in PBS for replating of EB outgrowths, filter-sterilized through a 0.22-μm filter.
3. Ethylenediamine tetraacetic acid (EDTA) solution: 0.02% EDTA (Sigma, St Louis, MO) in PBS for routine passaging, 0.08% EDTA in PBS for replating of EB outgrowths, filter-sterilized through a 0.22-μm filter.
4. Trypsin/EDTA: mix trypsin solution and EDTA solution at 1/1.
5. MC solution: dissolve 2 mg of mitomycin C (Serva) in 10 mL of PBS and filter-sterilize through a 0.22-μm filter. From this stock solution, dilute 300 μL into 6 mL of PBS (final concentration is 0.01 mg/mL). MC stock solution should be freshly prepared at weekly intervals and stored at 4°C. **Caution:** MC is carcinogenic.

2.1.2. Solutions Supporting Cell Adhesion

1. Gelatin solution: 1% gelatin (Fluka, Steinheim, Germany) in triple-distilled or Milli-Q water, autoclaved and diluted 1/10 with PBS. Incubate tissue culture dishes with 0.1% gelatin solution for 1 to 24 h at 4°C before use.

2. Poly-L-ornithine solution: 0.1 mg/mL poly-L-ornithine (Sigma) in 10 m*M* sodium borate buffer, pH 8.4. Filter-sterilize through a 0.22-μm filter. Incubate tissue culture dishes at 37°C for 3 h. Wash three times with Milli-Q water and incubate at room temperature for 12 h. Wash three times with Milli-Q water and dry at 40°C.
3. Laminin solution: 0.001 mg/mL laminin (Sigma) in PBS, filter-sterilized. Incubate poly-L-ornithine-coated tissue culture dishes with laminin solution at 37°C for 3 h. Wash twice with PBS before use. Store at room temperature.
4. Collagen type I solution: Dilute collagen type I (BD Biosciences, Bedford, MA) in sterile 0.02 *N* acetic acid solution. Incubate tissue culture dishes with 10 μg/cm² collagen at room temperature for 1 h. Wash twice with PBS and use immediately, or store for a maximum of 1 wk at 2 to 8°C.

2.1.3. Additives and Growth Factors (see **Note 1**)

1. Progesterone (Sigma): prepare a 1 m*M* stock solution in PBS and filter-sterilize through a 0.22-μm filter. Make fresh at weekly intervals and store at 4°C.
2. Putrescine (Sigma): prepare a 20 m*M* stock solution in PBS, filter-sterilize and store at 4°C.
3. Insulin (Sigma): prepare a stock solution from 100 mg of insulin into 10-mL Milli-Q water with 100 μL of glacial acetic acid and filter-sterilize. Make fresh every week and store at 4°C.
4. Sodium selenite (Sigma): prepare a 1 m*M* stock solution in PBS, filter-sterilize, and store at 4°C.
5. Fibronectin (Invitrogen, Karlsruhe, Germany): dissolve 1 mg of fibronectin in 1 mL of sterile Milli-Q water. Do not store in dilute solution and do not freeze/thaw repeatedly.
6. Transferrin (Sigma): prepare a stock solution with 4 mg/mL transferrin in Milli-Q water and filter-sterilize.
7. Nicotinamide (Sigma): prepare a 5 *M* stock solution in Milli-Q water and filter-sterilize.
8. Monothioglycerol (3-mercapto-1, 2-propanediol, MTG; Sigma): prepare a stock solution from 13 μL of MTG into 1 mL of Iscove's modification of Dulbecco's modification of Eagle's medium (IMDM; Invitrogen) and filter-sterilize through a 0.22-μm filter. Make fresh before use.
9. β-Mercaptoethanol (β-ME, Serva): prepare a stock solution from 7 μL of β-ME into 10 mL of PBS (stock concentration is 10 m*M*). Make fresh at weekly intervals and store at 4°C.
10. Additives I: to 100 mL of medium, add 1 mL of 200 m*M* L-glutamine stock (100X; Invitrogen), 1 mL of β-ME stock, 1 mL of nonessential amino acids stock (100X; Invitrogen) and 1 mL of penicillin–streptomycin stock (100X; Invitrogen).
11. Additives II: to 100 mL of medium, add 1 mL of 200 m*M* L-glutamine stock (100X), 300 μL of MTG of stock solution (final concentration is 450 μ*M*), 1 mL of nonessential amino acids stock (100X), and 1 mL of penicillin–streptomycin stock (100X).

12. Additives III: to 200 mL of medium, add 100 μL of insulin stock (final concentration is 5 μg/mL), 6 μL of sodium selenite stock (final concentration is 30 n*M*), 2.5 mL of transferrin stock (final concentration is 50 μg/mL), 1 mg of fibronectin (final concentration is 5 μg/mL), and 2 mL of penicillin–streptomycin stock (100X).

13. Additives IV: to 400 mL of medium, add 1 mL of insulin stock (final concentration is 25 μg/mL), 12 μL of sodium selenite stock, 5 mL of transferrin stock, 8 μL of progesterone stock (final concentration is 20 n*M*), 2 mL of putrescine stock (final concentration is 100 μ*M*), 400 μL of laminin (final concentration is 1 μg/mL), and 4 mL of penicillin–streptomycin stock (100X).

14. Basic fibroblast growth factor (bFGF; Strathmann Biotec AG, Hamburg, Germany): prepare a 10 μg/mL stock solution in sterile PBS with 0.1% bovine serum albumin (BSA; Invitrogen). Store aliquots in silanized tubes at –20°C (*see* **Note 1**).

15. Epidermal growth factor (EGF; Strathmann Biotec AG): prepare a 10 μg/mL stock solution in sterile PBS with 0.1% BSA. Store aliquots in silanized tubes at –80°C (*see* **Note 1**).

16. Leukemia inhibitory factor (LIF): 10 ng/mL (Chemicon International Inc., Temecula, CA).

2.1.4. Cultivation and Differentiation Media

1. Cultivation medium I: (DMEM; 4.5 g/L glucose; Invitrogen) supplemented with 15% heat-inactivated fetal calf serum (FCS; selected batches) and additives I for cultivation of MEF feeder layer cells.

2. Cultivation medium II: DMEM supplemented with 15% heat-inactivated FCS, LIF, and additives I for ES cell cultivation.

3. Differentiation medium I: IMDM supplemented with 20% heat-inactivated FCS and additives II for EB formation.

4. Differentiation medium II: DMEM/F12 (Invitrogen) supplemented with additives III for EB differentiation into neuronal cells.

5. Differentiation medium III: DMEM/F12 supplemented with additives IV, and daily addition of 10 ng/mL bFGF and 20 ng/mL EGF for differentiation of neuronal cells.

6. Differentiation medium IV: "Neurobasal" medium supplemented with 2% B27 (commercially available supplement containing neurotrophic factors, both Invitrogen) and 10% FCS for differentiation into neurons.

7. Differentiation medium V: DMEM/F12 supplemented with additives IV, 2% B27, 10 m*M* nicotinamide for differentiation into pancreatic cells.

8. Differentiation medium VI: hepatocyte culture medium composed of 500 mL of hepatocyte basal medium (Modified Williams'E), 0.5 mL of ascorbic acid, 10 mL of BSA–FAF (fatty acid-free), 0.5 mL of hydrocortisone, 0.5 mL of transferrin, 0.5 mL of insulin, 0.5 mL of human EGF, and 0.5 mL of gentamycin–amphothericin (GA-1000; all from Clonetics, Bio-Whittaker, Verviers, Belgium) supplemented with 10% FCS and 5 mL of penicillin–streptomycin stock (100X), prepared immediately before use.

2.2. Reverse Transcription Polymerase Chain Reaction (see Notes 2 and 3)

1. Diethyl pyrocarbonate-treated water (DEPC-H$_2$O): add 1 mL of DEPC (Invitrogen) to 1 L of Milli-Q water and stir overnight. DEPC is inactivated by heating to 100°C for 15 min or autoclaving for 15 min.

2. RNA lysis buffer: add 23.6 g of guanidinium thiocyanate to 5 mL of 250 m*M* Na-citrate, pH 7.0, 2.5 mL of 10% sarcosyl, and add DEPC–H$_2$O to a total volume of 49.5 mL and mix carefully. Make fresh at monthly intervals. Add 1% β-ME before use.

3. 2 *M* Na–acetate, pH 4.0: dissolve 27.2 g of Na–acetate·3 H$_2$O in 0.1% DEPC–H$_2$O, adjust the pH to 4.0 with glacial acetic acid, and adjust to 100 mL with DEPC–H$_2$O. Treat the buffer with 0.1% DEPC-H$_2$O at 37°C for at least 1 h and heat to 100°C or autoclave for 15 min.

4. Acidic phenol: phenol is saturated with DEPC–H$_2$O instead of Tris. The saturated acidic phenol contains 0.1% hydroxyquinoline (antioxidant, partial inhibitor of RNase, and a weak chelator of metal ions; its yellow color provides a convenient way to identify the organic phase). Store at 4°C for up to 2 mo.

5. Chloroform: isoamylalcohol (24/1).

6. 75% Ethanol: prepare in DEPC–H$_2$O.

7. 25 m*M* MgCl$_2$ (MBI Fermentas, St. Leon-Rot, Germany).

8. 5X Reaction buffer for reverse transcription (MBI Fermentas): 250 m*M* Tris-HCl, pH 8.3, 250 m*M* KCl, 20 m*M* MgCl$_2$, 50 m*M* dithiothreitol.

9. 10X polymerase chain reaction (PCR) buffer without MgCl$_2$ (MBI Fermentas): 750 m*M* Tris-HCl, pH 8.8; 200 m*M* (NH$_4$)$_2$SO$_4$, 0.1% Tween-20.

10. 40 U/μL RNase inhibitor (MBI Fermentas).

11. Oligo d(T)$_{18}$ (MBI Fermentas): 50 μ*M* in 10 m*M* Tris-HCl, pH 8.3.

12. 200 U/μL RevertAis™ M-MuLV reverse transcriptase (MBI Fermentas); 5 U/μL recombinant Taq DNA polymerase (MBI Fermentas).

13. 10 m*M* dNTP mix: dNTP (dGTP, dATP, dCTP, dTTP; MBI Fermentas) dilute to 100 m*M* with DEPC–H$_2$O and freeze at –20°C. 10 m*M* dNTP mix is freshly made by mixing the equal volumes of 100 m*M* of each dNTP before use.

14. Select PCR primer pairs: Dilute synthetic oligonucleotides to 10 μ*M* with DEPC–H$_2$O and freeze at –20°C.

15. 20 mg/mL Glycogen(MBI Fermentas).

16. 5 *M* NaCl: dissolve 29.2 g of NaCl in Milli-Q water; adjust to 100 mL with water and autoclave.

17. TE buffer: 10 m*M* Tris-HCl, 1 m*M* EDTA, pH 7.5, filter-sterilized through a 0.22-μm filter.

18. 10X Loening solution: 1.8 *M* Tris-HCl, pH 7.7, 1.5 *M* NaH$_2$PO$_4$·H$_2$O, 50 m*M* EDTA, and autoclave.

19. 6X Loading buffer: 10 mL of 10X Loening solution, 30 mL of glycerine, 10 mL of 10% sodium dodecyl sulfate, 20 mL of 10% *N*-lauroyl-sarcosine, 100 mg of bromcresol green (Sigma), and 30 mL of Milli-Q water.

20. 5X TBE: dissolve 54 g of Tris-base and 27.5 g of boric acid in Milli-Q water, add 20 mL of 0.5 M EDTA, pH 8.0, and adjust to 1 L with Milli-Q water.
21. Ethidium bromide aqueous solution (Serva): 1% w/v = 10 mg/mL. Agarose gels: melt electrophoresis grade agarose (Invitrogen) in 1X TBE by gentle boiling in a microwave oven. Cool to approx 60°C and pour into an agarose gel mold. Run small gels at approx 80 to 100 V by using bromcresol green in the stop mix as an indicator of migration.

Caution: Ethidium bromide is carcinogenic. Use nitrile gloves and dispose all contaminated tips, agarose gels, and buffers separately.

2.3. Immunohistochemical Analysis

1. 3.7% Paraformaldehyde (PFA; e.g., Fluka): dissolve 3.7 g of PFA in PBS and adjust to 100 mL with PBS, heat the mixture to 95°C, stir until the solution becomes clear, and cool to room temperature. **Caution:** PFA is toxic. Work under the hood and use gloves.
2. Methanol: acetone (7/3) fixative.
3. 10% goat serum (Invitrogen) or 1% BSA in PBS for blocking nonspecific binding of antibodies.
4. Hoechst 33342 (5 µg/mL in PBS).
5. Mounting medium: Vectashield (Linaris, Wertheim, Germany).
6. 0.5% BSA in PBS for dilution of secondary antibodies.

2.4. Enzyme-Linked Immunosorbent Assay

1. Krebs' Ringer bicarbonate HEPES (KRBH) buffer is used in sample preparation for insulin enzyme-linked immunosorbent assay (ELISA): combine 118 mM sodium chloride, 4.7 mM potassium chloride, 1.1 mM potassium dihydrogen phosphate, 25 mM sodium hydrogen carbonate (all from Roth, Karlsruhe, Germany), 3.4 mM calcium chloride (Sigma), 2.5 mM magnesium sulfate (Merck Biosciences, Schwalbach, Germany), 10 mM HEPES (Sigma), and 2 mg/mL BSA.
2. For insulin ELISA, prepare solutions of 2.5, 5.5, and 27.7 mM glucose and 5.5 mM glucose with 10 µM tolbutamide dissolved in KRBH buffer (all from Sigma).
3. Acid ethanol: 1 M hydrochloric acid/absolute ethanol (1/9).
4. Insulin ELISA (Mercodia, Uppsala, Sweden).
5. Albumin ELISA (Bethyl Laboratories, Montgomery, TX).
6. Bradford assay (Bio-Rad Laboratories, Munich, Germany).

2.5. Equipment

1. Tissue culture plates: 35 mm, 60 mm, and 100 mm (Nunc, Wiesbaden, Germany or BD Biosciences Falcon, Heidelberg, Germany).
2. Bacteriological petri dishes (Greiner, Frickenhausen, Germany): 60 mm for EB mass culture, 100 mm for EB hanging drop culture.
3. Pasteur pipets and 2-, 5-, 10-, 25-mL pipets.
4. Counting chamber (e.g., Thoma).

5. Tissue culture incubator with 37°C and 5% CO_2 atmosphere.
6. For feeder layer culture: sterile dissecting instruments, screen or sieve (~0.5–1-mm diameter), Erlenmeyer flasks with stir bars, centrifuge tubes.
7. Smart Spec 3000 (Bio-Rad Laboratories).
8. PCR apparatus: Mastercycler gradient (Eppendorf, Hamburg, Germany).
9. Electrophoresis equipment (Bio-Rad Laboratories)
10. 0.5- and 1.5-mL microtubes (Eppendorf) and 20-, 100-, 1000-µL filtertips (Biozym, Hess. Oldendorf, Germany).
11. Silanized 1.5-mL microtubes (Porex Bio Products, Inc., Petaluma, CA) and silanized 10-, 100-, and 1000-µL tips (Porex Bio Products, Inc.).
12. Cover slips (16 × 16 mm and 12 mm diameter) and slides.
13. ELISA Reader (Bio-Rad Laboratories).
14. LUCIA HEART imaging system (Nikon, Düsseldorf, Germany).

3. Methods

3.1. Cultivation of Undifferentiated ES Cells on Feeder Layer

3.1.1. Feeder Layer Culture (see **Note 4**)

1. Remove embryos from a mouse pregnant for 15 to 17 d (e.g., NMRI or CD-1 outbred strains), rinse in PBS, and remove placenta and fetal membranes, head, liver, and heart. Rinse the carcasses in trypsin solution.
2. Mince the tissue in 5 mL of fresh trypsin solution and transfer to an Erlenmeyer flask containing a stir bar.
3. Stir on a magnetic stirrer for 25 to 45 min (use a longer incubation time if the embryos are older), filter the suspension through a sieve or a screen, add 10 mL of culture medium I, and spin down (5 min at 1000g).
4. Resuspend the pellet in approx 3 mL of culture medium I and plate on 100-mm tissue culture plates (~2 × 10^6 cells per 100-mm dish) containing 10 mL of cultivation medium I, and incubate at 37°C and 5% CO_2 for 24 h.
5. Change the medium to remove debris, erythrocytes and unattached cellular aggregates, and cultivate for an additional 1 to 2 d.
6. Passage the primary culture of mouse embryonic fibroblasts: split 1/2 to 1/3 on 100-mm tissue culture plates and grow in cultivation medium I for 1 to 3 d. The cells in passages 2 to 4 are most suitable as feeder layer for undifferentiated ES cells.
7. Incubate feeder layer cells with MC buffer for 2 to 3 h, aspirate the MC solution, wash 3X with PBS, trypsinize the feeder cells, and replate to new gelatin (0.1%)-treated microwell plates or Petri dishes. Feeder layer cells prepared one day before ES cell subculture are optimal!

3.1.2. Culture of Undifferentiated ES Cells (see **Notes 5 and 6**)

It is important to passage ES cells every 24 or 48 h. Do not cultivate longer than 48 h without passaging or the cells may differentiate and be unsuitable for differentiation studies. Selected batches of FCS have to be used for ES cell culture (*see* **Note 6**).

1. Change the medium 1 to 2 h before passaging.
2. Aspirate the medium, add 2 mL of trypsin–EDTA, and incubate at room temperature for 30 to 60 s.
3. Carefully remove the trypsin–EDTA mixture and add 2 mL of fresh cultivation medium II.
4. Resuspend the cell population with a 2-mL glass pipet into a single cell suspension and split 1/3 to 1/10 to freshly prepared (60 mm) feeder layer plates.

3.2. In Vitro Differentiation Protocols

For the development of ES cells into differentiated phenotypes, cells are cultivated in three-dimensional aggregates or EBs. The differentiation of cardiac muscle, neuronal, pancreatic and hepatic cells requires different conditions. In this chapter, differentiation protocols utilizing the "hanging drop" method are described (*see* **Note 7**, **Fig. 1**).

3.2.1. Preparation of EBs

1. Prepare a single cell suspension containing a defined ES cell number of 200, 400, or 600 (*see* **Subheading 3.2.2.**) cells in 20 µL of differentiation medium I.
2. Place 20-µL drops (*n* = 50–60) of the ES cell suspension on the lids of 100-mm bacteriological Petri dishes containing 10 mL of PBS.
3. Cultivate the ES cells in hanging drops for 2 d. The cells will aggregate and form one EB per drop.
4. Rinse the aggregates carefully from the lids with 2 mL of medium, transfer into a 60-mm bacteriological Petri dish with 5 mL of differentiation medium I, and continue cultivation in suspension for 2 to 3 d until the time of plating.

3.2.2. Differentiation of EB Outgrowths (see **Note 8**)

3.2.2.1. Cardiac Muscle Cell Differentiation *(24)*

1. Use of 400 ES cells for preparation of EBs is optimal for cardiac differentiation.
2. Culture with differentiation medium I.
3. Plate 20 to 25 EBs onto gelatin-coated 60-mm tissue culture plates at day 5. The first beating clusters in EBs could already be seen 2 d after plating (–5 + 2 d).
4. Analyze cardiac muscle cells after EB plating by reverse transcription (RT)-PCR and immunofluorescence.

3.2.2.2. Neuronal Cell Differentiation *(56)*

Because the spontaneous differentiation into neuronal cell types is rather limited *(38)*, specific protocols were applied to increase the differentiation in vitro by (1) differentiation induction with retinoic acid *(50)*, (2) lineage selection *(57)*, (3) neuronal induction by stromal cell-derived inducing activity *(26)*, and (4) lineage-restricted differentiation *(39,40,56)*. In the following, the lineage-restricted differentiation protocol is described (*see* **Note 9**).

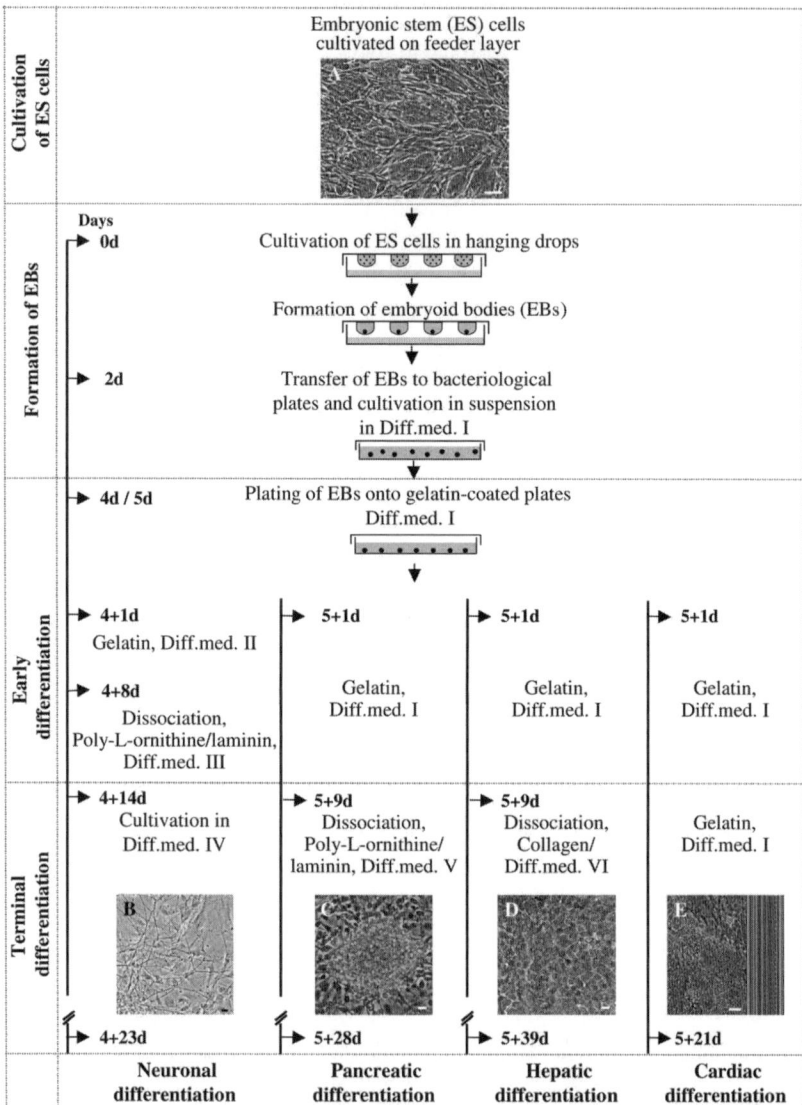

Fig. 1. Schematic diagram of differentiation of mouse ES cells into neuronal, pancreatic, hepatic and cardiac muscle cells in vitro. The protocols are based on the formation of embryoid bodies (EBs) by the "hanging drop" technique, followed by early and terminal differentiation induction after EB plating. (**A**) Undifferentiated ES cells cultivated 2 d on mitomycin-inactivated mouse embryonic fibroblasts as feeder layer. (**B–E**) Morphology of ES-derived neurons (B), pancreatic β-like cells (C), hepatocyte-like cells (D), and beating cardiomyocytes (E, left). Beating frequency is measured by the LUCIA HEART imaging system (E, right). Diff. med., Differentiation medium, Bar, 20 μm.

1. Neuronal phenotypes are obtained by using 200 ES cells for EB formation.
2. Culture with differentiation medium I.
3. Plate 20 to 30 EBs onto gelatin-coated 60-mm tissue culture dishes at day 4 and cultivate in differentiation medium I.
4. Exchange the medium with differentiation medium II, 1 d after plating (4 + 1 d).
5. Replenish medium every 2 d.
6. At days 4 + 8, dissociate EBs by incubation with trypsin (0.1%):EDTA (0.08%) 1/1 solution for 1 min, collect the cells by centrifugation (maximal 1000g, 5 min), and replate onto poly-L-ornithine/laminin-coated tissue culture dishes into differentiation medium III (*see* **Note 10**).
7. Change the medium every 2 d.
8. At day 4 + 14, induce neuronal cell differentiation by further cultivation with differentiation medium IV.
9. Analyze cultures for neuronal cells by RT-PCR and immunofluorescence.

3.2.2.3. PANCREATIC AND HEPATIC CELL DIFFERENTIATION (*SEE* **NOTE 11**)

1. Pancreatic and hepatic cell types are differentiated using 600 ES cells for preparation of EBs.
2. Culture EBs in differentiation medium I.
3. Plate 20 to 30 EBs onto gelatin-coated 60-mm tissue culture dishes at day 5 in differentiation medium I.
4. Replenish the medium every 2 d
5. At day 5 + 9, dissociate the EBs by incubation with trypsin (0.1%): EDTA (0.08%) 1/1 solution for 1 min, collect the cells by centrifugation (maximal 1000 rpm, 5 min), and replate onto poly-L-ornithine/laminin (pancreatic)- and collagen I (hepatic)-coated tissue culture dishes into differentiation medium V (pancreatic cells) and differentiation medium VI (hepatic cells), respectively.
6. Change the medium every 2 d.
7. Analyze cultures for 7 to 19 d (pancreatic cells) and 10 to 30 d (hepatic cells), respectively, after replating by RT-PCR, immunofluorescence, and ELISA.

3.3. Characterization of Differentiated Phenotypes

3.3.1. RT-PCR Analysis (see **Note 12**)

3.3.1.1. PREPARATION OF CELL SAMPLES (*SEE* **NOTES 13** AND **14**)

The transcripts of genes, which are specifically expressed during ES cell differentiation, are analyzed by RT-PCR with primers for tissue-specific genes (**Table 1**). The following steps are used to harvest differentiated cells:

1. Discard the medium and wash twice with PBS.
2. Add 400 µL of RNA lysis buffer per 60-mm culture dish. Allow the lysis buffer to spread across the surface of the dish and transfer the lysate into a 1.5-mL microtube.
3. Store samples at –20 or –80°C.

Table 1
Primers Used for RT-PCR

Genes	Primer sequences (forward/reverse)	Annealing temperature	Product size
Cardiac genes			
Nkx 2.5	5'-CGACGGAAGCCACGCGTGCT 5'-CCGCTGTCGCTTGCACTTG	60°C	181 bp
Cardiac α-MHC	5'-CTGCTGGAGAGGTTATTCCTCG 5'-GGAAGAGTGAGCGGCGCATCAAGG	64°C	301 bp
Cardiac β-MHC	5'-TGCAAAGGCTCCAGGTCTGAGGGC 5'-GCCAACACCAACCTGTCCAAGTTC	64°C	205 bp
ANF	5'-TGATAGATGAAGGCAGGAAGCCGC 5'-AGGATTGGAGCCCAGAGTGGACTAGG	64°C	203 bp
MLC-2V	5'-TGTGGGTCACCTGAGGCTGTGGTTCAG 5'-GAAGGCTGACTATGTCCGGGAGATGC	60°C	504 bp
Neural genes			
Nestin	5'-CTACCAGGAGCGCGTGGC 5'-TCCACAGCCAGCTGGAACTT	60°C	220 bp
En-1	5'-TGGTCAAGACTGACTCACAGCA 5'-TCTCGTCTTTGTCCTGAACCGT	61°C	390 bp
NFM	5'-TATTGTGACTGAGGGCTGTCGG 5'-GAGGCACTAAGGAGTCCCTG	60°C	297 bp
Tyrosine hydroxylase (TH)	5'-TGTCAGAGCAGCCCGAGGTC 5'-CCAAGAGCAGCCCATCAAAG	64°C	412 bp
Synaptophysin	5'-TACCGAGAGAACAACAAAGGGC 5'-GCCTGTCTCCTTGAACACGAAC	60°C	287 bp
GFAP	5'-TCGGAGTTGAAAGTTACAGG 5'-AGGATGGTTGTGGATTCTTC	50°C	234 bp
Pancreatic genes			
Glut-2	5'-TTCGGCTATGACATCGGTGTG 5'-AGCTGAGGCCAGCAATCTGAC	60°C	556 bp
IAPP	5'-TGATATTGCTGCCTCGGACC 5'-GGAGGACTGGACCAAGGTTG	65°C	233 bp
Insulin I	5'-TAGTGACCAGCTATAATCAGAG 5'-CGCCAAGGTCTGAAGGTCC	60°C	288 bp
Isl-1	5'-GTTTGTACGGGATCAAATGC 5'-ATGCTGCGTTTCTTGTCCTT	60°C	514 bp
Ngn3	5'-TGGCGCCTCATCCCTTGGATG 5'-AGTCACCCACTTCTGCTTCG	60°C	157 bp
Pax4	5'-ACCAGAGCTTGCACTGGACT 5'-CCCATTTCAGCTTCTCTTGC	60°C	300 bp
Pax6	5'-TCACAGCGGAGTGAATCAG 5'-CCCAAGCAAAGATGGAAG	58°C	332 bp
Pdx1	5'-CTTTCCCGTGGATGAAATCC 5'-GTCAAGTTCAACATCACTGCC	60°C	230 bp

(continued

able 1 *(Continued)*
'rimers Used for RT-PCR

ienes	Primer sequences (forward/reverse)	Annealing temperature	Product size
lepatic genes			
Albumin	5'-GTCTTAGTGAGGTGGAGCAT 5'-ACTACAGCACTTGGTAACAT	58°C	569 bp
Alpha-1- antitrypsin	5'-CAATGGCTCTTTGCTCAACA 5'-AGTGGACCTGGGCTAACCTT	63°C	518 bp
Alpha- fetoprotein	5'-CACTGCTGCAACTCTTCGTA 5'-CTTTGGACCCTCTTCTGTGA	58°C	301 bp
Cyp2β9	5'-GATGATGTTGGCTGTGATGC 5'-CTGGCCACCATGAAAGAGTT	53°C	153 bp
Cyp2β13	5'-CTGCATCAGTGTATGGCATTTT 5'-TTTGCTGGAACTGAGACTACCA	65°C	166 bp
HNF3 β	5'-GCGGGTGCGGCCAGTAG 5'-GCTGTGGTGATGTTGCTGCTCG	63°C	378 bp
Transthyretin	5'-CTCACCACAGATGAGAAG 5'-GGCTGAGTCTCTCAATTC	55°C	225 bp
Tyrosine amino- transferase	5'-ACCTTCAATCCCATCCGA 5'-TCCCGACTGGATAGGTAG	50°C	206 bp
lousekeeping gene			
β-tubulin	5'-TCACTGTGCCTGAACTTACC 5'-GGAACATAGCCGTAAACTGC	60°C	317 bp

MHC, myosin heavy chain.

3.3.1.2. Isolation of Total RNA (*see* Note 15)

The method described here is based on the use of a chaotropic agent (guanidine salt) for the disruption of cells and inactivation of ribonucleases *(58)*.

1. Thaw the lysate (400 μL) and vortex for 15 s.
2. Add 40 μL (1/10 vol) of 2 *M* Na–acetate, pH 4.0. Mix carefully.
3. Add 400 μL of acidic phenol and vortex vigorously.
4. Add 80 μL of chloroform:isoamylalcohol (24/1) and vortex again.
5. Store for 15 min on ice.
6. Separate the organic and aqueous phases by centrifugation at 16,000*g* for 10 min at room temperature.
7. Carefully transfer the upper aqueous phase to a fresh tube, add an equal volume of isopropanol, and mix well. Store for 1 h at –20°C (*see* Note 15).
8. Centrifuge at 16,000*g* for 10 min at room temperature. Carefully discard the supernatant.
9. Dissolve the pellet in 300 μL of lysis buffer. If the pellet is difficult to dissolve, heat to 65°C for several minutes. Add an equal volume (300 μL) of isopropanol and mix well. Store at –20°C for 1 h.

10. Centrifuge at 16,000g for 10 min at room temperature. Carefully discard the supernatant.
11. Wash the pellet with 500 μL of 75% ice-cold ethanol (made with DEPC–H$_2$O), vortex briefly, centrifuge at 16,000g for 10 min, discard the supernatant, and allow the pellet of nucleic acid to dry in the air.
12. Dissolve the RNA pellet in 30 μL of DEPC–H$_2$O and freeze at –80°C.
13. Dilute 1 μL of RNA with 100 μL of DEPC–H$_2$O, measure OD$_{260}$ and the concentration of RNA using the Smart Spec 3000 or a suitable spectrophotometer, adjust all samples to the same RNA concentration (e.g., 0.2 μg/μL) with DEPC–H$_2$O and measure again to ensure the same RNA concentration of all samples. The yield of RNA from EBs ($n = 20$) is in the range of 20 to 100 μg.

3.3.1.3. RT REACTIONS (*SEE* **NOTE 16**)

All RT and PCR solutions are available from commercial suppliers in ready-to-use form. Reverse transcription reactions are performed in 20-μL reaction volumes using 0.5-mL microcentrifuge tubes.

1. Label one PCR tube for each sample and appropriate controls. Add the same amount of RNA (0.5–1.0 μg in 3 μL) to each tube.
2. Prepare the following RT-Mastermix for 25 reactions (or a smaller quantity as required) containing: 261.25 μL of DEPC–H$_2$O, 100 μL of 5X reaction buffer, 20 μL of 10 mM dNTPs mix, 12.5 μL of RNase inhibitor, 25 μL of Oligo d(T)$_{18}$, and 6.25 μL of RevertAis™ M-MuLV RT to a total volume of 425 μL.
3. Add 17 μL of RT-Mastermix to each tube, mix carefully, and centrifuge briefly.
4. Transfer tubes to the thermal cycler and perform reverse transcription reactions for 1 h at 42°C and then heat to 99°C for 5 min.
5. Cool the samples to 4°C or store at –20°C until use.

3.3.1.4. PCR (*SEE* **NOTE 16**)

1. Prepare a PCR-Mastermix for 25 reactions (or a smaller quantity as required) containing: 798.75 μL of DEPC–H$_2$O, 125 μL of 10X PCR buffer, 90 μL of 25 mM MgCl$_2$, 80 μL of 10 mM dNTPs mix, 50 μL of 10 μM 5' sense primer of the target gene, 50 μL of 10 μM 3' antisense primer of the target gene, 6.25 μL of Taq DNA polymerase, to a total volume of 1200 μL.
2. Label new PCR tubes and add 2.0 μL of RT reaction product to each tube as template DNA.
3. Add 48 μL of PCR-Mastermix to each tube, mix by vortexing, and centrifuge briefly.
4. Transfer the tubes to the thermal cycler. Amplify the cDNA through 25 to 40 thermal cycles. Standard conditions are denaturation at 95°C for 40 s, annealing at 50 to 65°C for 40 s, and extension at 72°C for 40 s. The conditions depend on the primers and thermal cycler used.

5. Run a parallel reaction containing 2.0 µL of RT reaction product and 48 µL of PCR-Mastermix with primers of the internal standard gene (i.e., β-tubulin) instead of the target gene.
6. Cool the samples to 4°C and store at –20°C.

3.3.1.5. POST-PCR TREATMENT OF SAMPLES

1. Transfer the PCR products to 1.5-mL microtubes.
2. Add 2.5 µL of a 1/4 mixture of glycogen: 5 *M* NaCl and 150 µL of ice-cold ethanol to each tube.
3. Incubate at –20°C for at least 1 h and centrifuge at 16,000*g* for 15 min.
4. Dissolve the pellet in 25 µL of TE buffer, add 5 µL of 6X loading buffer, and store at 4°C.

3.3.1.6. ELECTROPHORESIS

1. Separate one third of each PCR reaction (10 µL) by electrophoresis on a 2% agarose gel in 1X TBE containing 0.35 µg/mL of ethidium bromide at 5 to 10 V/cm for 70 to 100 min.
2. Illuminate the gel by ultraviolet light and obtain a digital image.

3.3.2. Immunofluorescence Analysis (see **Notes 17** and **18**)

The formation of tissue-restricted proteins in the EB outgrowths is analyzed by immunofluorescence with a normally equipped fluorescence microscope or with a confocal laser scanning microscope (*see* **Note 17**). For the characterization of ES-derived cardiac, neuronal, pancreatic, and hepatic phenotypes (*see* **Fig. 2**, suitable primary antibodies are summarized in **Table 2**.

1. Rinse cover slips containing EB outgrowths twice with PBS.
2. Fix the cells onto cover slips with methanol: acetone (7/3) at –20°C for 10 min, or alternatively, with 3.7% paraformaldehyde in PBS at room temperature for 20 min (depending on the antibody used).
3. Rinse the cover slips twice with PBS at room temperature for 5 min.
4. Incubate the cells on cover slips with 10% goat serum or 1% BSA in PBS in a humidified chamber at room temperature for 30 min to prevent nonspecific immunostaining (*see* **Note 18**).
5. Incubate the cover slips with the primary antibody at 37°C for 60 min, or at 4°C overnight (final concentration according to manufacturer's instructions).
6. Rinse the cover slips with PBS three times at room temperature for 5 min.
7. Incubate the cover slips with fluorescently labelled specific secondary antibody (depending on the primary antibody) diluted in PBS with 0.5% BSA in a humidified chamber at 37°C for 45 to 60 min.
8. Incubate the cover slips with 5 µg/mL Hoechst 33342 in PBS for 10 min at room temperature to label the cell nuclei.

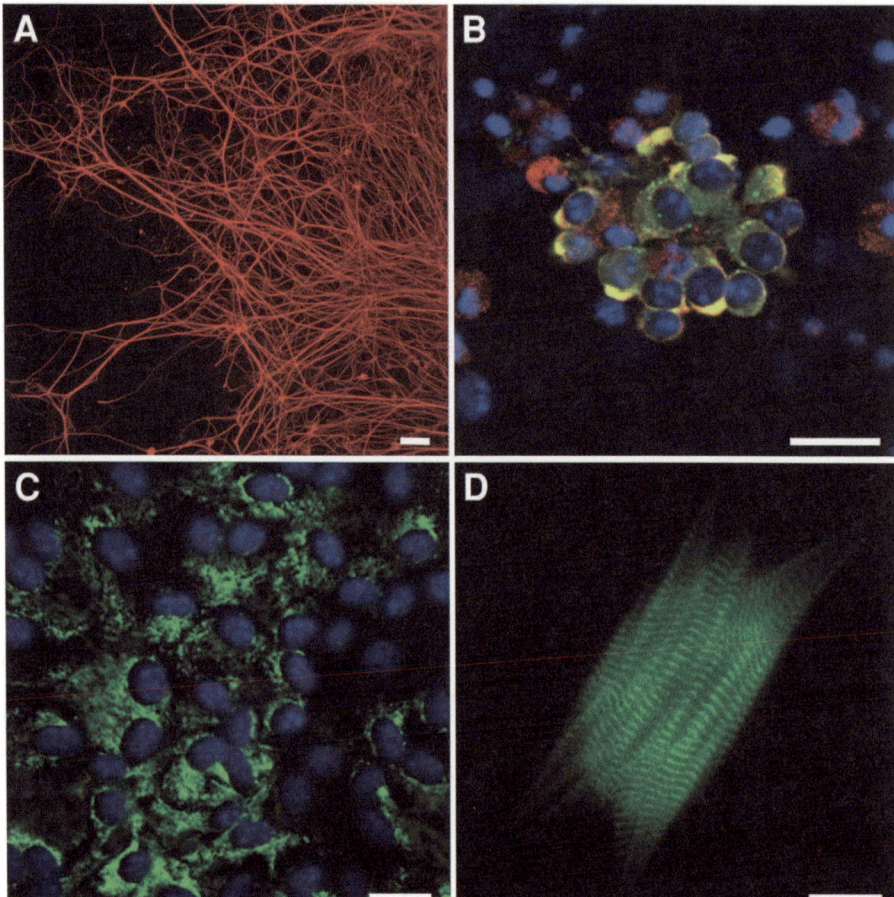

Fig. 2. Immunofluorescence analysis of differentiated ES-derived neuronal, pancreatic, hepatic and cardiac muscle cells. (**A**) Neuronal network positive for β-III-tubulin (stage 4 + 23d), (**B**) islet-like cluster of pancreatic endocrine cells expressing C-peptide (green) and insulin (red) (stage 5 + 28d), (**C**) hepatocyte-like cells expressing albumin (stage 5 + 39d), (**D**) cardiac muscle cell showing a well organized sarcomeric staining pattern of Z-disk epitopes of titin (stage 5 + 21 d). Cell nuclei were visualized by Hoechst 33342 staining (blue). Bar, 20 μm.

9. Rinse the cover slips twice with PBS at room temperature for 5 min.
10. Rinse the cover slips quickly with distilled water at room temperature.
11. Embed the cover slips in mounting medium and analyse the immunolabeled cells with a conventional fluorescence or confocal laser scanning microscope.

Table 2
Primary Antibodies for the Analysis of Cell-Specific Proteins by Immunofluorescence

Antigen	Antibody	Working dilution	Source
Cardiac muscle markers			
Titin	T12	1/100	*(64)*
α-Actinin	653	1/20	*(64)*
Myomesin	MyBB78	1/2	*(65)*
Sarcomeric MHC	MF-20	Undiluted	Developmental Studies Hybridoma Bank, IA
α-Sarcomeric actin	5C5	1/500	Sigma
α-Cardiac MHC	BA-G5	1/500	*(66)*
Neural markers			
Nestin	Rat 401	1/3	Developmental Studies Hybridoma Bank
β III-Tubulin	clone TU-20	1/200	Chemicon
Synaptophysin		1/150	Merck Biosciences
Tyrosine hydroxylase (TH)		1/150	Merck Biosciences
GABA		1/100	Sigma
Serotonin		1/100	Sigma
Glial fibrillary acidic protein (GFAP)	clone G-A-5	1/20	Chemicon
Common pancreatic and hepatic markers			
Cytokeratin 18	clone KS-B17.2	1/100	Sigma
Cytokeratin 19	clone LAS86	1/100	Chemicon
Pancreatic markers			
Carbonic anhydrase II		1/200	Abcam, Cambridgeshire, UK
C-peptide		1/100	Acris, Hiddenhausen, Germany
Insulin	hybridoma K36 aC10	1/40	Sigma
Isl-1		1/200	Abcam
Hepatic markers			
Albumin		1/100	Serotec, Oxford, UK
α-1-antitrypsin		1/100	Sigma
α-fetoprotein		1/100	Santa Cruz, Santa Cruz, CA
Amylase		1/100	Santa Cruz
Cytokeratin 14	clone CKB1	1/100	Sigma
Dipeptidyl peptidase IV		1/100	Santa Cruz

3.3.3. ELISA

3.3.3.1. INSULIN ELISA (*SEE* NOTE 19)

1. Wash the cells with PBS (5X) and preincubate in freshly prepared KRBH buffer with 2.5 mM glucose for 90 min at 37°C.
2. For a control, replace the KRBH buffer with 2.5 mM glucose by KRBH buffer with 5.5 mM glucose. For induction of insulin release, replace KRBH buffer with 2.5 mM glucose by KRBH buffer containing 27.7 mM glucose, or KRBH buffer containing 5.5 mM glucose and 10 µM tolbutamide.
3. Collect the supernatant for immediate determination of insulin release or store the supernatant at –20°C.
4. Dissociate the cells by treatment with 0.2% trypsin/0.02% EDTA 1/1 solution for 3 min and centrifuge.
5. Extract the proteins from the cells with acid ethanol overnight at 4°C, followed by cell sonication. Store samples at –20°C for determination of total cellular insulin and protein content.
6. Perform ELISA according to the manufacturer's recommendations.
7. Determine total protein content with a protein Bradford assay according to the manufacturer's recommendations.

3.3.3.2. ALBUMIN ELISA

1. Wash the cells with PBS (5X) and incubate in differentiation medium VI in the absence of BSA and FCS at 37°C for 24 h.
2. Collect the supernatant for immediate measurement of albumin release or store the supernatant at –20°C.
3. Dissociate the cells by treatment with 0.2% trypsin/0.02% EDTA 1/1 solution for 3 min, centrifuge, and add PBS for the determination of total cell number.
4. Perform the albumin ELISA according to the manufacturer's recommendations.

4. Notes

1. Repeated thawing and freezing of growth factors must be avoided; otherwise the biological activity will be decreased.
2. DEPC is a suspected carcinogen and should be handled with care.
3. If possible, the solutions should be treated with 0.1% DEPC at 37°C for 1 h and then heated to 100°C for 15 min or autoclaved for 15 min. DEPC reacts rapidly with amines and cannot be used to treat solutions containing buffers, such as Tris.
4. Feeder layer cells used for ES cell cultivation should be plated at an appropriate density and cell cycle-inactivated through mitomycin C treatment or gamma irradiation. They provide LIF and other still-unknown components essential for the prevention of mouse ES cell differentiation.
5. ES cells of mouse and human origin need feeder layer cells for growth in the undifferentiated state. In the case of mouse ES cells, feeder layer cells can be replaced by conditioned medium containing LIF. LIF is commercially available (e.g., Chemicon International, Inc.), or may be prepared from LIF expression vectors (*59,60*).

6. Good-quality FCS is critical for the long-term culture of ES cells, and failure to acquire good-quality serum may be one reason why ES cells fail to differentiate appropriately. Extensive serum testing is therefore necessary to achieve good results. The most sensitive tests for sera include: (1) comparative plating efficiencies at 10, 15, and 30% serum concentrations using ES and embryonic carcinoma cells, (2) alkaline phosphatase activity in undifferentiated ES cells (e.g., Vector Blue Substrate Kit, Vector Laboratories, Burlingame, CA, USA), and (3) test of in vitro differentiation capacity after three to five passages in selected serum batches *(61)*.

7. For preparation of EBs, three different protocols may be used: (1) the "hanging drop" method *(21–24)*, (2) the "mass culture" *(5)*; or (3) the "methylcellulose" technique *(25)*. The hanging drop method, as described here, generates EBs of a defined cell number (and size). This technique is especially used for developmental studies, because the differentiation pattern is dependent on the number of ES cells that differentiate within the EBs. A precondition for the use of this method is the dissociation of ES cells. (Unfortunately, hES cells could not yet be dissociated so far to get single cell suspension; for hES cells, "mass culture" can be used for EB preparation). Plate 5×10^5 to 2×10^6 cells into 60-mm bacteriological Petri dishes containing 5 mL differentiation medium. After 2 d, let the aggregates settle in a centrifuge tube, remove the medium and carefully transfer the aggregates with 5 mL of fresh differentiation medium into a new bacteriological dish. Change the medium every second day. Mass cultures of EBs also may be used for differentiation of a large number of cells. For hematopoietic differentiation, the "methylcellulose" method is used *(25)*. Methylcellulose (e.g., MethoCult H4100; Stem Cell Technologies, Vancouver, BC, Canada) is added to the differentiation medium at a final concentration of 0.9%.

8. To achieve efficient differentiation, cells should be plated on appropriate extracellular matrix proteins. Generally, poly-L-ornithine/laminin is used for neuronal and pancreatic differentiation, whereas collagen I is more suitable for cardiac and hepatic differentiation.

9. An efficient protocol for growth factor-mediated lineage selection of neuronal cells is described here. The protocol includes: (1) formation of cells of all three primary germ layers in the EBs, (2) selective differentiation of neuroectodermal cells by growth factor removal (serum depletion), (3) proliferation and maintenance of nestin-positive neural precursor cells in the presence of bFGF and EGF, and (4) the differentiation induction of functional neurons by withdrawal of bFGF/EGF, and the addition of neuronal differentiation factors *(39,56)*.

10. Replating of differentiating cells is a critical step. The cell density after replating should be optimal to prevent either overgrowth resulting in metabolic starvation, necrosis and cell death or poor differentiation efficiency because of reduced cell-to-cell contacts.

11. During pancreatic and hepatic differentiation, the cells show significant morphological changes. After induction of pancreatic differentiation, bipolar progenitor cells form islet-like cell clusters; after hepatic differentiation, binucleated large cuboidal, epithelial-like cells similar to primary hepatocytes develop.

12. Use gloves and filtertips throughout the whole procedure!
13. Do not leave RNA lysis buffer in culture dishes longer than 5 min because polystyrene is not resistant to lysis buffer.
14. mRNA isolation from small samples of cells can be performed using the Dynabeads® mRNA DIRECT™ Micro kit (Dynal Biotech, Oslo, Norway).
15. Never mix and disturb the organic and the aqueous phases!
16. For RT-PCR, rTth DNA polymerase can also be used as both reverse transcriptase and DNA polymerase *(62)*. In this case, the components of both RT- and PCR-Mastermix are different from when using MuLV reverse transcriptase and Taq DNA polymerase.
17. Confocal laser scanning microscope (CLSM) analysis can be used to study EBs or EB outgrowths. For immunofluorescence analysis of EBs, this method is used because the 3D EBs require an extended depth of focus. Some EBs are as much as 400 μm in diameter. EBs are scanned in thin sections (0.5–10 μm) using the appropriate filter combinations depending on the fluorescent dyes used.
18. Nonspecific binding of primary and especially secondary antibodies to the cells should be blocked by incubation with serum proteins of animal species, which were not used for the generation of the primary antibody. Specificity of immunostaining could be demonstrated by the absence of signals after incubation with PBS or with control antibodies instead of the specific primary antibody.
19. The analysis of differentiated pancreatic endocrine cells should include the determination of insulin production as a functional assay. The intracellular insulin content can be measured by the commercially available insulin ELISA. Bearing in mind recent publications showing the uptake of insulin from culture medium *(63)*, additionally glucose responsiveness should be tested. For this purpose, insulin release in the presence of low (5.5 mM, control) and high (27.7 mM) glucose concentration is determined. Tolbutamide (10 μM), a sulfonylurea known to stimulate insulin secretion together with 5.5 mM glucose, is used alternatively for the measurement of insulin release after induction. However, failure of the glucose response may be dependent on insufficient maturation during differentiation. Such effects were described during pancreatic differentiation of mES cells, where insulin was secreted in response to glucose at an advanced stage of 32 d of differentiation, but not at day 28 *(49)*.

Acknowledgments

We are grateful to Sabine Sommerfeld, Oda Weiss and Karla Meier, for her expert help in the establishment of culture and differentiation conditions. We thank the German Research Foundation (DFG), the Ministry of Education and Research (BMBF), Funds of the Chemical Industry (FCI) and the DeveloGen AG, Göttingen, Germany, for funding our stem cell projects.

References

1. Czyz, J., Wiese, C., Rolletschek, A., Blyszczuk, P., Cross, M., and Wobus, A. M. (2003) Potential of embryonic and adult stem cells in vitro. *Biol. Chem.* **384,** 1391–1409.

2. Evans, M. J. and Kaufman, M. H. (1981) Establishment in culture of pluripotential cells from mouse embryos. *Nature* **292**, 154–156.
3. Martin, G. R. (1981) Isolation of a pluripotent cell line from early mouse embryos cultured in medium conditioned by teratocarcinoma stem cells. *Proc. Natl. Acad. Sci. USA* **78**, 7634–7638.
4. Wobus, A. M., Holzhausen, H., Jäkel, P., and Schöneich, J. (1984) Characterization of a pluripotent stem cell line derived from a mouse embryo. *Exp. Cell Res.* **152**, 212–219.
5. Doetschman, T. C., Eistetter, H., Katz, M., Schmidt, W., and Kemler, R. (1985) The in vitro development of blastocyst-derived embryonic stem cell lines: formation of visceral yolk sac, blood islands and myocardium. *J. Embryol. Exp. Morphol.* **87**, 27–45.
6. Stewart, C. L., Gadi, I., and Bhatt, H. (1994) Stem cells from primordial germ cells can reenter the germ line. *Dev. Biol.* **161**, 626–628.
7. Prelle, K., Vassiliev, I. M., Vassilieva, S. G., Wolf, E., and Wobus, A. M. (1999) Establishment of pluripotent cell lines from vertebrate species—present status and future prospects. *Cells Tissues. Organs* **165**, 220–236.
8. Thomson, J. A., Itskovitz-Eldor, J., Shapiro, S. S., Waknitz, M. A., Swiergiel, J. J., Marshall, V. S., et al. (1998) Embryonic stem cell lines derived from human blastocysts. *Science* **282**, 1145–1147.
9. Williams, R. L., Hilton, D. J., Pease, S., Willson, T. A., Stewart, C. L., Gearing, D. P., et al. (1988) Myeloid leukaemia inhibitory factor maintains the developmental potential of embryonic stem cells. *Nature* **336**, 684–687.
10. Burdon, T., Chambers, I., Stracey, C., Niwa, H., and Smith, A. (1999) Signaling mechanisms regulating self-renewal and differentiation of pluripotent embryonic stem cells. *Cells Tissues Organs* **165**, 131–143.
11. Rohwedel, J., Sehlmeyer, U., Shan, J., Meister, A., and Wobus, A. M. (1996) Primordial germ cell-derived mouse embryonic germ (EG) cells in vitro resemble undifferentiated stem cells with respect to differentiation capacity and cell cycle distribution. *Cell Biol. Int.* **20**, 579–587.
12. Solter, D. and Knowles, B. B (1978) Monoclonal antibody defining a stage-specific mouse embryonic antigen (SSEA-1). *Proc. Natl. Acad. Sci. USA* **75**, 5565–5569.
13. Armstrong, L., Lako, M., Lincoln, J., Cairns, P. M., and Hole, N. (2000) mTert expression correlates with telomerase activity during the differentiation of murine embryonic stem cells. *Mech. Dev.* **97**, 109–116.
14. Schöler, H. R., Hatzopoulos, A. K., Balling, R., Suzuki, N., and Gruss, P. (1989) A family of octamer-specific proteins present during mouse embryogenesis: evidence for germline-specific expression of an Oct factor. *EMBO J.* **8**, 2543–2550.
15. Chambers, I., Colby, D., Robertson, M., Nichols, J., Lee, S., Tweedie, S., et al. (2003) Functional expression cloning of nanog, a pluripotency sustaining factor in embryonic stem cells. *Cell* **113**, 643–655.
16. Mitsui, K., Tokuzawa, Y., Itoh, H., Segawa, K., Murakami, M., Takahashi, K., et al. (2003) The homeoprotein nanog is required for maintenance of pluripotency in mouse epiblast and ES cells. *Cell* **113**, 631–642.

17. Wobus, A. M., and Boheler, K. R. (2005) Embryonic stem cells - prospects for developmental biology and cell therapy. *Physiol. Rev.* **85,** 635–678.
18. Niwa, H., Miyazaki, J., and Smith, A. G. (2000) Quantitative expression of Oct-3/ 4 defines differentiation, dedifferentiation or self-renewal of ES cells. *Nat. Genet.* **24,** 372–376.
19. Ying, Q. L., Nichols, J., Chambers, I., and Smith, A. (2003) BMP induction of Id proteins suppresses differentiation and sustains embryonic stem cell self-renewal in collaboration with STAT3. *Cell* **115,** 281–292.
20. Bradley, A., Evans, M., Kaufman, M. H., and Robertson, E. (1984) Formation of germ-line chimaeras from embryo-derived teratocarcinoma cell lines. *Nature* **309,** 255–256.
21. Wobus, A. M., Wallukat, G., and Hescheler, J. (1991) Pluripotent mouse embryonic stem cells are able to differentiate into cardiomyocytes expressing chronotropic responses to adrenergic and cholinergic agents and Ca2+ channel blockers. *Differentiation* **48,** 173–182.
22. Wobus, A. M., Kaomei, G., Shan, J., Wellner, M. C., Rohwedel, J., Ji, G., et al. (1997) Retinoic acid accelerates embryonic stem cell-derived cardiac differentiation and enhances development of ventricular cardiomyocytes. *J. Mol. Cell Cardiol.* **29,** 1525–1539.
23. Boheler, K. R., Czyz, J., Tweedie, D., Yang, H. T., Anisimov, S. V., and Wobus, A. M. (2002) Differentiation of pluripotent embryonic stem cells into cardiomyocytes. *Circ. Res.* **91,** 189–201.
24. Wobus, A. M., Guan, K., Yang, H. T., and Boheler, K. R. (2002) Embryonic stem cells as a model to study cardiac, skeletal muscle, and vascular smooth muscle cell differentiation. *Methods Mol. Biol.* **185,** 127–156.
25. Wiles, M. V. and Keller, G. (1991) Multiple hematopoietic lineages develop from embryonic stem (ES) cells in culture. *Development* **111,** 259–267.
26. Kawasaki, H., Mizuseki, K., Nishikawa, S., Kaneko, S., Kuwana, Y., Nakanishi, S., et al. (2000) Induction of midbrain dopaminergic neurons from ES cells by stromal cell-derived inducing activity. *Neuron* **28,** 31–40.
27. Ying, Q. L., Stavridis, M., Griffiths, D., Li, M., and Smith, A. (2003) Conversion of embryonic stem cells into neuroectodermal precursors in adherent monoculture. *Nat. Biotechnol.* **21,** 183–186.
28. Maltsev, V. A., Rohwedel, J., Hescheler, J., and Wobus, A. M. (1993) Embryonic stem cells differentiate in vitro into cardiomyocytes representing sinusnodal, atrial and ventricular cell types. *Mech. Dev.* **44,** 41–50.
29. Drab, M., Haller, H., Bychkov, R., Erdmann, B., Lindschau, C., Haase, H., et al. (1997) From totipotent embryonic stem cells to spontaneously contracting smooth muscle cells: a retinoic acid and db-cAMP in vitro differentiation model. *FASEB J.* **11,** 905–915.
30. Yamashita, J., Itoh, H., Hirashima, M., Ogawa, M., Nishikawa, S., Yurugi, T., et al. (2000) Flk1-positive cells derived from embryonic stem cells serve as vascular progenitors. *Nature* **408,** 92–96.

31. Rohwedel, J., Maltsev, V., Bober, E., Arnold, H. H., Hescheler, J., and Wobus, A. M. (1994) Muscle cell differentiation of embryonic stem cells reflects myogenesis in vivo: developmentally regulated expression of myogenic determination genes and functional expression of ionic currents. *Dev. Biol.* **164,** 87–101.
32. Nishikawa, S. I., Nishikawa, S., Hirashima, M., Matsuyoshi, N., and Kodama, H. (1998) Progressive lineage analysis by cell sorting and culture identifies FLK1+VE-cadherin+ cells at a diverging point of endothelial and hemopoietic lineages. *Development* **125,** 1747–1757.
33. Dani, C., Smith, A. G., Dessolin, S., Leroy, P., Staccini, L., Villageois, P., Darimont, C., and Ailhaud, G. (1997) Differentiation of embryonic stem cells into adipocytes in vitro. *J. Cell Sci.* **110,** 1279–1285.
34. Kramer, J., Hegert, C., Guan, K., Wobus, A. M., Müller, P. K., and Rohwedel, J. (2000) Embryonic stem cell-derived chondrogenic differentiation in vitro: activation by BMP-2 and BMP-4. *Mech. Dev.* **92,** 193–205.
35. Risau, W., Sariola, H., Zerwes, H. G., Sasse, J., Ekblom, P., Kemler, R., and Doetschman, T. (1988) Vasculogenesis and angiogenesis in embryonic-stem-cell-derived embryoid bodies. *Development* **102,** 471–478.
36. Bain, G., Kitchens, D., Yao, M., Huettner, J. E., and Gottlieb, D. I. (1995) Embryonic stem cells express neuronal properties in vitro. *Dev. Biol.* **168,** 342–357.
37. Fraichard, A., Chassande, O., Bilbaut, G., Dehay, C., Savatier, P., and Samarut, J. (1995) In vitro differentiation of embryonic stem cells into glial cells and functional neurons. *J. Cell Sci.* **108,** 3181–3188.
38. Strübing, C., Ahnert-Hilger, G., Shan, J., Wiedenmann, B., Hescheler, J., and Wobus, A. M. (1995) Differentiation of pluripotent embryonic stem cells into the neuronal lineage in vitro gives rise to mature inhibitory and excitatory neurons. *Mech. Dev.* **53,** 275–287.
39. Okabe, S., Forsberg-Nilsson, K., Spiro, A. C., Segal, M., and McKay, R. D. (1996) Development of neuronal precursor cells and functional postmitotic neurons from embryonic stem cells in vitro. *Mech. Dev.* **59,** 89–102.
40. Lee, S. H., Lumelsky, N., Studer, L., Auerbach, J. M., and McKay, R. D. (2000) Efficient generation of midbrain and hindbrain neurons from mouse embryonic stem cells. *Nat. Biotechnol.* **18,** 675–679.
41. Bagutti, C., Wobus, A. M., Fässler, R., and Watt, F. M. (1996) Differentiation of embryonal stem cells into keratinocytes: comparison of wild-type and beta 1 integrin-deficient cells. *Dev. Biol.* **179,** 184–196.
42 Hamazaki, T., Iiboshi, Y., Oka, M., Papst, P. J., Meacham, A. M., Zon, L. I., and Terada, N. (2001) Hepatic maturation in differentiating embryonic stem cells in vitro. *FEBS Lett.* **497,** 15–19.
43. Chinzei, R., Tanaka, Y., Shimizu-Saito, K., Hara, Y., Kakinuma, S., Watanabe, M., et al. (2002) Embryoid-body cells derived from a mouse embryonic stem cell line show differentiation into functional hepatocytes. *Hepatology* **36,** 22–29.
44. Jones, E. A., Tosh, D., Wilson, D. I., Lindsay, S., and Forrester, L. M. (2002) Hepatic differentiation of murine embryonic stem cells. *Exp. Cell Res.* **272,** 15–22.

45. Yamada, T., Yoshikawa, M., Kanda, S., Kato, Y., Nakajima, Y., Ishizaka, S., et al. (2002) In vitro differentiation of embryonic stem cells into hepatocyte-like cells identified by cellular uptake of indocyanine green. *Stem Cells* **20**, 146–154.

46. Kania, G., Blyszczuk, P., Czyz, J., Navarrete-Santos, A., and Wobus, A. M. (2003) Differentiation of mouse embryonic stem cells into pancreatic and hepatic cells. *Methods Enzymol.* **365**, 287–303.

47. Lumelsky, N., Blondel, O., Laeng, P., Velasco, I., Ravin, R., and McKay, R. (2001) Differentiation of embryonic stem cells to insulin-secreting structures similar to pancreatic islets. *Science* **292**, 1389–1394.

48. Hori, Y., Rulifson, I. C., Tsai, B. C., Heit, J. J., Cahoy, J. D., and Kim, S. K. (2002) Growth inhibitors promote differentiation of insulin-producing tissue from embryonic stem cells. *Proc. Natl. Acad. Sci. USA* **99**, 16105–16110.

49. Blyszczuk, P., Czyz, J., Kania, G., Wagner, M., Roll, U., St Onge, L., and Wobus, A. M. (2003) Expression of Pax4 in embryonic stem cells promotes differentiation of nestin-positive progenitor and insulin-producing cells. *Proc. Natl. Acad. Sci. USA* **100**, 998–1003.

50. Rohwedel, J., Guan, K., and Wobus, A. M. (1999) Induction of cellular differentiation by retinoic acid in vitro. *Cells Tissues Organs* **165**, 190–202.

51. Leahy, A., Xiong, J. W., Kuhnert, F., and Stuhlmann, H. (1999) Use of developmental marker genes to define temporal and spatial patterns of differentiation during embryoid body formation. *J. Exp. Zool.* **284**, 67–81.

52. Maltsev, V. A., Wobus, A. M., Rohwedel, J., Bader, M., and Hescheler, J. (1994) Cardiomyocytes differentiated in vitro from embryonic stem cells developmentally express cardiac-specific genes and ionic currents. *Circ. Res.* **75**, 233–244.

53. Rohwedel, J., Kleppisch, T., Pich, U., Guan, K., Jin, S., Zuschratter, W., et al. (1998) Formation of postsynaptic-like membranes during differentiation of embryonic stem cells in vitro. *Exp. Cell Res.* **239**, 214–225.

54. Rohwedel, J., Horak, V., Hebrok, M., Fuchtbauer, E. M., and Wobus, A. M. (1995) M-twist expression inhibits mouse embryonic stem cell-derived myogenic differentiation in vitro. *Exp. Cell Res.* **220**, 92–100.

55. Prelle, K., Wobus, A. M., Krebs, O., Blum, W. F., and Wolf, E. (2000) Overexpression of insulin-like growth factor-II in mouse embryonic stem cells promotes myogenic differentiation. *Biochem. Biophys. Res. Commun.* **277**, 631–638.

56. Rolletschek, A., Chang, H., Guan, K., Czyz, J., Meyer, M., and Wobus, A. M. (2001) Differentiation of embryonic stem cell-derived dopaminergic neurons is enhanced by survival-promoting factors. *Mech. Dev.* **105**, 93–104.

57. Li, M., Pevny, L., Lovell-Badge, R., and Smith, A. (1998) Generation of purified neural precursors from embryonic stem cells by lineage selection. *Curr. Biol.* **8**, 971–974.

58. Chomczynski, P. and Sacchi, N. (1987) Single-step method of RNA isolation by acid guanidinium thiocyanate- phenol-chloroform extraction. *Anal. Biochem.* **162**, 156–159.

59. Smith, D. B. and Johnson, K. S. (1988) Single-step purification of polypeptides expressed in Escherichia coli as fusions with glutathione S-transferase. *Gene* **67**, 31–40.

60. Gearing, D. P., Nicola, N. A., Metcalf, D., Foote, S., Willson, T. A., Gough, N. M., et al. (1989) Production of leukemia inhibitory factor in *Escherichia coli* by a novel procedure and its use in maintaining embryonic stem cells in culture. *BioTechnology* **7**, 1157–1161.

61. Robertson, E. J. (1987) Embryo-derived stem cell lines, in *Teratocarcinoma and Embryonic Stem Cells: A Practical Approach* (Robertson, E. J., ed.), IRL Press, Oxford, pp. 71–112.

62. Myers, T. W. and Gelfand, D. H. (1991) Reverse transcription and DNA amplification by a Thermus thermophilus DNA polymerase. *Biochemistry* **30**, 7661–7666.

63. Rajagopal, J., Anderson, W. J., Kume, S., Martinez, O. I., and Melton, D. A. (2003) Insulin staining of ES cell progeny from insulin uptake. *Science* **299**, 363.

64. Fürst, D. O., Osborn, M., Nave, R., and Weber, K. (1988) The organization of titin filaments in the half-sarcomere revealed by monoclonal antibodies in immunoelectron microscopy: a map of ten nonrepetitive epitopes starting at the Z line extends close to the M line. *J. Cell Biol.* **106**, 1563–1572.

65. Obermann, W. M., Gautel, M., Steiner, F., van der Ven, P. F., Weber, K., and Fürst, D. O. (1996) The structure of the sarcomeric M band: localization of defined domains of myomesin, M-protein, and the 250-kD carboxy-terminal region of titin by immunoelectron microscopy. *J. Cell Biol.* **134**, 1441–1453.

66. Rudnicki, M. A., Jackowski, G., Saggin, L., and McBurney, M. W. (1990) Actin and myosin expression during development of cardiac muscle from cultured embryonal carcinoma cells. *Dev. Biol.* **138**, 348–358.

15

Staining Embryonic Stem Cells Using Monoclonal Antibodies to Stage-Specific Embryonic Antigens

Bruce A. Fenderson, Maria P. De Miguel, April D. Pyle, and Peter J. Donovan

Summary

Stage-specific embryonic antigens (SSEAs) are cell-surface molecules that exhibit lineage-restricted patterns of expression during development. These antigens provide useful markers for identifying embryonic stem cells and their differentiated derivatives. SSEA-1 provides a surface marker for mouse and human primordial germ cells and mouse embryonic stem cells. SSEA-3 and SSEA-4 provide surface markers for human embryonic stem cells. Here, we describe methods for staining embryonic stem cells in culture and suspension using specific primary antibodies and fluorescent dye- or enzyme-conjugated secondary antibodies. Methods for alkaline phosphatase staining and immunohistochemical localization of antigens in paraffin sections are also described. Monoclonal antibodies directed to SSEAs are valuable research tools that can be used to: (1) determine the percentage of stem cells in mixed cultures; (2) examine the spatial distribution of stem cells and differentiated derivatives in monolayer cultures and embryoid bodies; and (3) isolate populations of differentiated and undifferentiated cells by immunomagnetic or fluorescence-activated cell sorting for cellular and molecular analyses.

Key Words: Embryonic stem cells; embryonic germ cells; embryonal carcinoma cells; primordial germ cells; monoclonal antibodies; stage-specific embryonic antigens; immunofluorescence; alkaline phosphatase; immunohistochemistry.

1. Introduction

The inner cell mass (ICM) of the peri-implantation embryo gives rise to the primary germ layers and a vast array of differentiated cells and tissues. When the ICM is cultured in vitro on an appropriate feeder layer, epiblast cells can be propagated indefinitely as a self-renewing population of undifferentiated embryonic stem (ES) cells (reviewed in **ref. 1**). ES cells are capable of extensive differentiation in vitro, including the formation of primordial germ cells (PGCs)

From: *Methods in Molecular Biology, vol. 325: Nuclear Reprogramming: Methods and Protocols*
Edited by: S. Pells © Humana Press Inc., Totowa, NJ

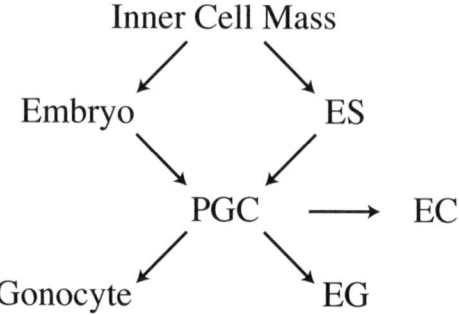

Fig. 1. Derivation of embryonic stem (ES) cells from the mammalian embryo. When the inner cell mass (ICM) is cultured in vitro on a fibroblast feeder layer, epiblast cells can be propagated indefinitely as ES cells. When primordial germ cells (PGCs) are explanted in vitro on a fibroblast feeder layer in the presence of specific growth factors they give rise to self-renewing embryonic germ (EG) cells. Embryonal carcinoma (EC) cells are the undifferentiated stem cells of teratocarcinomas that are themselves derived from PGCs in vivo.

and gametes *(2,3)*. In the early embryo, PGCs arise at the boundary of the embryonic and extra-embryonic ectoderm and migrate to the genital ridge by way of the gut endoderm (reviewed in **ref. *4***). When these migratory PGCs are placed in culture on an appropriate feeder layer in the presence of specific growth factors they give rise to colonies of self-renewing, undifferentiated stem cells termed embryonic germ (EG) cells *(5–7)*. PGCs give rise to another type of pluripotent stem cell in vivo termed embryonal carcinoma (EC). EC cells are the malignant stem cells of teratocarcinomas. These germ cell tumors arise spontaneously in the ovaries and testes of both humans and experimental animals. They can also be induced experimentally by grafting early embryos and PGCs to ectopic sites *(8)*. Teratocarcinomas contain a haphazard arrangement of somatic cell derivatives representing all three primary germ layers. ES and EG cells have a normal diploid karyotype, whereas malignant EC cells typically have an aneuploid karyotype.

Although ES, EG, and EC cells are derived from embryonic cells in different ways (**Fig. 1**), they share common patterns of gene expression *(9)* and common morphological features (high nuclear-to-cytoplasmic ratio, prominent nucleoli, and growth in colonies with indistinct cell borders). They also share a propensity for rapid, spontaneous differentiation. For this reason, markers are essential for distinguishing between undifferentiated stem cells and differentiated derivatives. Among the first set of markers useful for this purpose were a series of developmentally regulated antigens defined by monoclonal antibodies. Davor Solter and Barbara Knowles referred to these cell-surface antigens in the late 1970s as "stage-specific embryonic antigens" (SSEAs) *(10)*. SSEA-1 appears on cleavage-stage mouse embryos and its expression becomes restricted to the ICM, embryonic ectoderm, and migrating PGCs *(10,11)*. SSEA-1 also is highly expressed by mouse EC, ES, and EG cells. SSEA-3 and SSEA-4 are different cell-surface antigens that are present on mouse oocytes, eggs, and cleavage-stage embryos *(12,13)*. SSEA-3 provides a useful marker for visceral endoderm in the postimplantation mouse embryo *(14)*. Many such developmentally regulated antigens have been described; however, most laboratories continue to use SSEA-1, -3, and -4 as useful markers of pluripotency and differentiation.

Patterns of SSEA expression in the mouse and human are different. In contrast to mouse embryonic stem cells, which express SSEA-1, human EC, ES, and EG cells express SSEA-3 and SSEA-4 *(7,12,13,15,16)*. SSEA-3 and SSEA-4 also are expressed by the ICM of the human blastocyst *(16)*. Differentiation of human EC and ES cells is associated with a decline in the expression of SSEA-3 and SSEA-4 and a reciprocal increase in the expression of SSEA-1 *(17,18)*. One must note, however, that SSEA-1 is expressed by human PGCs *(19)* and human EG cells *(7)*. Despite these species differences, SSEA-1, -3, and -4 provide reliable markers for identifying mammalian embryonic stem cells and their differentiated derivatives. The profile of SSEA expression on mouse and human stem cells, PGCs, and early embryos is summarized in **Table 1** *(20–22)*.

In this chapter, we describe methods for staining embryonic stem cells using primary antibodies and conjugated secondary antibodies. This indirect staining method serves to amplify the primary signal and improve the sensitivity of antigen detection. The exquisite specificity of secondary antibodies makes it possible to detect different classes of primary antibodies (hence different antigens) simultaneously using multicolor immunolabeling techniques. Fluorescent dyes such as fluorescein (green) and rhodamine (red) are visualized using a fluorescent microscope. Enzyme-conjugates such as horseradish peroxidase or alkaline phosphatase are visualized easily by light microscopy upon addition of a substrate that forms a colored precipitate. A comprehensive reference guide for immunological methods is provided by Harlow and Lane *(23)*.

2. Materials
2.1. Immunological Reagents

1. Monoclonal antibodies directed to SSEAs can be obtained as hybridoma supernatants or ascites fluids (*see* **Note 1**) from commercial sources (e.g., Chemicon International). They can also be obtained from the "Developmental Studies Hybridoma Bank" of the National Institutes of Health, which is maintained at the University of Iowa (http://www.uiowa.edu/~dshbwww/info.html). Culture super-

Table 1
Markers for Identifying Pluripotent Stem Cells[a]

Marker	Mouse embryo		Mouse cells	Human embryo		Human cells	
	(ICM)	(PGC)	(ES, EG, EC)	(ICM)	(PGC)	(ES, EC)	(EG)
SSEA-1	•	•	•		•		•
SSEA-3			•			•	•
SSEA-4			•			•	•
TRA-1–60			•			•	•
Alkaline phosphatase	•	•	•	•	•	•	•
Oct4	•	•	•	•	•	•	•

[a]Positive labeling is indicated by a symbol (•). In our experience, SSEA-4 provides a stronger and more reliable marker for human EC cells than does SSEA-3 *(17)*. TRA-1–60 monoclonal antibody *(20)*, alkaline phosphatase *(21,22)*, and the Oct4 transcription factor (reviewed in **ref.** *9*) provide additional markers for embryonic stem cells. SSEA-1 is a useful marker for human PGCs *(19)* and human EG cells *(7)*.

EC, embryonal carcinoma; ES, embryonic stem; EG, embryonic germ; PGC, primordial germ cell; ICM, inner cell mass.

natant is a better source of primary antibody because it is close to the desired working concentration (1–10 µg/mL) and not contaminated with host immunoglobulins that are present in mouse ascites (*see* **Note 2**). Primary antibodies should be diluted serially and then tested for binding activity to determine the optimum working dilution (this process is termed "antibody titration"). Antibodies can be diluted in 10 mM phosphate-buffered 0.9% saline (PBS, pH 7.2) and stored at 4°C for years. Bovine serum albumin (Sigma, St. Louis, MO; fraction V, 0.1% w/v) commonly is added as a "carrier protein" to prevent nonspecific absorption of antibody onto glass or plastic. For long-term storage, sodium azide (Sigma, 0.1% w/v) is added to prevent bacterial growth (*see* **Note 3**). It is critical to avoid repeated freeze–thaw cycles to avoid protein denaturation and loss of antibody binding activity. Characteristics and specificity of the hybridomas producing anti-SSEA monoclonal antibodies are summarized in **Table 2** *(24)*.

2. Secondary antibodies conjugated with fluorescent dyes or enzymes can be obtained from a variety of commercial sources, including Sigma, Vector Laboratories (Burlingame, CA) and Accurate Chemical and Scientific (Westbury, NY). Mouse IgM antibodies, such as anti-SSEA-1, can be detected using either a heavy chain-specific anti-mouse IgM or an anti-mouse polyvalent Ig that reacts with the heavy chains of IgG, IgA, and IgM. Secondary antibodies should be diluted according to the manufacturer's instructions (typically 1/200–1/1000) and tested for binding activity to determine the optimum working dilution. Fluorescent antibodies are light sensitive (*see* **Note 4**) and should be stored in the dark at 4°C in PBS (0.1% sodium azide can be added for long-term storage).

able 2
lonoclonal Antibodies Defining Stage-Specific Embryonic Antigens[a]

ame	Hybridoma (antibody class)	Carbohydrate antigen
SEA-1	MC480 (Mouse IgM)	Galβ1→4GlcNAcβ1→3Galβ1→4Glc
		3
		↑
		Fucα1
SEA-3	MC631 (Rat IgM)	Galβ1→3GalNAcβ1→3Galα1→4Galβ1→4Glc
SEA-4	MC813-70 (Mouse IgG)	NeuAcα2→3Galβ1→3GalNAcβ1→3Galα1→4Galβ1→4Glc

[a]Biochemical studies have shown that these antigens are carbohydrates carried on glycolipids and ycoproteins. Simple sugar additions or deletions generate a variety of oligosaccharides, some of which ovide markers for mouse and human stem cells. Glycolipid carriers can be identified by thin-layer romatography immunostaining *(15,17)*. The relative distribution of carbohydrate antigens carried on ycolipids and glycoproteins can be determined using a pharmacological agent (PDMP) that blocks ycolipid biosynthesis *(24)*. It is important to remember that anti-SSEA-3 is a rat monoclonal antibody. SEA-4 is a neuraminidase-sensitive antigen; neuraminidase-treatment of pluripotent stem cells can mask SSEA-3.

3. Avidin–biotin (ABC) peroxidase or ABC alkaline phosphatase staining kits are obtained from Vector Laboratories. Detailed protocol sheets accompany these products. Store kits at 4°C.

2.2. Buffers and Solutions

1. Dulbecco's-modified Eagle's medium (DMEM; Invitrogen Life Technologies, Carlsbad, CA) containing 15% heat-inactivated fetal bovine serum (FBS; Hyclone, Logan, UT), 4.5 g/L glucose (high glucose), 1 m*M* glutamine, 1 m*M* pyruvate, 2-mercaptoethanol (5 μL/500 mL), and 1% (v/v) penicillin–streptomycin (Invitrogen Life Technologies) is used for mouse cell culture.

2. Trypsin (0.05%) and ethylenediaminetetraacetic acid (0.53 m*M*) in Hank's balanced salt solution (Invitrogen Life Technologies) is used for dissociating cells for single cell (suspension) assays. Store at 4°C.

3. Dulbecco's phosphate-buffered saline without calcium or magnesium (10 m*M* PBS 0.9% saline, pH 7.2; Invitrogen Life Technologies) is used for most immunoassay washing steps. When working with live cells, 5% FBS (v/v) and 0.1% sodium azide are commonly added to maintain cell viability and inhibit internalization of cross-linked antigens. Store at 4°C.

4. Porcine skin gelatin (cat. no. 1890, Sigma) is used to coat plastic dishes and glass cover slips to promote cell substrate attachment. Make a 0.1% (w/v) solution in distilled water and autoclave on slow exhaust.

5. Goat serum (Sigma) is used as a nonspecific protein blocking solution for assays in which the secondary antibody was produced in goat (e.g., goat anti-mouse IgM). Make a 10% (v/v) solution in PBS with 0.1% (w/v) sodium azide added to prevent bacterial growth. Store at 4°C.

6. PBS containing 0.1% (v/v) Triton X-100 (Sigma) is used for permeabilization of adherent cells attached to plastic dishes or glass cover slips. Store at 4°C.

2.3. Enzyme Substrates

1. Diaminobenzidine (DAB) is used for detection of horseradish peroxidase-conjugated secondary antibodies. Use preweighed DAB tablets (Sigma) to prepare 0.05% solution in 50 mM Tris buffer pH 7.5, containing 0.3% H_2O_2 (1:100 dilution of peroxide stock solution). DAB is a known carcinogen and should be used with caution (*see* **Note 5**).

2. Fast Red/naphthol phosphate staining solution is used for detecting membrane-associated alkaline phosphatase and alkaline phosphatase-conjugated secondary antibodies. Prepare a 1 mg/mL solution of Fast Red TR salt in distilled water (cat no. F8764, Sigma; store stock at –20°C). Add 40 μL/mL naphthol AS-MX phosphate solution (cat. no. 85-5, Sigma; store stock at 4°C). Both reagents should be made fresh and used immediately.

3. Nitroblue tetrazolium (NBT)/5-bromo-4-chloro-3-indolyl phosphate (BCIP) staining solution is also used for detecting alkaline phosphatase. NBT (Sigma) stock solution is 0.5 g of NBT in 10 mL of 70% dimethylformamide (store at 4°C). BCIP (Sigma) stock solution is 0.5 g BCIP in 10 mL of 100% dimethylformamide (store at 4°C). Alkaline phosphatase buffer is 100 mM sodium chloride and 5 mM magnesium chloride in 100 mM Tris buffer, pH 9.5. For each assay, add 66 μL of NBT stock to 10 mL of alkaline phosphatase buffer, mix well, add 33 μL of BCIP stock, and use immediately.

2.4. Fixatives, Dyes, and Mounting Media

1. Paraformaldehyde (Sigma) in PBS (4% w/v final concentration) is used for fixing cells for immunoassays. Prepare a 10% w/v stock solution in distilled water, preheated to 60°C in a fume hood with stirring. Titrate 0.1 N NaOH carefully until the solution becomes clear. Cool before using. Freeze aliquots at –20°C. To make a fresh 4% working solution: mix 4 mL of 10% stock paraformaldehyde, 1 mL of 10X PBS salts (Invitrogen Life Technologies), and 5 mL of distilled water. Store at 4°C. Cells can also be fixed using a 1:10 dilution of 37% formaldehyde solution (Fisher Scientific, Pittsburgh, PA) in PBS.

2. Glycerol (70% v/v) in 50 mM Tris, pH 9.6, provides a mounting medium for microscopic examination of air-dried or fixed cells attached to glass slides or cover slips. Vectashield (Vector Laboratories) contains an antiphotobleaching agent that is useful for preserving fluorescence.

3. CrystalMount (Biomedia, Foster City, CA) provides an excellent water-based mounting medium for immunoperoxidase and immunoalkaline phosphatase-labeled cells and tissues.

4. Bouin's solution (Fisher Scientific) is often used for fixing embryos or embryoid bodies before paraffin embedding and immunohistochemical assays.

5. Xylene and ethanol (Fisher Scientific) are used for deparaffinizing tissue sections.

6. Methylene Blue (Sigma) is used as a counterstain for immunohistochemical assays. Make a 0.1% (w/v) solution in distilled water and pass through filter paper before use.
7. Permount (Fisher Scientific) is used for permanent mounting of dehydrated tissue sections after immunohistochemical staining.
8. Silane-coated glass microscope slides (Sigma) provide an improved substrate for attachment of cells and paraffin sections.

3. Methods

3.1. General Considerations

The success of an immunoassay depends on the quality of antigen preservation, dilution and specificity of the antibodies, and sensitivity of the detection system. In this connection, it is important to note that the SSEAs described here are trypsin-resistant cell-surface carbohydrates. Trypsin-dissociation of mouse and human EC cells enhances staining with anti-SSEA monoclonal antibodies (unpublished observations), and may provide a useful technique for studies of human ES and EG cells. SSEAs withstand air drying and paraformaldehyde fixation. Acetone fixation and paraffin embedding should be avoided when staining for SSEA-3 and SSEA-4, because these antigens are carried in part on membrane glycolipids *(13,15,17)*.

Immunoassays are sensitive and there is a significant potential for misinterpretation of results. Thus, it is important to include a positive and a negative control in all immunoassays. The best negative control is one that includes all incubation steps, including the addition of secondary antibody and detection substrate, but that uses an isotype-matched primary antibody with no known cellular reactivity. For example, mouse IgM antimeiotic male germ cell antibody (FE-J1) provides a negative control for mouse IgM anti-SSEA-1.

In the remainder of this section, we introduce methods for staining embryonic stem cells using immunofluorescence and immunoperoxidase labeling techniques (**Subheadings 3.2.–3.7.**). We further describe a method for identifying embryonic stem cells based on their high level of cell-surface alkaline phosphatase activity (**Subheading 3.8.**) and a method for immunohistochemical localization of antigens in paraffin sections (**Subheading 3.9.**). The advantages and disadvantages of these staining methods are summarized in **Table 3**.

3.2. Immunofluorescence Analysis of Trypsin-Dissociated Single Cells

Indirect immunofluorescence can be used to localize cells of interest in normal embryos (**Fig. 2** and **ref.** *25*) or rapidly determine the percentage of labeled cells in culture. Staining assays involving cultured stem cells are typically performed in round bottomed 96-well microtiter plates.

Table 3
Methods for Staining Embryonic Stem Cells

Staining method	Advantages/disadvantages
IF single cells: Trypsin-dissociated cells are labeled in suspension and observed using a fluorescent microscope	• Rapid identification of stem cells (% positive) • Trypsin enhances SSEA expression • Fixation not required • Subjective analysis
IF two color: Cells labeled in suspension then reattached to glass slides for cytoplasmic double-labeling assays	• Trypsin enhances SSEA expression • Restaining permits simultaneous analysis of cell-surface and cytoplasmic antigens • Limited to visual inspection of small numbers of cells
FACS: Trypsin-dissociated cells are labeled in suspension then analyzed by flow cytofluorometry	• Quantitative, multicolor analysis • Records viability and cell cycle information • Positive and negative cells can be sorted • Cell viability after sorting often is poor
IF adherent cells: Cells growing on glass cover slips are fixed *in situ* and permeabilized before antibody labeling	• High-resolution antigen localization • Spatial relationships of undifferentiated stem cells and differentiated derivatives are preserved • Glass slides or cover slips must be pretreated with adhesion molecules to permit cell substrate attachment
IP adherent cells: Cells growing on plastic dishes are fixed for immunoperoxidase or immunoalkaline phosphatase	• Cells can be analyzed directly on plastic dishes • Analysis of colonies by light microscopy • Low-resolution staining • Not useful for double-labeling assays
AP adherent cells: Cells growing on plastic dishes are fixed for alkaline phosphatase immunocytochemistry	• Cells can be analyzed directly on plastic dishes • Analysis of colonies by light microscopy • Low-resolution staining • Not useful for double-labeling assays
IH paraffin sections: Tissues are fixed, embedded, and sectioned for ABC immunoperoxidase	• High-sensitivity localization of antigens in tissues • Low background • Some antigens do not survive paraffin embedding • Difficult to detect mouse IgG labeling patterns due to presence of endogenous (host) IgG

IF, immunofluorescence; IP, immunoperoxidase; AP, alkaline phosphatase; IH, immunohistochemistry; FACS, fluorescence-activated cell sorting.

1. Dissociate the cells completely using trypsin–ethlyenediaminetetraacetic acid at 37°C; add an equal volume of complete medium (DMEM–15% FBS) and centrifuge at 175g (1000 rpm) for 5 min; resuspend the cells in PBS–5% FBS and adjust the concentration to approx 2 × 10^7 cells/mL. Hold the cells on ice before use (*see* **Note 6**).

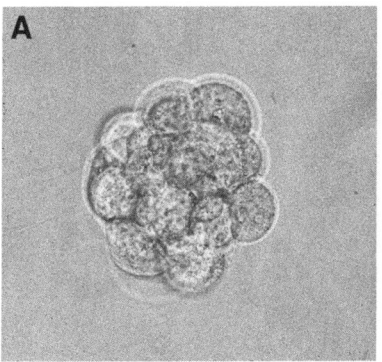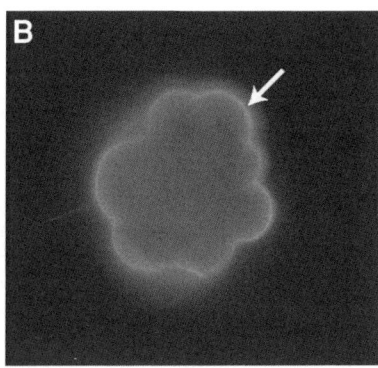

Fig. 2. Indirect immunofluorescence staining of a mouse embryo with anti-SSEA-1 monoclonal antibody. Morula-stage embryos were collected from the uteri of pregnant mice on day 3.5 postcoitum and labeled with anti-SSEA-1 hybridoma supernatant, followed by FITC-conjugated anti-mouse IgM as described previously *(25)*. Embryos were washed and observed under a fluorescent microscope using phase **(A)** or fluorescent **(B)** microscopy. Specific labeling is evidenced by the presence of a membrane-bound "ring" stain (arrow). Original magnification ×250.

2. Add 50 µL of cells per well to round-bottomed microtiter plates (10⁶ cells/well). Skip a row between antibodies to avoid cross-contamination. Add 150 µL of primary antibody (e.g., undiluted hybridoma supernatant). Mix cells and antibody by tapping the plate placed on a solid surface and incubate for 40 min at 4°C (*see* **Note 6**). Resuspend the cells after 20 min using a multichannel pipet or by gently tapping the microtiter plate.
3. Place the microtiter plate in a swinging plate holder and centrifuge at 175*g* (1000 rpm) for 3 min. Gently invert the plate and shake once (one motion) to remove the supernatant. Immediately add 200 µL of PBS-5% FBS; tap to resuspend the cells, centrifuge and shake.
4. Add 150 µL of fluorescein–isothiocyanate (FITC)-conjugated secondary antibody diluted in PBS–5% FBS. Mix the cells and antibody by gently tapping the plate; incubate for 40 min at 4°C and resuspend the cells after 20 min as before.
5. Add 100 µL of PBS–5% FBS and wash the cells by centrifugation as described in **Subheading 3.2.3.** Resuspend the labeled cells in 50 µL of PBS–5% FBS. Transfer 20-µL aliquots to a glass slide and overlay with a 22 × 22-mm glass cover slip.
6. Observe the cells using a fluorescent microscope with epi-illumination. Positive cells will show a bright "ring" stain (*see* blastomeres in **Fig. 2**). Cytoplasmic fluorescence typically represents nonspecific uptake of secondary antibody.

3.3. Immunofluorescence Analysis of Labeled Cells Reattached to Glass Slides

Single cells that are surface-labeled in suspension (*see* **Subheading 3.2.**) can be double-labeled for cytoplasmic markers, provided that the cells are securely attached to a glass slide or cover slip, fixed, and permeabilized. This technique provides important information on co-expression of cell-surface and cytoplasmic markers (e.g., %A, %B, and %A+B). If the first label is green (e.g., fluorescein), then the second color should be red (e.g., rhodamine). It is critical to remember that the second round of labeling must not employ immunological reagents that detect the same primary or secondary antibody that was used to label cells in suspension (*see* **Note 7**).

1. Label the cells as described previously (**Subheading 3.2.**). Transfer 20-µL aliquots to spots on silane-coated glass slides. Allow the cells to dry rapidly and thoroughly on a slide warming tray at 37°C (*see* **Note 8**).
2. Fix the cells with 4% paraformaldehyde for 30 min at room temp. Permeabilize the cells with PBS–0.1% Triton X-100 for 30 min at room temp; wash four times with PBS (*see* **Note 9**).
3. Dry cell-free margins of the slide with tissue paper, leaving a hydrated island over the cells and incubate with 100 µL of PBS-10 % goat serum for 30 min at room temp. Rinse once with PBS.
4. Add 100 µL of primary antibody and place in humidified chamber (e.g., 150-mm plastic tissue culture dish containing a wet paper towel) for 40 min at 4°C.
5. Wash three times with PBS-5% FBS. Add 100 µL of isotype-specific secondary antibody and place in a humidified chamber for 40 min at 4°C. Do not allow the slides to dry during the wash and incubation steps.
6. Wash the cells four times with PBS; add 20 µL of glycerol mounting medium (Vectashield) and a glass cover slip. Examine the cells using a fluorescent microscope with epi-illumination. Use appropriate filters to monitor specific green and red fluorescence.

3.4. Fluorescence-Activated Cell Sorting

Fluorescence-activated cell sorting (FACS) is a powerful system for quantitatively measuring fluorescence intensity in a population of cells. Fluorescence is measured from single cells passing in small droplets through a narrow gauge nozzle. This technique (flow cytofluorometry) permits the detection of multiple cell-surface markers simultaneously. Positive and negative cells can also be diverted to collection tubes when an electric charge is applied to the droplet surface. Propidium iodide (PI) and other dyes such as Hoechst and diamidino-2-phenylindole (DAPI) can be used to exclude dead cells and measure DNA content. Perhaps the most useful feature of FACS analysis is the ability to sort viable cells for molecular and cellular assays. For viable cell sorting, solutions should be free of sodium azide and passed through a 0.22-µm filter to remove debris.

1. Dissociate the cells completely and stain by indirect immunofluorescence as described in **Subheading 3.2.** Viability can be improved by reducing the incubation time to 20 min and by limiting centrifugation washing steps.

2. Resuspend the cells in PBS–5% FBS at a concentration of about 5×10^6/mL (total volume less than 1 mL) and transfer to a 12×75-mm polystyrene tube. Hold the labeled cells on ice before FACS analysis.
3. Based on "light scatter" (a measure of cell size and viability) and fluorescence intensity in the negative control sample, set electronic "gates" so that specific fluorescence emission can be analyzed in experimental samples.
4. For cell sorting experiments, embryonic stem cells should be collected in complete culture medium containing 50% FBS.

3.5. Immunofluorescence Analysis of Viable Cells in Culture

Antigen expression by cells in culture can be monitored for short periods of time (minutes to hours) while maintaining cell viability. Thus, embryonic stem cells can be examined by fluorescence microscopy for antigen expression *in situ* and then returned to the incubator for continued growth. This procedure requires the use of an inverted phase/fluorescence microscope.

1. Aspirate the culture medium and replace with medium diluted 1:1 with sterile hybridoma culture supernatant (sodium azide-free) for 40 min at 37°C.
2. Wash the culture dish gently with complete medium three times. Add FITC-conjugated second antibody prepared in sterile PBS–5% FBS for 30 min at 37°C.
3. Wash the culture dish with PBS–5% FBS and observe adherent cells using an inverted phase/fluorescent microscope. After viewing, replace the PBS buffer with complete culture medium, and return the plate to the incubator for continued cell growth.

3.6. Immunofluorescence Analysis of Adherent Cells

High-resolution studies of cell-surface and cytoplasmic antigen expression are performed using stem cells cultured on glass cover slips. To achieve cell attachment to glass, it is necessary to pretreat cover slips with a strong adhesion molecule (e.g., collagen or fibronectin). For double-labeling assays involving cell-surface antigens, primary and secondary antibodies can be mixed together to shorten incubation times, provided that the primary antibodies are of a different class or species, and that the secondary antibodies are isotype specific (*see* **Note 7**). For double-labeling assays involving cell-surface and cytoplasmic markers, cells are fixed with 4% paraformaldehyde and permeabilized with 0.1%Triton X-100 before antibody labeling (*see* **Subheading 3.3.**).

1. Place sterile 22×22-mm cover slips on the bottom of 60-mm or 100-mm plastic culture dishes and overlay with sterile 0.1% gelatin solution and incubate at room temp for 1 h. Remove excess fluid, add complete medium, and allow cells to attach (*see* **Note 10**).
2. Carefully remove the cover slips from the dish using forceps and rinse with PBS. Fix with 4% paraformaldehyde for 30 min at room temp.

3. Rinse the cover slips with PBS three times. Block nonspecific protein binding sites with PBS–10% goat serum for 20 min at room temp.
4. Place the cover slips (cell side up) on 15-mL centrifuge tube caps placed inside a humidified chamber (e.g., 150-mm plastic culture dish with wet paper towel) and add 50 µL each of the primary antibodies (e.g., anti-SSEA-4 and anti-SSEA-1 hybridoma supernatants); incubate for 40 min at 4°C.
5. Wash the cover slips with PBS three times and add secondary antibodies (e.g., FITC anti-mouse IgM µ chain specific and RITC anti-mouse IgG γ chain specific) for 40 min at 4°C.
6. Rinse the cover slips in PBS three times. Invert on a glass slide using 20 µL of glycerol mounting medium (Vectashield) and examine the cells using a fluorescent microscope with epi-illumination using appropriate filters for red and green fluorescence.

3.7. Immunoperoxidase Analysis of Adherent Cells

Indirect immunoperoxidase analysis is useful for identifying stem cells cultured on plastic dishes by routine light microscopy. It is useful for low-resolution studies and for comparison with results obtained by staining for alkaline phosphatase (*see* **Subheading 3.8.**). In double-staining assays, immunoperoxidase labeling must be performed before staining for alkaline phosphatase (*see* **Note 11**).

1. Rinse adherent cells with PBS and fix with 4% paraformaldehyde for 30 min at room temp.
2. Wash with PBS three times; add primary antibody and incubate for 40 min at 4°C.
3. Wash with PBS three times; add horseradish peroxidase-conjugated secondary antibody and incubate for 40 min at 4°C.
4. Wash the plates with PBS three times and overlay with DAB enzyme substrate (*see* **Note 5**). Stop the enzyme reaction by rinsing with tap water.
5. Observe the staining reaction by light microscopy using bright-field optics. Results can be preserved by removing excess water and adding a drop of water-soluble Crystal Mount followed by a glass cover slip. For long-term storage, the edges of the cover slip can be sealed with Permount to prevent evaporation.

3.8. Alkaline Phosphatase Histochemistry

Germ cells and pluripotent stem cells of the early mammalian embryo are known to express high levels of alkaline phosphatase *(21,22)*. Although this enzyme is not unique to stem cells, it provides a rapid assay that can be used in parallel with more specific immunological and molecular assays (*see* **Note 11**). Alkaline phosphatase staining can be used to analyze the overall distribution of stem cell colonies or primordial germ cells growing on plastic dishes, using routine light microscopy (**Fig. 3**).

1. Wash the dish with PBS and fix with 4% paraformaldehyde in PBS for 30 min at room temp.

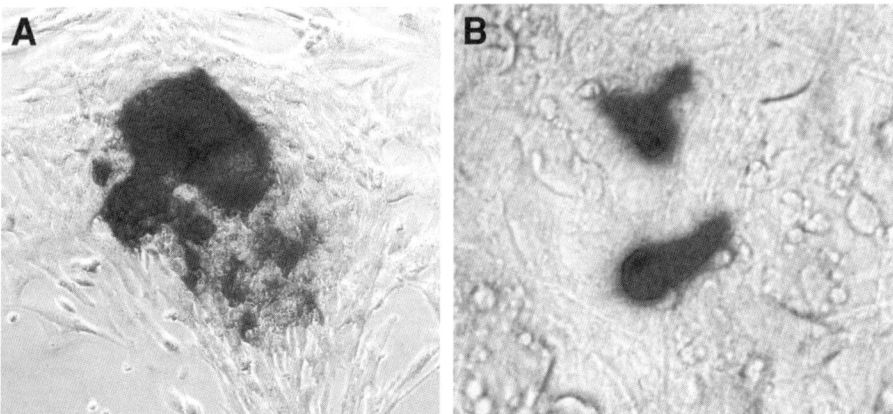

Fig. 3. Identification of pluripotent cells using alkaline phosphatase cytochemistry. Mouse ES cells and PGCs were cultured on plastic tissue culture plates, fixed with paraformaldehyde, washed, and incubated with Fast Red/naphthol phosphate as described *(22)*. As is shown, undifferentiated mouse ES cells (**A**) and migrating PGCs (**B**) stain bright red. Results are observed using light microscopy with bright field optics. In these experiments, PGCs were explanted from E10.5 mouse embryos and cultured on irradiated STO fibroblasts for 24 h. Mouse ES cells were plated on bacteriological dishes in the absence of leukemia inhibitory factor (LIF) and in the presence of 10^{-6} M all-*trans* retinoic acid (Sigma) for 6 d to promote differentiation. Embryoid bodies were harvested and allowed to "implant" on tissue culture plastic for 24 h. A large colony of undifferentiated ES cells (red cells) is surrounded by differentiated derivatives. Original magnification ×200 (A) and ×400 (B).

2. Wash the plate with PBS three times and rinse once with distilled water.
3. Add Fast Red/napthol phosphate staining solution for 40 min at room temp. Rinse with tap water and observe the labeled cells using bright field optics. The red color reaction begins to fade after 2 or 3 d, even if the cells are kept at 4°C.
4. A blue color reaction product can be obtained using the NBT/BCIP enzyme substrate (*see* **Subheading 2.3.**, **item 3**). Results can be preserved by removing excess water and adding a drop of water-soluble Crystal Mount followed by a glass cover slip. Seal the edges of the cover slip with Permount for long-term storage.

3.9. Immunohistochemical Analysis Using Avidin–Biotin Immunoperoxidase

The avidin–biotin complex (ABC) immunoperoxidase technique commonly is used to identify SSEAs in paraffin sections of embryos and embryoid bodies (**Fig. 4**). Tissues are fixed with Bouin's solution at 4°C overnight, embedded in paraffin, and sectioned at 6 μm. After deparaffinization through xylene and a graded series of alcohols, tissue sections are stained with monoclonal antibody as follows (*see* **Note 12**).

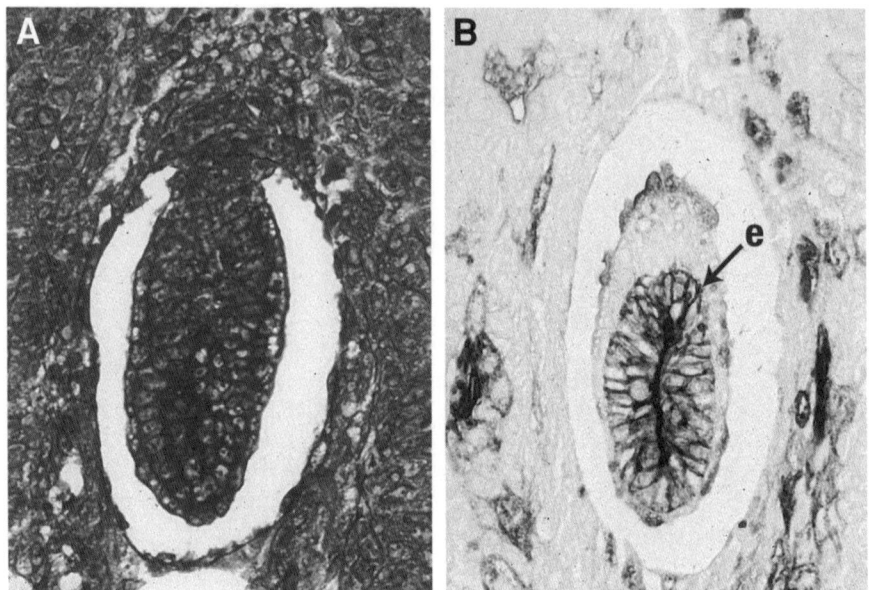

Fig. 4. Immunohistochemical staining of postimplantation mouse embryos for SSEA-1 using the ABC immunoperoxidase labeling technique. This paraffin section of an E6.5 mouse embryo was stained with Methylene Blue (**A**) or labeled sequentially with anti-SSEA-1 antibody, biotinylated goat anti-mouse IgM, and peroxidase-conjugated avidin–biotin complex (**B**) as described *(25)*. Anti-SSEA-1 antibody labels embryonic ectodermal cells (e), but not extra-embryonic ectodermal cells. This labeling technique also can be used to identify stem cells in paraffin sections of embryoid bodies and in experimentally induced teratomas. Original magnification ×250. (B, reprinted from **ref. 25**, with permission from Elsevier.)

1. Block endogenous peroxidase activity by incubating deparaffinized sections in 0.3% hydrogen peroxide for 1 h at room temp. Rinse the slides with tap water and dry the area around the section, creating a hydrated island over the tissue section.
2. Overlay the tissue with 100 μL of PBS–10% goat serum for 30 min at room temp in a humidified chamber (e.g., 150-mm plastic culture dish with wet paper towel).
3. Decant the blocking solution and dry the cell-free margins with tissue paper. Add 100 μL of primary antibody, forming a hydrated bubble over the tissue and incubate in a humidified chamber for 2 h (or overnight) at 4°C.
4. Wash with PBS three times and dry the surrounding glass. Add 100 μL of biotinylated goat anti-IgM (follow kit instructions for dilution) and incubate in a humidified chamber for 1 h at room temp. Do not allow the slides to dry during the labeling and washing steps.
5. Wash the slides with PBS three times and dry the surrounding glass. Add peroxidase-conjugated ABC (follow the kit mixing instructions carefully) and incubate in a humidified chamber for 1 h at room temp.

6. Wash the slides with PBS three times and add DAB enzyme substrate (*see* **Subheading 2.3., item 1**).
7. Stop the enzyme reaction by gently rinsing the slides in several changes of tap water. Slides may be counter-stained if desired (e.g., 0.1% aqueous Methylene Blue for 30 s) and then rinse with tap water. Dehydrate sections using a graded series of alcohols and xylene.
8. Place a drop of Permount on each section and overlay with 22 × 22-mm glass cover slip. Allow the Permount to dry thoroughly and observe using bright field optics.

4. Notes

1. Monoclonal antibodies are secretory products of hybrids between an activated B-cell and a myeloma cell. After selection and cloning, "hybridoma" cell lines produce a monoclonal antibody recognizing a single antigenic epitope. This monoclonal antibody is secreted into the culture medium where it is collected. Hybridomas can also be cultured within the peritoneal cavity of a syngeneic mouse. The ascites fluid that accumulates contains a high concentration of monoclonal antibody. Care must be taken to ensure that the host animal does not suffer undue pain and suffering.
2. The concentration of immunoglobulin in culture supernatant is close to the desired working concentration (~1–10 µg/mL). The concentration of immunoglobulin in ascites fluid is commonly in excess of 1 mg/mL. Most hybridoma culture supernatants can be diluted 1/20 without appreciable loss of antibody binding activity. Most hybridoma ascites can be diluted 1/1000 without appreciable loss of antibody binding activity.
3. Sodium azide is added to many staining solutions to prevent bacterial contamination. It also is used to prevent capping and internalization of surface antigens on live cells. It is our experience that sodium azide neither kills cells nor inhibits their long-term propagation, provided that treated cells are washed and returned to complete culture medium.
4. For most immunofluorescence assays, it is unnecessary to take precautions regarding ambient light. Photobleaching under these conditions is minimal and will not measurably affect the outcome.
5. Care must be taken when handling DAB, because it is a known carcinogen. Treat DAB solutions with bleach before discarding down the sink.
6. Holding live cells on ice during staining reactions slows cellular metabolism and helps to prevent cell clumping, as well as cell-surface receptor cross-linking. Holding live cells on ice also reduces endocytosis of labeled secondary antibodies, a common source of cytoplasmic nonspecific labeling. Human ES cells are extremely sensitive to changes in medium and lack of serum. To improve cell viability during staining, live cells can be resuspended in complete ES medium.
7. It is essential to use class and isotype-specific, affinity-purified secondary antibodies when staining for two or more antigens. Polyclonal anti-IgG secondary antibodies cross-react with IgM (due to common light chain determinants) and xenogeneic IgG (because of common heavy chain determinants). Thus, if the

first round of labeling involves mouse IgM, the primary antibody used for the second round of labeling must either be an IgG, or derived from a different species (e.g., rat or goat).

8. A cytocentrifuge is also used for loading cells on glass slides. Air-dried samples on glass slides can be stored at 4°C or frozen at –20°C for several weeks prior to use.

9. Once a labeled cell is dried or fixed with paraformaldehyde, immune complexes on the plasma membrane cannot be removed by subsequent mild detergent extraction.

10. Allow at least 6 h for feeder layer cells to spread on protein-coated glass cover slips. Human fibronectin (Sigma; 10 µg/mL in sterile PBS) provides an alternative to the use of gelatin for promoting cell attachment to glass slides and cover slips.

11. Cells that are initially stained for alkaline phosphatase cannot be stained subsequently using immunofluorescence or immunoperoxidase labeling techniques; however, the reverse is feasible.

12. Double labeling using ABC immunoperoxidase generally is not advised. A second concern with immunohistochemical assays of embryos or teratomas is the presence of endogenous (host) immunoglobulin, particularly IgG. This problem can be avoided by selecting biotinylated secondary antibodies that do not react with IgG (e.g., goat anti-mouse IgM). Always include a negative control.

Acknowledgments

This work was supported by research grants RO1-HD38252 and RO1-HD41553 to P. J. Donovan from the NIH.

References

1. Donovan, P. J. and Gearhart, J. (2001) The end of the beginning for pluripotent stem cells. *Nature* **414,** 92–97.
2. Hubner, K., Fuhrmann, G., Christenson, L. K., Kehler, J., Reinbold, R., De La Fuente, R., et al. (2003) Derivation of oocytes from mouse embryonic stem cells. *Science* **300,** 1251–1256.
3. Geijsen, N., Horoschak, M., Kim, K., Gribnau, J., Eggan K., and Daley, G. Q. (2004) Derivation of embryonic germ cells and male gametes from embryonic stem cells. *Nature* **427,** 148–154.
4. De Miguel, M. P., Federspiel, M. J., and Donovan, P. J. (2000) Regulation of growth and survival in the mammalian germline, in *The Testis: From Stem Cell to Sperm Function* (Goldberg, E., ed). Serono Symposium. Springer-Verlag, New York, NY, pp. 55–70.
5. Resnick, J. L., Bixler, L. S., Cheng, L., and Donovan, P. J. (1992) Long-term proliferation of mouse primordial germ cells in culture. *Nature* **359,** 550–551.
6. Matsui, Y., Zsebo, K., and Hogan, B. L. (1992) Derivation of pluripotential embryonic stem cells from murine primordial germ cells in culture. *Cell* **70,** 841–847.
7. Shamblott, M. J., Axelman, J., Wang, S. P., Bugg, E. M., Littlefield, J. W., Donovan, P. J., et al. (1998) Derivation of pluripotent stem cells from cultured human primordial germ cells. *Proc. Nat. Acad. Sci. USA* **95,** 13726–13731.

8. Stevens, L. C. (1980) Testicular, ovarian and embryo-derived teratomas. *Cancer Surveys* **2,** 75–91.

9. Donovan PJ. (2001) High Oct-ane fuel powers the stem cell. *Nat. Genet.* **2,** 246–247.

10. Solter, D. and Knowles, B. B. (1978) Monoclonal antibody defining a stage-specific mouse embryonic antigen (SSEA-1) *Proc. Nat. Acad. Sci. USA* **75,** 5565–5569.

11. Fox, N., Damjanov, I., Martinez-Hernandez, A., Knowles, B. B., and Solter, D. (1981) Immunohistochemical localization of the early embryonic antigen (SSEA-1) in postimplantation mouse embryos and fetal and adult tissues. *Dev. Biol.* **83,** 391–398.

12. Shevinsky, L. H., Knowles, B. B., Damjanov, I., and Solter, D. (1982) Monoclonal antibody to murine embryos defines a stage-specific embryonic antigen on mouse embryos and human teratocarcinoma cells. *Cell* **30,** 697–705.

13. Kannagi, R., Cochran, N. A., Ishigami, F., Hakomori, S., Andrews, P. W., Knowles, B. B., et al (1983) Stage-specific embryonic antigens (SSEA-3 and -4) are epitopes of a unique globo-series ganglioside isolated from human teratocarcinoma cells. *EMBO J.* **2,** 2355–2361.

14. Fox, N. W., Damjanov, I., Knowles, B. B., and Solter, D. (1984) Stage-specific embryonic antigen 3 as a marker of visceral extraembryonic endoderm. *Dev. Biol.* **103,** 263–266.

15. Wenk, J., Andrews, P. W., Casper, J., Hata, J., Pera, M. F., Von Keitz, A., et al. (1994) Glycolipids of germ cell tumors: Extended globo-series glycolipids are a hallmark of human embryonal carcinoma cells. *Int. J. Cancer* **58,** 108–115.

16. Henderson, J. K., Draper, J. S., Baillie, H. S., Fishel, S., Thomson, J. A., et al. (2002) Preimplantation human embryos and embryonic stem cells show comparable expression of stage-specific embryonic antigens. *Stem Cells* **20,** 329–37.

17. Fenderson, B. A., Andrews, P. W., Nudelman, E., Clausen, H., and Hakomori, S. (1987) Glycolipid core structure switching from globo- to lacto- and ganglio-series during retinoic acid-induced differentiation of TERA-2-derived human embryonal carcinoma cells. *Dev. Biol.* **122,** 21–34.

18. Draper, J. S., Pigott, C., Thomson, J. A., and Andrews, P. W. (2002) Surface antigens of human embryonic stem cells. changes upon differentiation in culture. *J. Anat.* **200,** 249–258.

19. Turnpenny, L., Brickwood, S., Spalluto, C. M., Piper, K., Cameron, I. T., Wilson, D. I., et al. (2003) Derivation of human embryonic germ cells: an alternative source of pluripotent stem cells. *Stem Cells* **21,** 598–609.

20. Badcock, G., Pigott, C., Goepel, J., and Andrews, P. W. (1999) The human embryonal carcinoma antigen TRA-1-60 is a sialylated keratan sulphate proteoglycan. *Cancer Res.* **59,** 4715–4719.

21. Berstine, E. G., Hopper, M. L., Grandchamp, L., and Ephrussi, B. (1973) Alkaline phosphatase activity in mouse teratoma. *Proc. Natl. Acad. Sci. USA* **70,** 3899–3903.

22. De Miguel, M. P. and Donovan, P. J. (2000) Isolation and culture of mouse germ cells, in *Developmental Biology Protocols* (Tuan, R. S. and Lo, C. W., eds.) Humana Press, Totowa, NJ, pp. 403–408.

23. Harlow, E. and Lane, D. (1999) *Using Antibodies. A Laboratory Manual.* Cold Spring Harbor Laboratory Press, Cold Spring Harbor, NY.
24. Fenderson, B. A., Radin, N. S., and Andrews, P. W. (1993) Differentiation antigens of human germ cell tumors: distribution of carbohydrate epitopes on glycolipids and glycoproteins analyzed using PDMP, an inhibitor of glycolipid synthesis. *Eur. Urol.* **23,** 30–37.
25. Fenderson, B. A., Holmes, E. H., Fukushi, Y., and Hakomori, S. (1986) Coordinate expression of X and Y haptens during murine embyogenesis. *Dev. Biol.* **114,** 12–21.

16

Analysis of the Nucleolar Compartment of the Nucleus as an Indicator of Nuclear Reprogramming After Nuclear Transfer

Jacques-E. Fléchon

Summary

When analyzing reprogramming after nuclear transfer, it is interesting to focus on the nucleolar compartment, which is the most morphologically well-defined compartment in the nucleus. As with many messenger RNA-encoding genes, the ribosomal RNA genes are expressed in the nuclei of cells used for nuclear transfer. We suppose that a successful passage from the expression of genes specific to somatic cells to those characteristic of an early embryo implies the transient arrest of any expression under the effect of the oocyte cytoplasm. After nuclear transfer, it is possible to observe using electron microscopy the changes in nucleoli that reflect their activity. In successful cases, the nucleoli are deactivated effectively before the end of the one-cell stage. Sometimes however, incomplete changes or delays in the process may be observed that eventually are associated with abnormal development. It is possible to confirm the diagnosis using other techniques, three examples of which are given: the loss of reticulated structure of nucleoli may be quickly detected by immunofluorescence; proteins that are specific for a nucleolar component can be tracked during nucleolar changes by immunochemistry on thin sections; and the presence of deoxyribonucleic acid inside a nucleolus (indispensable for its activity) can be verified by ultrastructural cytochemistry.

Key Words: Cloning; nucleolus; reprogramming; electron microscopy; immunocytochemistry.

1. Introduction

Cloning is not widely used as a technique to multiply (i.e., "copy") exceptional farm animals (cattle). However cloning, or more precisely, embryonic development controlled by the nucleus of a differentiated cell, appears to offer a promising new field of study for developmental biology. The development of an embryo operated by the genome of a somatic cell means that the somatic

From: *Methods in Molecular Biology, vol. 325: Nuclear Reprogramming: Methods and Protocols*
Edited by: S. Pells © Humana Press Inc., Totowa, NJ

genes expressed by its nucleus are repressed and also that the genes normally expressed at the onset of development are activated. This complex nuclear re-arrangement is called reprogramming. The differences between embryos after in vitro fertilization and nuclear transfer embryos in a panel of expressed genes may be very illustrative of early abnormal reprogramming *(1)*. In fact the genomes of zygotes and even cleaving embryos may not be expressed initially *(2)*; therefore, we propose the hypothesis that all genes of a transferred nucleus should first be completely repressed to give rise to normal progeny. Not only messenger RNAs (mRNA), but also other types of RNAs, such as ribosomal RNAs (rRNAs), are synthesized by early embryos. The latter are produced in the nucleolar compartment of the nucleus. What is interesting for practical purposes is that the morphology of a nucleolus reflects its activity, as described herein. Essentially, the method described in this chapter is the observation, by electron microscopy, of the nucleoli of transferred nuclei at each early cleavage stage of cloned embryos that were obtained under standardized conditions. By comparison with normal control embryos, it is possible to follow the steps by which nucleoli of transferred nuclei are deactivated and eventually reactivated. Cytochemical techniques using light and/or electron microscopy also help in this analysis.

1.1. Ultrastructure and Role of Nucleoli

Ribosomal genes of eukaryotes are distributed in tandem clusters on different chromosomes, forming nucleolus-organizing regions (NORs) in which nucleoli are formed *(3)*. Active nucleoli are composed of three morphologically distinct elements, the fibrillar centers (FCs), which are surrounded by the dense fibrillar component (DFC), and the granular component (GC), which contain preribosomal particles (**Fig. 1**).

Schematically, the nucleolus is a compartmentalized machine in which the ribosomal deoxyribonucleic acid (rDNA, the code) is located in the FCs, and the rRNA is synthesized by RNA polymerase I in the FCs at the boundary with the DFC *(4)*. Preribosomal RNA is elongated in the DFC and processed into 5S, 18S, and 28S RNAs in the GC *(5)*. In actuality, the nucleolus looks like a 3D network in which all the processes are ongoing co-locally and in which sharp functional boundaries cannot be absolutely established.

Specific macromolecular complexes of enzymes and cofactors are found in the sites of polymerization and splicing. Some of these multifunctional proteins are listed in **Table 1** *(6–10)*, where their distributions and roles are briefly indicated.

At each M phase of the somatic mitotic cycle, the nucleoli become inactive and disassemble; most of their components except those associated with the NOR, such as RNA polymerase I *(11)*, disaggregate and disperse. Metaphase

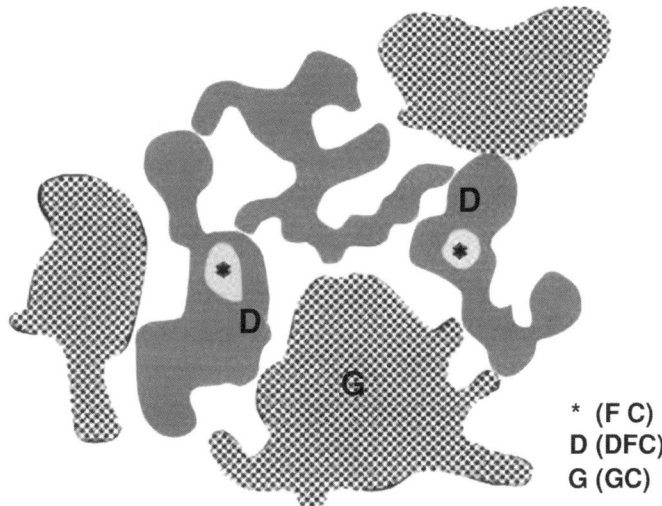

* (F C)
D (DFC)
G (GC)

Fig 1. Schematic structure of an active nucleolus. The FCs (*) are surrounded by the DFC (D), forming a reticulated structure sometimes called nucleolonema. The GC (G) is made of preribosomal particles.

Table 1
List of Proteins Useful as Markers for Nucleoli

Name	Nature and role	Localization
Fibrillarin	Protein of a ribonucleoprotein complex involved in early processing of pre-rRNA (6)	DFC (7)
Nucleolin (C23)	Multifunctional phosphoprotein involved in rRNA transcription and/or processing (8)	DFC and GC (9)
Nucleophosmin (B23)	Multifunctional phosphoprotein involved in late preribosomal assembly (10)	DFC and GC (9)

If oocytes are an exception (12). After diakinesis, nucleoli are rebuilt by association of the NOR with prenucleolar bodies containing proteins of the DFC (3,11).

The activity of interphase nucleoli is reflected in their size and complexity: a reticulated appearance is seen in proliferating hepatocytes (13) and lymphocytes (14); it is much less so in terminally differentiated cells (15).

1.2. Nucleologenesis in Control Embryos

The synthesis of rRNA does not start immediately in zygotes, and complete nucleologenesis generally is achieved only after a few cell cycles. The inactive nucleoli are called nucleolar precursor bodies (NPBs), not to be confused with the prenucleolar bodies of postmitotic somatic cells. According to Fléchon and

Kopečný *(16)*, two types of NPBs and of nucleogenesis are observed in mammals. The cow represents one model in which rRNA synthesis begins inside the NPB, whereas in the mouse model, the activity starts at the periphery, leaving the center of the NPB temporarily unchanged. We will take two examples of early embryonic nucleologenesis, the rabbit (mouse type), a laboratory mammal, and the cow itself, a farm mammal, both used for cloning.

1.2.1. Cleaving Rabbit Embryo

In the cleaving rabbit embryo, nucleologenesis was observed to be completed at the 16-cell stage *(2,17)*. At the two-cell stage, the NPBs were still compact spheres of fibrillar material (**Fig. 2A**). At the four-cell stage, caps were adjoined onto the NPBs (**Fig. 2B**). DNA, when localized by ultrastructural cytochemistry, was bound only to the cavitated surface of some NPBs at the 8-cell stage, whereas it was not found inside. The last step was observed at the 16-cell stage, when the nucleoli were clearly activated: the inner residual body disappeared, the DNA-containing areas (FCs) were surrounded by the DFC, and the GC accumulated (**Fig. 2C**). Markers such as fibrillarin could be followed, by the immunogold technique, from the periphery of the NPB to the reticulated structure of the DFC at later stages.

1.2.2. Cleaving Bovine Embryo

In the bovine embryo, three steps have been described before the fully reticulated active nucleoli form *(18)*. At first, the NPBs are spherical homogeneous masses of fibrils. They then acquire a large, unique vacuole and finally secondary vacuoles before the FCs are formed (a sign of reactivation). In the last step, the nucleolus, which now contains the GC, is fully reticulated and activated. These steps can be recognized morphologically, and nucleolin may be used as a marker to follow the DFC *(19)*.

Fig. 2. *(opposite page)* Rabbit nucleologenesis observed by TEM in control rabbit embryos (**A–C**) and after nuclear transfer (**D–F**) according to Kaňka et al. *(2)*. (Original magnification ×15,000 for all figures that do not give information on the actual size of the structures as sections may be either equatorial or tangential). (**A**) At the two-cell stage, each NPB is still a compact fibrillar mass. (**B**) At the four-cell stage, a "cap" is visible on the NPB surface, to which is attached perinucleolar chromatin. (**C**) At the 16-cell stage, the nucleoli are well-reticulated and contain GC. (**D**) In a cloned one-cell embryo, this nucleolus shows a remnant of DFC and a central NPB in formation. (**E**) At the four-cell stage, a NPB looking like control (**B**). (**F**) At the 32-cell stage, reticulated nucleoli are lacking GC. (Reproduced from Kaňka et al. *(2)* with permission.)

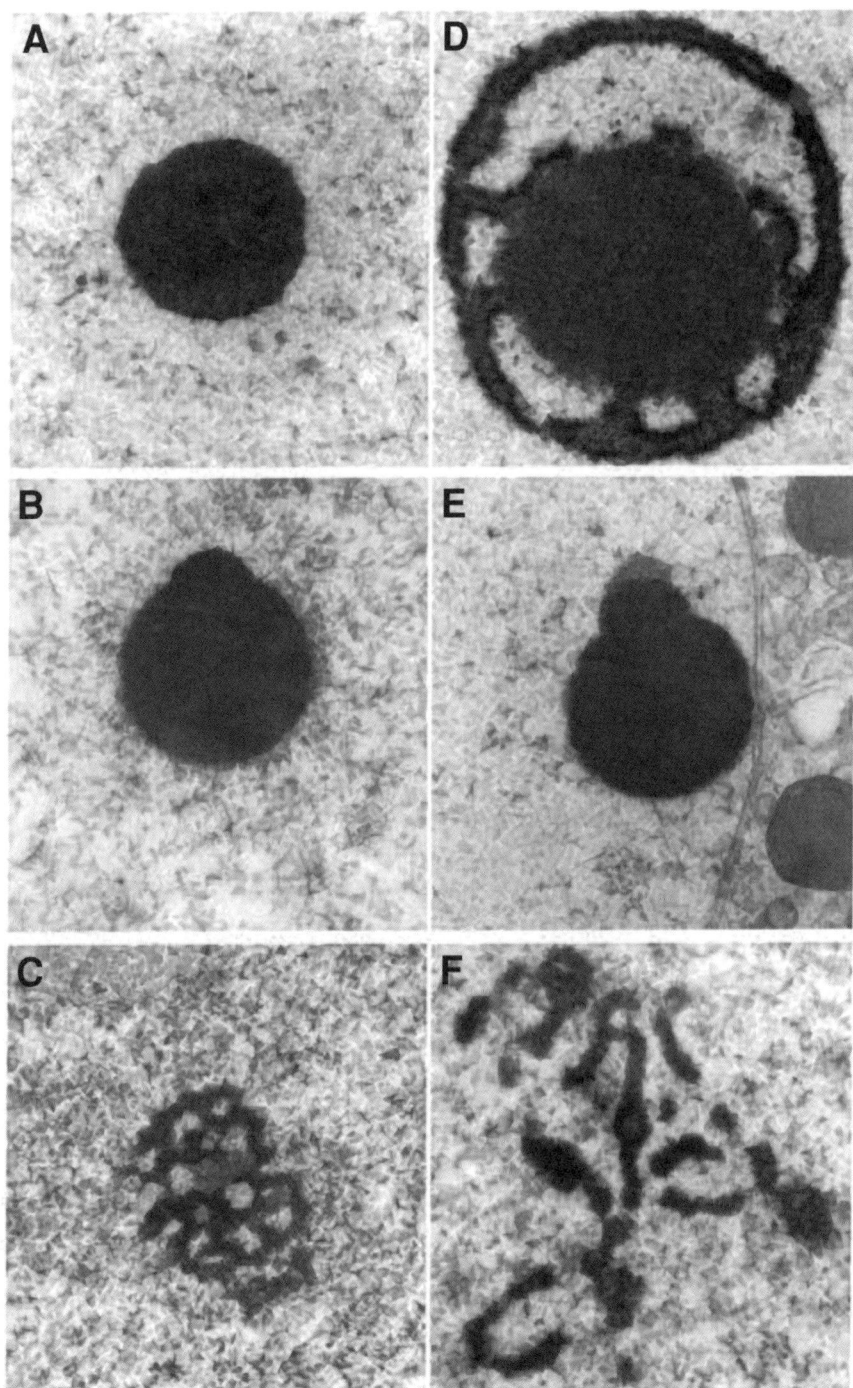

Nucleologenesis is not synchronous even in the same blastomere in both rabbit and bovine embryos. Nevertheless, nucleoli incompletely rebuilt during the first cell cycles are essentially fully reactivated by the 8-cell stage in the bovine species (*18,20,21*) and by the 16-cell stage in the rabbit species (*2*).

1.3. Nucleologenesis in Cloned Embryos

Whatever the cells habitually used as donors for nuclear transfer, they contain active nucleoli in interphase and are capable of rebuilding them after mitosis. The purpose of the method described here is to follow the morphology and activity of nucleoli in embryonic nuclei obtained by micromanipulation. If rRNA synthesis is transitorily repressed, it is possible to check whether the morphological involution of the nucleoli is analogous to the segregation of nucleolar components obtained artificially by treatment of cells with inhibitors of rRNA synthesis such as actinomycin D (*19*) or to a more physiological process such as the involution of nucleoli at the end of oocyte growth (*22*), which is a more or less exact reverse of the normal process of embryonic nucleologenesis. Nucleologenesis may then eventually start again.

1.3.1. Cloned Rabbit Embryos

The nuclei of 32-cell stage blastomeres contain active nucleoli. Upon transfer to enucleated oocytes (*2*), the nuclei no longer contain reticulated nucleoli, but rather regressing forms analogous to those of mature oocytes, which are never observed in normal zygotes. These forms contain no GC but instead only a few FCs and a reduced DFC surrounding vacuoles and/or a central NPB-like residual body (**Fig. 2D**). Less frequently, nucleoli were reduced to an NPB with cap. Nuclear incorporation of 5-³H-uridine could no longer be visualized by autoradiography before the end of the one-cell stage. At the 2-cell stage, the nucleoli looked like NPBs in normal zygotes or like NPBs with a cap as in normal 4-cell stage embryos (**Fig. 2E**). From this step, the nucleologenesis was completed almost as in control embryos, although even at the 32-cell stage, the GC did not clearly reappear (**Fig. 2F**).

1.3.2. Cloned Bovine Embryos

Activated or nonactivated cytoplasts were compared as recipients for nuclei of day 4 to 5 blastomeres (*21*). In both cases, the fully reticulated nucleoli of the transferred nuclei disappeared. However in nonactivated cytoplasts, the nucleoli regressed only to multivesiculated NPBs at the 2-cell stage and then development stopped, whereas in activated cytoplasts an almost-typical compact NPB was observed and normal development ensued, although accelerated one stage ahead for the rebuilding of active nucleoli.

During the 2- and 4-cell stages of bovine clones (*19*), the reticulated structure of nucleoli collapsed and looked like NPBs containing fewer and fewer vacuoles. In the best cases, a reassociation of DNA with the nucleoli was obtained at the 8-cell stage and a complete, reticulated structure was rebuilt at the 16-cell stage (a slight delay). Differences resulting from donor cell types could be observed, with somatic cells giving worse results (at least delayed de- and redifferentiation associated with lower cleaving rate) than embryonic cells. A different cytoplasmic milieu may therefore be required for reprogramming embryonic or adult nuclei *(23)*. More generally, the cell cycle stage of the donor cells and of the cytoplasts should be equivalent.

All the examples chosen and other similar studies *(24,25)* indicated that rRNA synthesis is generally repressed after nucleotransfer before it is reactivated at the right time in successful cases. Early defects (delay or arrest of cleavage) may be correlated with initial incomplete gene repression and/or reactivation.

2. Materials

2.1. Animals

The experiments can be performed in any species used for nuclear transfer (*see* **Note 1**).

2.2. Chemicals

2.2.1. Transmission Electron Microscopy (TEM)

1. Phosphate-buffered saline (PBS) Dulbecco A (OXOID, Basingstoke, UK); dilute as required.
2. Potassium phosphate, monobasic: KH_2PO_4.
3. Disodium phosphate, dibasic: Na_2HPO_4.
4. Glutaraldehyde (EM grade, 25% in water; Polyscience), store at 4°C, dilute as required.
5. Sodium cacodylate: $C_2H_6As\ NaO_2 \cdot 3H_2O$. *Warning:* Poison.
6. Osmium tetroxide: OsO_4, crystalline. *Warning:* Poison.
7. Polybed 8/2 embedding Kit BDMA (Polyscience).
8. Methylene Blue (Merck).
9. Azur Blue (Merck).
10. Uranyl acetate. Light sensitive.
11. Lead nitrate: $Pb(NO3)_2$.
12. Sodium citrate $Na_3(C_6H_5O_7) \cdot 2H_2O$.

2.2.2. Immunofluorescence

1. Pronase E (Sigma-Aldrich).
2. Paraformaldehyde EM grade (Polyscience).
3. Bovine serum albumin (BSA), fraction V, fatty acid free (Sigma-Aldrich), store at 4°C.
4. Triton X 100.
5. Ammonium chloride.
6. Moviol (Hoechst).

2.2.3. Immunogold

1. Agarose, low-melting-point (Polyscience).
2. Lowicryl K4 M Kit (Agar Scientific). Store at 4°C.
3. Normal goat serum (NGS), freeze.
4. Gold-coupled antibodies (Aurion, Wageningen, NL). Store at 4°C.

2.2.4. Terminal Nucleotidyltransferase (TdT) Technique

All chemicals are from Sigma.

1. 5 bromo-2'-deoxyuridine triphosphate (BrdUTP).
2. 5 bromo-uridine monophosphate.
3. $CoCl_2$.
4. β-mercaptoethanol.
5. Calf thymus terminal deoxynucleotidyl transferase (TdT).
6. Deoxynucleotides.
7. Monoclonal antibody anti-BrdU.

3. Methods

Electron microscopy is sufficient to show changes of nucleolar morphology (**Fig. 2**). However, to follow the fate of some nucleolar components, cytochemical techniques can be used, three of which are additionally described: immunofluorescence, ultrastructural immunocytochemistry, and nucleic acid cytochemistry.

3.1. TEM

1. Fix using the classical techniques (1% glutaraldehyde and 1% or 2% osmium tetroxide in cacodylate buffer for 1 h at room temperature; wash in cacodylate buffer; *see* **Note 2**).
2. Dehydrate in an ethanol series (50, 70, 90, 95, and two times 100%, 10 min each, at room temperature).
3. Embed in Epon (*see* **Note 3**).
4. Use an ultramicrotome to obtain semi-thin sections; observe by light microscopy after methylene blue staining to select the level where nuclei and nucleoli are present (*see* **Note 4**).
5. Cut thin sections and stain them with uranyl acetate and lead citrate (*see* **Note 5**).

3.2. Immunofluorescence

This technique can be used to rapidly explore the changes in morphology of the nucleoli after nuclear transfer. As shown in **Fig. 3A,B**, the use of antifibrillarin (characteristic of the DFC) allows one to observe an almost instant loss of the reticulated aspect of the nucleoli *(19)*.

All treatments are performed at room temperature except **step 1**.

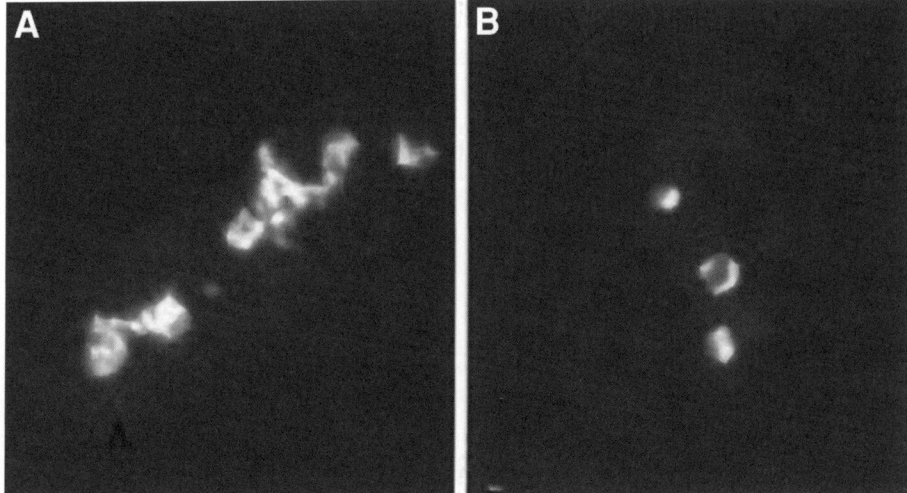

Fig. 3. Immunofluorescent distribution of fibrillarin in nucleoli in a cultured bovine fibroblast (**A**) and in nucleoli of the same cell type, 1 h after nuclear transfer (**B**). Note collapse of the reticulated structure *(2)*.

1. Wash embryos in protein/serum-free medium used for their collection or culture and treat with 0.5% pronase at 38°C for approx 30 s (depending to the species) to remove the zona pellucida. Wash in PBS with 0.2 % bovine serum albumin (BSA) for 15 min.
2. Fix them in 3.7% paraformaldehyde dissolved in PBS for ≤60 min (*see* **Note 6**).
3. Wash in the same medium with 0.2 % BSA (three times for 10 min).
4. Permeabilize with 0.5% (v/v) Triton X-100 in PBS for 15 min and wash with PBS.
5. Incubate with the primary antibody diluted in PBS containing 0.25% BSA for 1 h (*see* **Note 7**).
6. Wash and incubate with the fluorescently labeled secondary antibody diluted in the same medium.
7. Whole mounts of washed embryos are made on glass slides in Moviol.

Omit the primary antibody in controls.

3.3. Immunogold Technique

Anti-nucleolin was used *(19)* to follow the changes in the DFC (from reticulated structure to NPB and vice-versa) between the onset and the end of nucleolar metamorphosis after nuclear transfer (**Fig. 4A**).

All treatments are done at room temperature, except **steps 2** and **6**.

1. Wash embryos in medium without BSA.

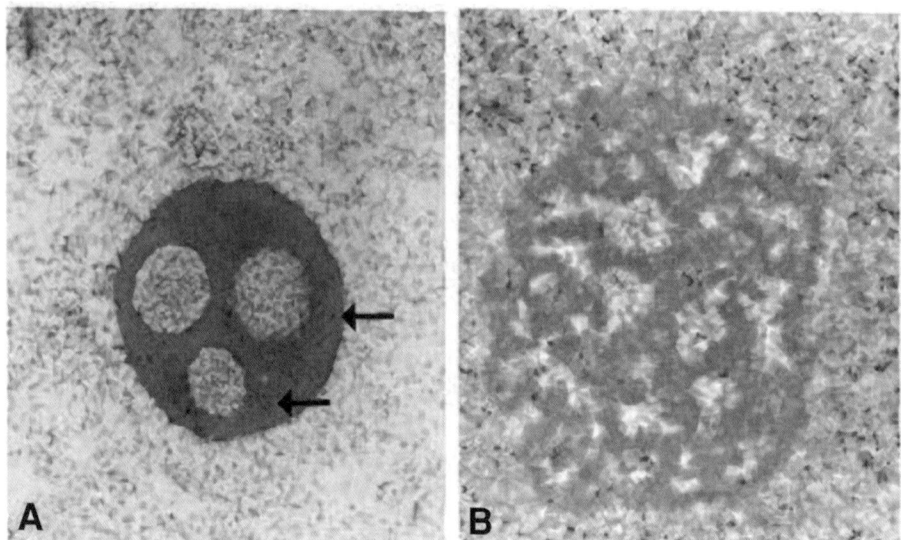

Fig. 4. Ultrastructural cytochemistry. **(A)** Immunogold ultrastructural localization of nucleophosmin in a remnant of DFC compacting into a NPB at the four-cell stage of a cloned bovine embryo. **(B)** Labeling of DNA in perinucleolar chromatin and inside the reticulated nucleolus at the eight-cell stage of a control bovine embryo. (Reproduced from Baran et al. *[19]* with permission.)

2. Fix them in 3.7% paraformaldehyde and 0.2% glutaraldehyde in Sörensen phosphate buffer pH 7.3 for 1 h at 4°C (*see* **Note 8**). Wash in the same buffer three times for 20 min.
3. Pre-embed the specimens in 2% (w/v) agar dissolved in PBS.
4. Dehydrate in an ethanol series as above and embed in Lowicryl K4M. Polymerize at –20°C under ultraviolet illumination for 2 d.
5. Preincubate ultrathin sections in PBS containing 5% (v/v) NGS for 60 min.
6. Incubate with the primary antibody diluted in PBS containing 2.5% NGS overnight at 4°C. Wash in PBS three times for 20 min.
7. Incubate the sections in the goat secondary antibody coupled with colloidal gold particles and wash again (*see* **Note 9**).
8. Stain the sections for 10 min with 4% uranyl acetate in water.

For controls, the primary antibody is omitted or the grids are incubated on antibody-free gold particles (*see* **Note 9**).

3.4. Nucleic Acid Cytochemistry

As a DFC is sometimes not obvious to recognize, it is important to show the presence of intranucleolar DNA, as proof of the rDNA gene activity (**Fig. 4B**). DNA can be labeled at the ultrastructural level by the terminal TdT technique

(26). For this purpose, ultrathin sections in Lowicryl are prepared as for ultra-structural immunocytochemistry.

1. Incubate the sections for 30 min at 37°C. in a solution containing 25 μM BrdUTP, 100 mM sodium cacodylate, pH 7.0, 2 mM CaCl$_2$, 10 mM β-mercaptoethanol, 50 μg/mL BSA, and 125 U/mL calf thymus TdT in PBS.
2. Incubate for 30 min at 37°C in the same medium with 4 mM each of dCTP, dGTP, dATP and rinse two times with distilled water. To decrease nonpecific binding, pretreat sections with 3% GNS and 1% BSA in PBS and rinse with PBS containing 1% BSA.
3. Incubate the sections for 4 h at room temperature with a monoclonal antibody to BrdU, diluted 1/50 in PBS supplemented with 0.2 % BSA and 1/50 GNS. Wash with PBS containing 1% BSA.
4. Incubate at room temperature for 1 h with goat anti-mouse IgG coupled to colloidal gold 5 to 10 nm in diameter, diluted in PBS, pH 8.2, with 0.2% BSA added.
5. Wash in PBS containing 1% BSA, rinse in distilled water, and mount the sections on nickel grids.
6. Stain as above (*see* **Subheading 3.3.**, **step 8**).

Controls may omit TdT or BrdUTP or replace the latter by 5-bromo-uridine monophosphate (*see* **Note 10**).

4. Notes

1 All the techniques for cloning should be mastered and, unfortunately, they are still very different between laboratories. However, once chosen, conditions should be invariable to obtain reproducible results with such an invasive technique, except in experiments for which one plans to study the effect of changing one parameter. The controls must also be obtained in well-defined conditions. In vitro fertilization would more closely match the conditions (medium, temperature, etc.) used for nuclear transfer; however, it should be remembered that embryos cultured in vitro may already express some genes differently from in vivo controls *(27)*. Although the pattern of nucleologenesis in vitro seems to match that in vivo, the occurrence of chromosome aberrations may increase in vitro *(28)*.
2. Cacodylate buffer 0.25 M at pH 7.3 (stock solution at 4°C) is used to prepare 1% glutaraldehyde fixative (from 25% glutaraldehyde stock solution) in 0.1 M cacodylate buffer solution and the 0.1 M cacodylate buffer washing solution. Osmium tetroxide (OsO4) is stored in glass ampullae, which can be broken inside a bottle containing the required amount of 0.1 M cacodylate buffer to prepare 1% or 2% fixative solution. The flask must be stopped up rapidly and left at room temperature in a hood. The solution can be used the next day after dissolution of crystals.
3. Embedding and sectioning: it is recommended that one follow the manufacturer's instructions in use of the embedding kit and the ultramicrotome. A period in an electron microscope laboratory is necessary for all beginners, including directions on the use of the electron microscope.

4. Azure bue and methylene blue staining: mix 1% methylene blue in aqueous borax and 1% azur blue in water (v/v); dilute with distilled water (1/2, v/v). Semithin sections are stained for 5 min at 40°C and rinsed in distilled water.

5. Uranyl acetate staining: prepare a 1 to 4% aqueous solution in a stopped, tinted bottle stored in a dark space. Drops of filtered solution are deposited on a piece of Parafilm. After rinsing in distilled water and drying, each grid with the sections downwards is placed on the surface of a drop for approx 10 min; the grids are then removed with tweezers, rinsed with a stream of distilled water flowed from a pipet, and dried on a filter paper. "Reynolds" lead citrate staining: Prepare an aqueous solution (30 mL) of 1.33 g of lead nitrate and 1.76 g of sodium citrate in a flask, close, and agitate. Dissolve the white precipitate with 8 mL of 1 N NaOH and make to 50 mL with distilled water. The solution can be stored for a few weeks in tubes protected from the atmosphere (CO_2) by paraffin oil.

6. To prepare fixative, add the required quantity of paraformaldehyde to a flask containing PBS and heat to 60°C with slow agitation for 10 min. Once cleared (if necessary, add some drops of 1 N NaOH), filtered, and cooled, the solution is ready to use.

7. To avoid background staining, incubation in 50 mM ammonium chloride in PBS for 1 h at room temperature can be applied after fixation to block free aldehyde groups. Nonspecific binding may be diminished by preincubation in 1% to 2% BSA or NGS in PBS for 1 h at room temperature before antibody treatment.

8. 0.1 M Sörensen phosphate buffer, pH 7.4, is prepared by mixing 19.6 mL of 0.066 M KOH and 80.4 mL of 0.066 Na_2 solution.

9. For gold-coupled antibodies, follow the instructions of the supplier.

10. Additional controls are the same as for immunogold.

References

1. Daniels, R., Hall, V., and Trounson, A. O. (2000) Analysis of gene transcription in bovine nuclear transfer embryos reconstructed with granulosa cell nuclei. *Biol. Reprod.* **63,** 1034–1040.

2. Kaňka, J., Hozák, P., Heyman, Y., Chesné, P., Degrolard, J., Renard, J. P., et al. (1996) Transcriptional activity and nucleolar ultrastructure of embryonic rabbit nuclei after transplantation to enucleated oocytes. *Mol. Reprod. Dev.* **43,** 135–144.

3. Scheer, U., Thiry, M., and Goessens, G. (1993) Structure, function and assembly of the nucleolus. *Trends Cell Biol.* **3,** 236–241.

4. Cheutin, T., O'Donohue, M.-F., Beorchia, A., Vandelaer, M., Kaplan, H., Deféver, B., et al. (2002) Three-dimensional organization of active rRNA genes within the nucleolus. *J. Cell Sci.* **115,** 3297–3307.

5. Puvion-Dutilleul, F., Puvion, E., Bachellerie, J.-P. (1997) Early stages of pre-rRNA formation within the nucleolar ultrastructure of mouse cells studied by in situ hybridization with a 5'ETS leader probe. *Chromosoma* **105,** 496–505.

6. Azum-Gelade, M.-C., Noaillac-Depeyre, J., Caizergues-Ferrer, M. and Gas, N. (1994) Cell cycle redistribution of U3 snRNA and fibrillarin. *J. Cell Sci.* **107,** 463–475.

7. Ochs, R. L., Lischwe, M. A., Spohn, H., and Busch, H. (1985) Fibrillarin: a new protein of the nucleolus identified by autoimmune sera. *Biol. Cell* **54**, 123–124.

8. Lapeyre, B., Bourbon, H., and Amalric, F. (1987) Nucleolin, the major nucleolar protein of growing eukaryotic cells: an anusual protein structure revealed by the nucleotide sequence. *Proc. Natl. Acad. Sci. USA* **84**, 1472–1476.

9. Biggiogera, M., Bürki, K., Kaufmann, S. H., Shaper, J. H., Gas, N., Almaric, F., et al. (1990) Nucleolar distribution of proteins B23 and nucleolin in mouse preimplantation embryos as visualized by immunoelectron microscopy. *Development* **110**, 1263–1270.

10. Okuwaki, M., Tsujimoto, M., and Nagata, K. (2002) The RNA binding activity of a ribosome biogenesis factor, nucleophosmin/B23, is modulated by phosphorylation with a cell cycle-dependent kinase and byassociation with its subtype. *Mol. Biol. Cell* **13**, 2016–2030.

11. Dousset, T., Wang, C., Verheggen, C., Chen, D., Hernandez-Verdun, D., Huang, S. (2000) Initiation of nucleolar assembly is independent of RNA polymerase I transcription. *Mol. Biol. Cell* **11**, 2705–2717.

12. Zatsepina, O. V., Bouniol-Baly, C., Debey, P. (2000) Functional and molecular reorganization of the nucleolar apparatus in maturing mouse oocytes. *Dev. Biol.* **223**, 3543–3570.

13. Derenzini, M., Sirri, V., Pession, A., Trere, D., Roussel, P., Ochs, R. L., et al. (1995) Quantitative changes of the two major AgNOR proteins, nucleolin and protein B23, related to stimulation of rDNA transcription. *Exp. Cell Res.* **219**, 276–282.

14. Dergunova, N. N., Bulycheva, T. I., Artemenko, E. G., Shpakova, A. P., Pegova, A. N., Gemjian, E. G., et al. (2002) A major nucleolar protein B23 as a marker of proliferation activity of human peripheral lymphocytes. *Immunol. Lett.* **83**, 67–72.

15. Verheggen, C., Le Panse, S., Almouzni, G., and Hernandez-Verdun, D. (2001) Maintenance of nucleolar machineries and pre-rRNAs in nucleolus of erythrocyte nuclei and remodelling in Xenopus extracts. *Exp. Cell Res.* **269**, 23–34.

16. Fléchon, J. E. and Kopečný, V. (1998) The nature of the 'nucleolus precursor body' in early preimplantation embryos: a review of fine-structure cytochemical, immunocytochemical an autoradiographic data related to nucleolar function. *Zygote* **6**, 183–191.

17. Baran, V., Mercier, Y., Renard, J.-P., and Fléchon, J. E. (1997) Nucleolar substructures of rabbit cleaving embryos: an immunocytochemical study. *Mol. Reprod. Dev.* **48**, 34–44.

18. Kopečný, V. (1989) High-resolution autoradiographic studies of comparative nucleogenesis and genome reactivation during early embryogenesis in pig, man and cattle. *Reprod. Nutr. Dev.* **29**, 589–600.

19. Baran, V., Vignon, X., LeBourhis, D., Renard, J.-P., and Fléchon, J. E. (2002) Nucleolar changes in bovine nucleotransferred embryos. *Biol. Reprod.* **66**, 534–543.

20. Baran, V., Fléchon, J. E., and Pivko, J. (1996) Nucleologenesis in the cleaving bovine embryo : immunocytochemical aspects. *Mol. Reprod. Dev.* **44**, 63–70.

21. Kaňka, J., Smith, S. D., Soloy, E., Holm, P. and Callesen, H. (1999) Nucleolar ultrastructure in bovine nuclear transfer embryos. *Mol. Reprod. Dev.* **52**, 253–263.

22. Crozet, N., Kaňka, J., Motlik, J., and Fulka, J. (1986) Nucleolar fine structure and RNA synthesis of bovine oocytes from antral follicles. *Gamete Res.* **14,** 65–73.
23. Du, F., Sung, L.- Y., Tian, X. C., and Yang X. (2002) Differential cytoplast requirement for embryonic and somatic cell nuclear transfer in cattle. *Mol. Reprod. Dev.* **63,** 183–191.
24. Hyttel, P., Viuff, D., Fair, T., Laurincik, J., Thomsen, P. D., Callesen, H., et al. (2002) Ribosomal RNA gene expression and chromosome aberrations in bovine oocytes and preimplantation embryos. *Reproduction* **122,** 21–30.
25. Laurincik, J., Zakhartchenko, V., Stojkovic, M., Brem, G., Wolf, E., Müller, M., et al. (2002) Nucleolar protein allocation and ultrastructure in bovine embryos produced by nuclear transfer from granulosa cells. *Mol. Reprod. Dev.* **61,** 477–487.
26. Thiry, M., Ploton, D., Menager, M., and Goessens, G. (1993) Ultrastructural distribution of DNA within the nucleolus of various animal cell lines or tissues revealed by terminal deoxynucleotidyl transferase. *Cell Tissue Res.* **271,** 33–45.
27. Wrenzycki, C., Hermann, D., Carnwath, J. W., and Niemann, H. (1999) Alterations in the relative abundance of gene transcripts in preimplantation bovine embryos cultured in medium supplemented with either serum or PVA. *Mol. Reprod. Dev.* **53,** 8–18.
28. Hyttel, P., Laurincik, J., Zakhartchenko, V., Stojkovic, M., Wolf, E., Müller, M., et al. (2001) Nucleolar protein allocation and ultrastructure in bovine embryos produced by nuclear transfer from embryonic cells. *Cloning* **3,** 69–82.

17

Methylation-Sensitive Polymerase Chain Reaction

Hannah R. Moore, Richard R. Meehan, and Lorraine E. Young

Summary

Here, we describe a robust and reproducible methylation-sensitive polymerase chain reaction (MS-PCR) method to detect the percentage methylation in repeat sequences of individual pre-implantation ovine embryos produced by different embryo technologies. This method allows the comparison of embryos produced by nuclear transfer with other production and embryo culture methods, accounting for the heterogeneity between embryos within a single treatment. DNA extracted from single embryos is digested with a methylation-sensitive restriction enzyme to determine the percentage methylation after PCR amplification in comparison with an undigested control. The undigested control represents 100% methylation because methylation-sensitive enzymes do not cut methylated DNA, allowing the entire sample to be amplified by PCR. Image analysis quantification of the digested subsample PCR product on an ethidium bromide-stained agarose gel is proportional to the amount of methylated DNA in each embryo. By comparing quadruplicate values obtained for each embryo against a standard curve, we are able to ensure the validity of our results for each individual embryo. Compared with bisulphite sequencing methods, the method described is rapid, inexpensive, and relatively high-throughput.

Key Words: DNA methylation; pre-implantation embryo; sheep; somatic cell nuclear transfer; methylation-sensitive PCR; Satellite I; SINE; single embryo; blastocyst.

1. Introduction

In pre-implantation embryo studies, robust data on methylation status require the analysis of individual embryos, especially when applying treatments such as in vitro fertilization (IVF) and somatic cell nuclear transfer (SCNT), where the expected heterogeneity between embryos is likely to be significant *(1–3)*. By pooling embryos, the variations that exist between individual embryos are not measurable and therefore the data provide an inaccurate picture of the typical situation. With the heterogeneity between embryos we have detected using bisulphite sequencing *(4)* and 5-methylcytosine immunostaining

From: *Methods in Molecular Biology, vol. 325: Nuclear Reprogramming: Methods and Protocols*
Edited by: S. Pells © Humana Press Inc., Totowa, NJ

(3) of individual blastocysts, analyses of pools of, for instance, 10 embryos are likely to detect methylation differences equally high between different pools within treatments as between treatments. Thus, considerable numbers of embryos are required for adequate replication when analysing pooled embryos.

Methylation analysis of single embryos has recently been reported *(5)* utilizing the pre-treatment of DNA with sodium bisulfite, a process that converts unmethylated cytosine residues to uracil *(6,7)*. The methylation status of multiple specific CpG dinucleotides can then be assessed using a number of different methods. Bisulfite-treated DNA can be amplified by polymerase chain reaction (PCR) to obtain specific sequences of interest and the PCR products are then cloned and sequenced to determine the methylation status of the cytosine residues within the sequence *(8,9)*. This method is labor intensive and expensive as it requires the sequencing of many clones (ideally 20 per sample) to differentiate between the normal variability within a population of cells. An alternative method using bisulphite treatment involves the design of PCR primers that will only amplify "converted" (unmethylated) DNA, allowing the assessment of the methylation status for a number of CpG dinucleotides within the primer sequence. This method has recently been used in pig and bovine embryos, although ≥45 to 65 cycles of hemi-nested PCR were required for product detection *(2,10,11)*, a situation that can be prone to PCR-based anomalies. In addition, all bisulfite-based techniques have inherent problems. Attainment of complete conversion of the sequence of interest is problematic *(12)*, especially within CpG-rich sequences and DNA-containing unusual secondary structures. Quantification of methylation levels also can be problematic when using primers to amplify converted C-rich sequences. It has been demonstrated that primers for this form of methylation analysis do not always amplify methylated and unmethylated DNA proportionately *(13)*.

Here, we describe a robust, nonbisulfite-based method for the quantification of methylation status of individual sequences in single pre-implantation embryos. MS-PCR techniques were developed to analyze the DNA methylation status of repetitive DNA sequences in single oocytes and pre-implantation embryos by combining DNA digestion with the methylation sensitive restriction enzyme, *Hha*I, with single-embryo PCR. These methods have been used to compare the methylation status of DNA in ovine embryos derived in vivo, via IVF, or SCNT but should be applicable to other species. The method is described for analysis of the heterochromatic sheep Satellite I sequence that is found predominantly in centromeric chromosomal regions *(14–16)*. However, examination of other repetitive sequence compartments within the sheep genome also is possible with this method (*see* **Note 1**), allowing further insight into the global methylation events occurring during pre-implantation development.

2. Materials

1. Proteinase K (Sigma).
2. Pellet Paint® Co-precipitant (Novagen).
3. *Hha*I restriction endonuclease, NEBuffer 4, and bovine serum albumin (New England Biolabs).
4. *Sss* I methylase and *S*-adenosyl-methionine (New England Biolabs).
5. OVSAT1 Oligonucleotide primers : OVSAT1F (ATTCCTCTCCCGCTGATGCC) and OVSAT1R (CCTGAGAACCTTACCACCGTG; *see* **Note 2**).
6. Thermo-Start® PCR mastermix, (Abgene).
7. 0.5-mL Thin-walled PCR tubes.
8. Autoclaved distilled water (dH$_2$O).
9. Agarose (BDH).
10. 1X TBE (Tris-borate ethylenediaminetetraacetic acid).
11. Ethidium bromide (Sigma).
12. Gel loading solution (Sigma).
13. Multi-Analyst software (Bio-Rad).
14. QIAQuick gel extraction kit (Qiagen).
15. Topo-TA cloning kit (Invitrogen).
16. Embryos between one-cell and blastocyst stage, washed in culture medium and subsequently stored at –80°C in 1 µL of culture medium carefully placed at the bottom of an autoclaved 0.5-mL Eppendorf tube using a P10 Gilson pipet. Embryos can be stored in this manner for at least 2 yr.

3. Methods

Figure 1 outlines the complete method from DNA extraction to the quantification of the resultant PCR products.

3.1. Extraction of DNA From Single Oocytes or Embryos

1. Tubes containing single oocytes or embryos in 1 µL of culture medium are subjected to three freeze–thaw cycles before the extraction of DNA. Perform freezing at –80°C for periods of 30 min, defrosting the embryos at room temperature between freezes.
2. Digest embryos with Proteinase K (0.1 µg; in 10 µL volume Tric-acetate buffer [TAE]) at 50°C for 1 h, then incubate at 80°C for 20 min to destroy the proteinase K activity. Incubations are best performed on a reliable heating block or Thermal Cycler.
3. Ethanol-precipitate the DNA by adding 2 µL of Pellet Paint® (Novagen) and 1 µL of 3 *M* sodium acetate, pH 5.2 (supplied). Perform a 100% ethanol wash (2 vol, i.e., 20 µL), vortexing before centrifuging at 22,130 relative centrifugal force (RCF) in a benchtop refrigerated centrifuge for 1 h. Remove the supernatant and then add 100 µL of 70% ethanol, vortexing and centrifuging for 30 min as before. The Pellet Paint co-precipitates with the DNA to improve recovery and spins down to a visible pink pellet, facilitating resuspension. It does not interfere with restriction digests or PCR/agarose electrophoresis.

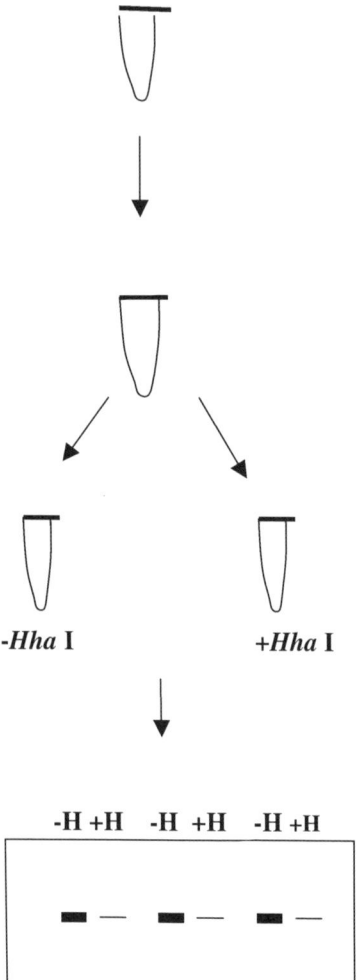

DNA extracted from single embryos by repeat freeze-thawing and proteinase K digestion.

DNA ethanol precipitated using Pellet Paint® (Novagen) and resuspended in 6μl dH₂0.

DNA divided into two aliquots; one digested with *Hha* I (+*Hha* I; 20μl volume), the second aliquot remains undigested (-*Hha* I; *Hha* I-free 20μl digest).

Sequence-specific MS-PCR: 4 replicate PCR reactions per digested/undigested single embryo DNA were performed, each using 2 μl template DNA. The intensity of the *Hha* I-digested DNA PCR product was directly compared to the undigested DNA PCR product (representing 100% methylation), to ascertain the % methylation of the *Hha* I site within the DNA sequence.

Fig. 1. Outline of methyl-sensitive PCR method.

4. Remove most of the supernatant with a pipet, leaving the sample in a fume hood for 5 min (or on the benchtop for 10 min) to remove residual ethanol.
5. Resuspend the DNA pellet in 6 μL of dH₂O.

3.2. Digestion of Single-Embryo DNA

1. Divide the DNA from single oocytes/embryos into two aliquots of 3 µL.
2. Digest one aliquot with *Hha*I restriction endonuclease in a 19 µL of digestion reaction containing 20 U of *Hha*I, 1X NEBuffer 4 (supplied), and 1X BSA for 2 h at 37°C. Add a further 20 U of *Hha*I after 2 h and incubate for an additional 2 h to ensure complete digestion of DNA.
3. Heat-kill the restriction enzyme activity by incubation at 65°C for 20 min.
4. Incubate the second DNA aliquot in a 20 µL of sham-digest reaction containing 1X New England Biolabs (NEB) Buffer 4 and 1X BSA for 4 h at 37°C, followed by a second 65°C incubation of 20 min.

3.3. Methylation-Sensitive PCR

1. The percentage methylation calculated for each embryo is obtained by comparing the band intensity of an undigested DNA PCR product (representing 100% methylation) with the *Hha*I-digested PCR product.
2. Perform quadruplicate PCRs in a 25-µL volume containing 12.5 µL of 2X Thermostart mastermix, 250 nM each primer (2.5 µL), and 2 µL of either *Hha*I-digested or -undigested DNA.
3. Perform PCR in a thermal cycler with a heated lid (to avoid using mineral oil) at 1 cycle of 15 min at 94°C, followed by repeated cycles of 94°C for 30 s, 62°C for 30 s, and 72°C for 45 s. For OVSAT1 primers (*see* **Note 2**) the optimal cycle number in our hands was 19 for blastocysts, 21 for morulae, 22 for 8- to 16-cell embryos, and 24 for oocytes. In all cases, methylation values were normalized against the relevant standard curve (*see* **Subheading 3.5.**).

3.4. Gel Electrophoresis and Image Quantification

1. Mix a 20-µL aliquot of PCRs with 3 µL of loading dye (Sigma) and electrophorese in a 1.5 to 2% agarose gels using 1X TBE containing 0.2 µg/mL ethidium bromide at 8 to 9 V/cm.
2. Visualize PCR products on a UV transilluminator and the record the image.
3. Quantify the band intensity of the PCR products (we used Multi-Analyst software with local background subtraction).

3.5. Generation of Standard Curves for MS-PCR Analysis

To define the linearity of the MS-PCR amplification within the PCR cycle used for quantification, standard curves were established to analyze DNA standards representing various levels of methylation. Standard curves were generated for each embryo stage analysed (*see* **Note 3**).

1. Use DNA extracted and pooled from 3 oocytes, 8 to 16 cell embryos, morulae, or blastocysts for standard curve generation.
2. Perform PCR at the same cycle number as optimized for embryo MS-PCR (*see* **Subheading 3.3.**).

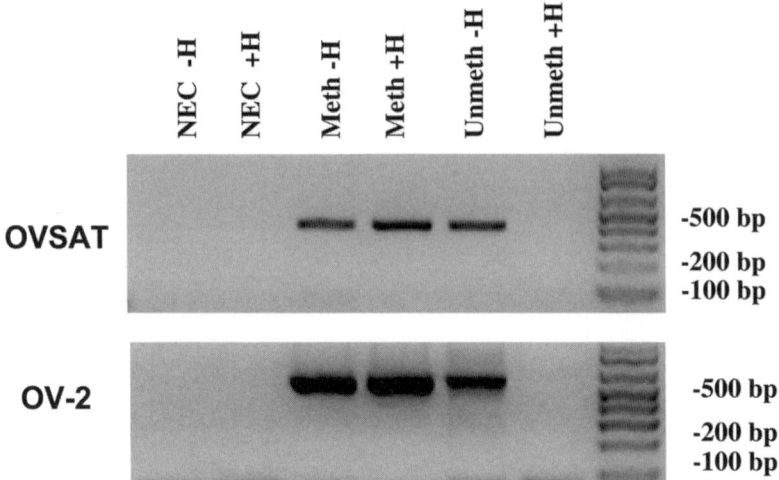

Fig. 2. PCR controls used for Satellite I (OVSAT) or OV-2 SINE DNA methylation analysis.

3. Standards consist of known quantities of undigested DNA ranging from 2 μL of embryo/oocyte DNA (the equivalent amount of template used in a MS-PCR reaction) down to 0.4 μL, at 0.4-μL intervals and with four replicates per specified DNA quantity. Because the DNA template is undigested, each DNA standard represents a known methylation level from 100% (2 μL of DNA) down to 20% methylated (0.4 μL of DNA). To predict the observed methylation level value from the known standard, divide the intensity of the PCR product by the "100% methylation" (i.e., 2 μL) standard PCR product intensity, thus replicating the method of quantification used for individual embryo analysis. Plot the observed values against the predicted methylation level values to generate standard curves specific to each embryo stage and primer set. Examples are shown in **Fig. 2A–C**.

3.6. MS-PCR Controls and Assessment of Reproducibility

Representative controls established for Satellite I MS-PCR are demonstrated in **Fig. 3**.

1. Initiate a "no embryo control" (NEC; 1 μL of dH$_2$O) at the DNA extraction step during each experiment. Add 1 μL of dH$_2$O to a 0.5-mL Eppendorf tube and carry through the entire method described in **Subheadings 3.1.** to **3.4.** to ensure the reagents and plastic consumables are not contaminated.
2. A control also is required to ensure that the *Hha*I enzyme (present in excess) completely digests the unmethylated DNA. Because there is no available methylation-insensitive isoschizomer for *Hha*I, an unmethylated PCR product amplified by the OVSAT1 primers and then gel extracted/purified using the QIAquick Gel Extraction kit as per manufacturers instructions is used for this particular control.

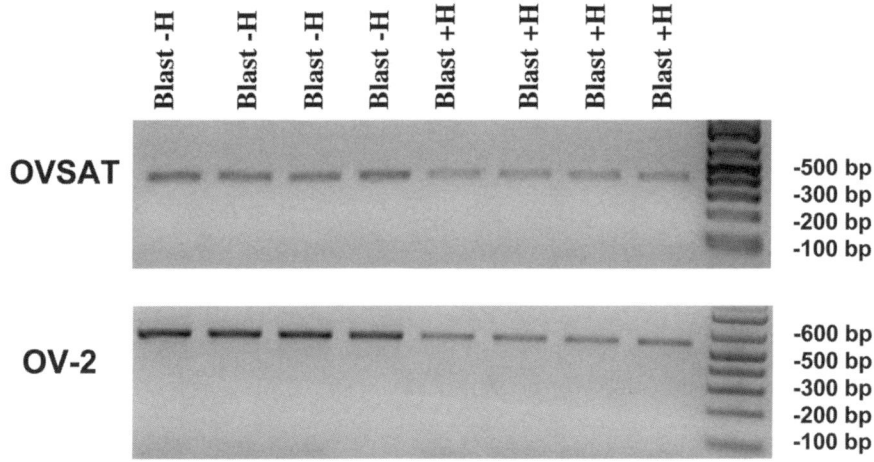

Fig. 3. PCR replicates for DNA derived from a single blastocyst using the Satellite I (OVSAT) or OV-2 SINE primers.

3. To ensure the *Hha*I enzyme does not digest methylated DNA, generate a pCR2.1-TOPO (Invitrogen) plasmid containing the Satellite I PCR product according to the manufacturers instructions. Incubate 1 µg of plasmid DNA with *Sss* I methylase for 1 to 2 h at 37°C in a reaction mixture containing 1X *S*-adenosyl methionine, 1X NEB Buffer 4, and 5 U *Sof ss*I methylase made up to 20 µL with dH₂O. Destroy *Sss*I methylase activity at 65°C for 20 min, before restriction enzyme digestion. Both the unmethylated and methylated controls are digested alongside the embryo/oocyte DNA and subsequently used as template in a PCR with the OVSAT1 primer pair (**Subheading 3.3.**).
4. Check the PCR products visually to confirm complete digestion of unmethylated controls and no digestion of methylated plasmid DNA.
5. Perform four PCR replicates routinely per digested/undigested single embryo DNA sample for each primer set. We have found the reproducibility of the MS-PCR method to be very high. **Figure 4** demonstrates the reproducibility for PCR replicates (performed at 19 cycles) for a single blastocyst undergoing methylation analysis of the Satellite I sequence. Data derived from single replicates are only removed when a clear difference was observed between quadruplicate values, for example, those attributable to pipetting errors, and data are only used if at least three replicates gave values ±10% (*see* **Note 4**).

3.7. Statistical Analysis

1. In our experience, the line of best fit for the standard curves described in **Subheading 3.5.** was found to be quadratic for all stages. The relevant quadratic equation was subsequently used to modify all % methylation values calculated from the MS-PCR experiments.

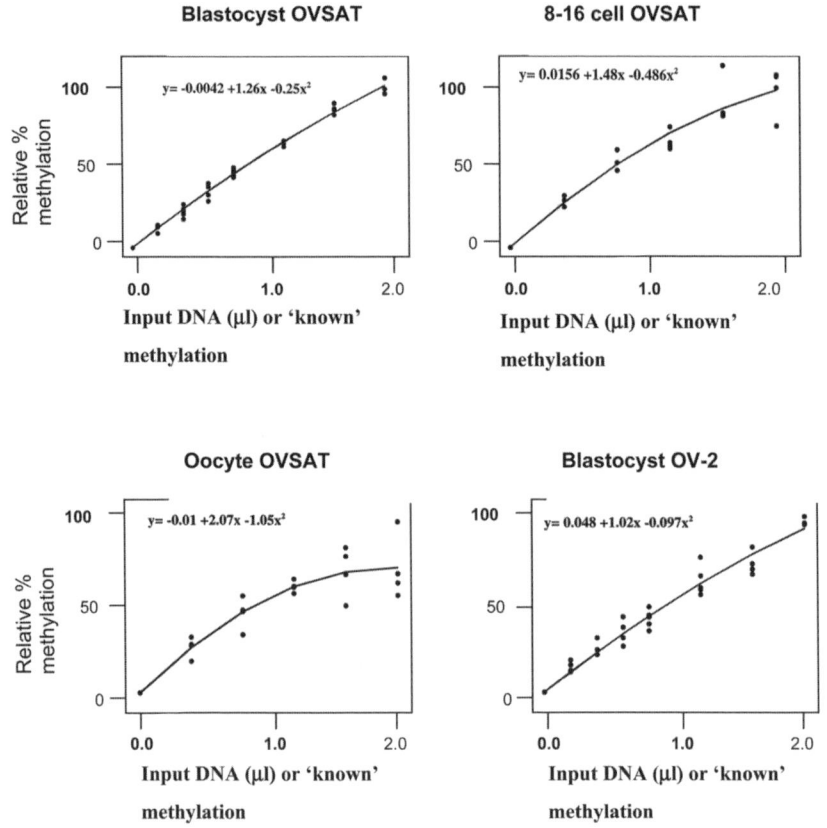

Fig. 4. Examples of standard curves for calculating the methylation status Satellite I (OVSAT) and OV-2 SINE (OV-2).

2. To assess whether the differences observed between different embryo treatment groups are statistically significant, perform one-way analysis of variance. Because the methylation data collected from individual embryos often do not demonstrate equal variance between the treatment groups, a square root transformation may first be required.

3. Pairwise comparisons of methylation level differences between treatment groups are analyzed using the Tukey simultaneous test (95% confidence intervals). *p* Values were calculated from t values generated from the Tukey tests, using a t-distribution table.

4. Notes

1. The use of MS-PCR with alternative ovine repeat sequences: the MS-PCR technique can be used successfully for repeat sequences other than ovine Satellite I. For example, we have analyzed the OV-2 SINE element, which unlike the Satel-

lite sequence is interspersed throughout the genome. The OV-2 SINE element sequence was isolated unintentionally in our laboratory during sheep lambda phage genomic library screening for an unrelated sequence. Sequencing of a 4 kb subclone isolated during this screen (using a CG-rich ovine *IGF2R* exon 1 probe) demonstrated the presence of a 556-bp sequence highly homologous to the previously reported bovine BOV-2 SINE element (accession no. X64125.1). The OV-2 sequence was restriction mapped using the "map" application of GCG10, identifying a single *Hha*I site for use in MS-PCR (nucleotides 177-180). The "pPrime" application of GCG10 was used to identify PCR primer positions flanking the *Hha*I site and the primers named OV-2F (GAAATTAAAAGATGCTGGCTCC) and OV-2R (AGTTCAGTCGCTCAGTCGTG). The MS-PCR assay is performed as described above for the Satellite I sequence, making sequence-specific standard curves (*see* **Fig. 2D**). For OV-2 SINE MS-PCR, sheep blastocyst DNA was amplified over the course of 24 cycles.

2. Selection of Satellite sequences used for methylation-sensitive PCR: the method chosen for analysis of the sheep Satellite I DNA sequence used the design of primers spanning a single *Hha*I site. The sheep Satellite I sequence (accession No. Z18540) was mapped for *Hha*I sites using the Map function on GCG10 and primers spanning a single *Hha*I site were designed using the Prime function of GCG10. The *Hha*I site used for methylation analysis was situated at nucleotides 480–483, and the positions of flanking MS-PCR primers were OVSAT1F, 329-348 and OVSAT1R, 717-697.

3. The use of the technique to analyze methylation levels in tissue or cultured cells: we also have measured DNA methylation levels in fetal and adult tissues and in cultured fetal fibroblasts by using the equivalent amount of DNA to a single blastocyst (1 ng) as starting material (results not shown). The methylation values obtained from direct comparison of PCR product intensities are standardized against the pooled blastocyst-generated standard curves. The mean of three values for each tissue sample is calculated to give a representative figure.

4. Potential limitations of technique associated with low copy number repeats: the method was found to be reproducible for Satellite I DNA methylation analysis for all embryo stages and oocytes, but not reproducible when analyzing the methylation status of the OV-2 SINE element at early embryo stages (oocyte and 8- to 16-cell; blastocyst results were reproducible and valid). This most likely reflects the very high copy number of satellite DNA within the genome. The BOV-2 SINE element is known to have approx 100,000 copies and is dispersed throughout the genome (*17*). Although similarly dispersed (occurring at an average interval of 20 kb in the sheep genome *[18]*), the OV-2 SINE has not been characterized in terms of copy number. However, our ability to amplify the OV-2 SINE element from small amounts of DNA at very low PCR cycle numbers (24 cycles when using 1/10th blastocyst DNA; data not shown) would indicate that copy number is relatively high. Generation of standard curves using OV-2 SINE primers for 8- to 16-cell stage embryos and oocytes failed to provide reproducible standard curves with simple linear or quadratic properties, suggesting the

OV-2 primers were inefficient when used to amplify template of decreased concentration. Therefore, only OV-2 MS-PCR data collected for the blastocyst stage were valid. The validity of the MS-PCR assay only in blastocysts for the SINE element suggests that we are approaching the limits of copy number sensitivity for single embryo analysis and demonstrates the need to optimize MS-PCR assays when using new primer sets. Whether this can be further improved by altering PCR conditions and/or reagents is the subject of current investigation in our laboratory

Acnowledgments

We would like to thank Jane Taylor and John Gardner for provision of embryos and David Waddington for statistical advice. This work was supported by BBRSC.

References

1. Dean, W., Santos, F., Stojkovic, M., Zakhartchenko, V., Walter, J., Wolf, E., et al. (2001) Conservation of methylation reprogramming in mammalian development: aberrant reprogramming in cloned embryos. *Proc. Natl. Acad. Sci. USA* **98,** 13734–13738.

2. Kang, Y. K., Koo, D. B., Park, J. S., Choi, Y. H., Chung, A. S., Lee, K. K., et al. (2001) Aberrant methylation of donor genome in cloned bovine embryos. *Nat. Genet.* **28,** 173–177.

3. Beaujean, N., Taylor, J., Gardner, J., Wilmut, I., Meehan, R., and Young, L. (2004) Effect of limited DNA methylation reprogramming in the normal sheep embryo on somatic cell nuclear transfer. *Biol. Reprod.* **71,** 158–193.

4. Taylor, J., Fairburn, H., Beaujean, N., Meehan, R., and Young, L. (2003) Gene expression in the developing embryo and fetus. *Reprod. Suppl.* **61,** 151–65.

5. Geuns, E., De Rycke, M., Van Steirteghem, A., and Liebaers, I. (2003) Methylation imprints of the imprint control region of the SNRPN-gene in human gametes and preimplantation embryos. *Hum. Mol. Genet.* **12,** 22,2873–22,2879.

6. Frommer, M., McDonald, L. E., Millar, S., Collis, C. M., Watt, F., Grigg, G. W., Molloy, P. L., et al. (1992) A genomic sequencing protocol that yields a positive display of 5-methylcytosine residues in individual DNA strands. *Proc. Natl. Acad. Sci. USA* **89,** 1827–1831.

7. Olek,, A., Oswald, J., and Walter, J. (1996) A modified and improved method for bisulphite based cytosine methylation analysis. *Nucleic Acids Res.* **24,** 24,5064–24,5066.

8. Oswald, J., Engemann, S., Lane, N., Mayer, W., Olek, A., Fundele, R., et al. (2000) Active demethylation of the paternal genome in the mouse zygote. *Curr. Biol.* **10,** 475–478.

9. Lane, N., Dean, W., Erhardt ,S., Hajkova, P., Surani, A., Walter, J., and Reik, W. (2003) Resistance of IAPs to methylation reprogramming may provide a mechanism for epigenetic inheritance in the mouse. *Genesis* **35,** 2, 88–2, 93.

10. Kang, Y. K., Koo, D. B., Park, J. S., Choi, Y. H., Kim, H. N., Chang, W. K., et al. (2001) Typical demethylation events in cloned pig embryos. clues on species-specific differences in epigenetic reprogramming of a cloned donor genome. *J. Biol. Chem.* **276**, 39980–39984.

11. Kang, Y.K, Koo, D. B., Park, J. S., Choi, Y. H., Lee, K. K., and Han, Y. M. (2001) Influence of oocyte nuclei on demethylation of donor genome in cloned bovine embryos. *FEBS Lett.* **499**, 55–58.

12. Rother, K. I., Silke, J., Georgiev, O., Schaffner, W., and Matsuo, K. (1995) Influence of DNA sequence and methylation status on bisulfite conversion of cytosine residues. *Anal. Biochem.* **231**, 263–265.

13. Warnecke, P. M., Stirzaker, C., Melki, J. R., Millar, D. S., Paul, C. L., and Clark, S. J. (1997) Detection and measurement of PCR bias in quantitative methylation analysis of bisulphite-treated DNA. *Nucleic Acids Res.* **25**, 21,4422–21,4226.

14. Kurnit, D. M., Brown, F. L., and Maio, J. J. (1978) Mammalian repetitive DNA sequences in a stable Robertsonian system. Characterization, in situ hybridizations, and cross-species hybridizations of repetitive DNAs in calf, sheep, and goat chromosomes. *Cytogenet. Cell. Genet.* **21**, 145–167.

15. Novak, U. (1984) Structure and properties of a highly repetitive DNA sequence in sheep. Nucleic Acids Res. 12, 2343–2350.

16. Burkin, D. J., Broad, T. E., and Jones, C. (1996) The chromosomal distribution and organization of sheep satellite I and II centromeric DNA using characterized sheep-hamster somatic cell hybrids. *Chromosome Res.* **4**, 49–55.

17. Lenstra, J.A,., van Boxtel, J. A., Zwaagstra, K. A., and Schwerin, M. (1993) Short interspersed nuclear element (SINE) sequences of the Bovidae. *Anim. Genet.* **24**, 33–39.

18. Buchanan, F. C., Littlejohn, R. P., Galloway, S. M., and Crawford, A. M. (1993) Microsatellites and associated repetitive elements in the sheep genome. *Mamm. Genome.* **4**, 5,258–5,264.

18

Analysis of DNA Methylation Profiles in Preimplantation Embryos Using Bisulfite Mutagenesis

Yong-Mahn Han, Seok-Ho Kim, and Yong-Kook Kang

Summary

For developmental competence of mammalian embryos, dynamic epigenetic changes to both the maternally and paternally derived contributions to the genome of the zygote should be brought about in the early cleavage stages. DNA methylation is a typical epigenetic mark modified during pre-implantation development. Here, we describe how to analyze DNA methylation profiles in early-stage embryos.

Key Words: Bisulfite mutagenesis; DNA methylation; nuclear transfer; pre-implantation.

1. Introduction

It has been known for several years that DNA methylation plays an important role in the regulation of gene expression in mammals. The DNA demethylation process is a unique event occurring during pre-implantation development. Aberrant methylation states were observed at various genomic loci in cloned bovine embryos, suggesting that developmental failures of cloned embryos may be caused by incomplete epigenetic reprogramming of donor genomic DNA *(1)*. This was the first report showing that methylation profiles of early stage embryos, or other situations in which amounts of DNA were limiting, could be analyzed by the bisulfite mutagenesis method.

A variety of different methods can be used to determine DNA methylation states *(2,3)*. The bisulfite conversion polymerase chain reaction (PCR) method is a powerful technique for investigating methylation states of genomic DNA in various species *(1,4,5)*. It has been used to detect 5-methylcytosine (5-MeC) and the status of methylation on CpG dinucleotides in DNA sequences. When DNA sequences are exposed to bisulfite, unmethylated cytosine is converted

From: *Methods in Molecular Biology, vol. 325: Nuclear Reprogramming: Methods and Protocols*
Edited by: S. Pells © Humana Press Inc., Totowa, NJ

to uracil, but 5-MeC is nonreactive *(6)*. Therefore, after PCR amplification, the unmethylated CpG dinucleotides are amplified as CpA and the methylated CpG dinucleotides as CpG. Thus, using the principle of bisulfite mutagenesis, we can determine whether genomic DNA sequences of a target region are methylated or not. This chapter describes the use of bisulfite mutagenesis to analyze the DNA methylation status of pre-implantation embryos, especially cloned embryos. Bovine satellite I DNA fragment (211 bp), which contains 12 CpG dinucleotides and 2 *Aci*I recognition sites (5'-ccgc-3') is exemplified as the target sequence *(1)*.

2. Materials

1. Sodium bisulfite (Sigma).
2. pGem T-easy vector system (Promega).
3. PCR premix (Bioneer, Korea).
4. Restriction enzymes (Boehringer Mannheim).
5. pSP70 Vector (Promega).
6. Embryo lysis buffer (ELB): 20 mM Tris-HCl, pH 8.0; 0.9% Tween-20; 0.9% Nonidet P40; 0.4 mg/mL proteinase K. Store at −70°C.
7. Proteinase K (Boehringer Mannheim).
8. TE: 10 mM Tris-HCl, pH 8.0; 0.1 M ethylenediaminetetraacetic acid.
9. Ampicillin (Sigma).
10. Isopropyl-β-D-thio-galactopyranoside (IPTG).
11. Phenol/chloroform (Sigma).
12. Extraction buffer: 10 mM Tris-HCl, pH 8.0, 0.1 M ethylenediaminetetraacetic acid, pH 8.0, 20 µg/mL pancreatic RNase, 0.5% sodium dodecyl sulfate.
13. Transfer ribonucleic acid (tRNA; carrier) or glycogen.
14. Hydroquinone (Sigma).
15. Sodium hydroxide (Sigma).
16. PCR purification kit (Qiagen).
17. Plasmid mini prep kit (Intron, Korea).
18. DNA Thermal Cycler 480 (Perkin Elmer).
19. α-[^{32}P] dCTP.
20. Tina 20 Image Analyzer (Fuji film, BAS-1500).
21. *Aci* I restriction enzyme (Promega).
22. Wizard DNA purification kit (Promega).
23. 5% Polyacrylamide gel.

3. Methods

The methylation status of CpG dinucleotides is distinguishable from cognate sequences by sequencing PCR products amplified from bisulfite-treated genomic DNA (*see* **Fig. 1**). The methylation state can also be determined by digesting the PCR products with restriction enzymes, which recognize only unconverted CpG dinucleotides with specific flanking-sequences (*see* **Fig. 2**).

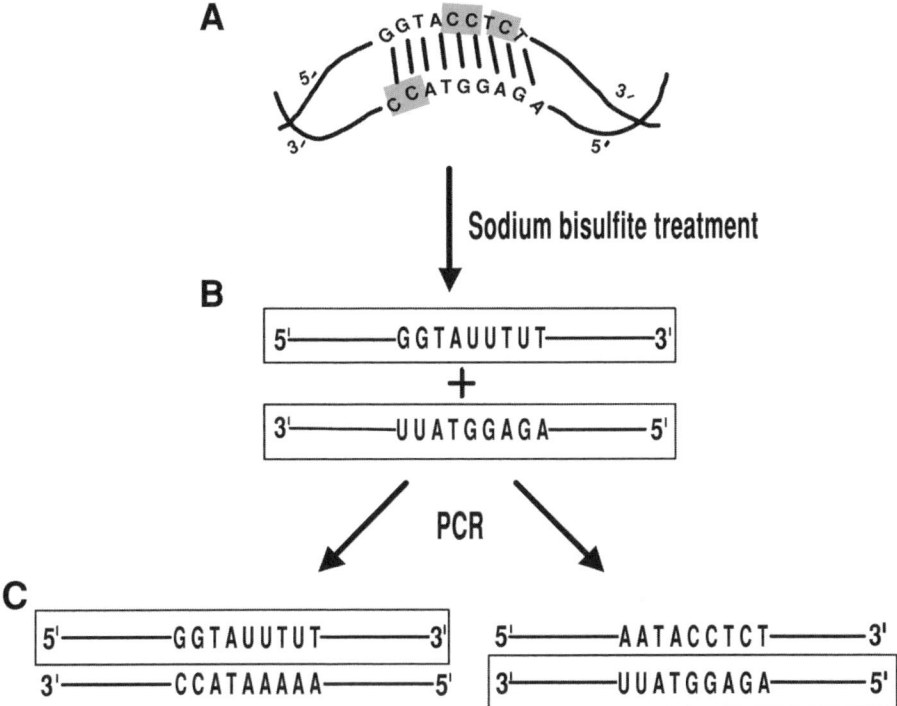

Fig. 1. Principle of bisulfite conversion PCR. (A) Partial sequence of the bovine satellite I region. (B) DNA sequence converted by sodium bisulfite treatment. (C) DNA sequence amplified from bisulfite-treated DNA sequence by PCR.

Here, two different methods are described in detail: (1) bisulfite conversion PCR sequencing, and (2) bisulfite conversion PCR digestion.

3.1. Bisulfite Conversion PCR Sequencing

The methods described below outline (1) designing primers, (2) isolation of genomic DNA from pre-implantation embryos, (3) bisulfite treatment, (4) PCR amplification, cloning and sequencing, and (5) identification of the DNA methylation state.

3.1.1. Designing Primers

The key element of the bisulfite conversion PCR method is to design accurate primers (*see* **Note 1**). The primer sequences are complementary to the oligonucleotide sequences converted by bisulfite treatment (*see* **Fig. 3**). Test the designed primers first using genomic DNA derived from tissue or cultured cells (*see* **Note 2**).

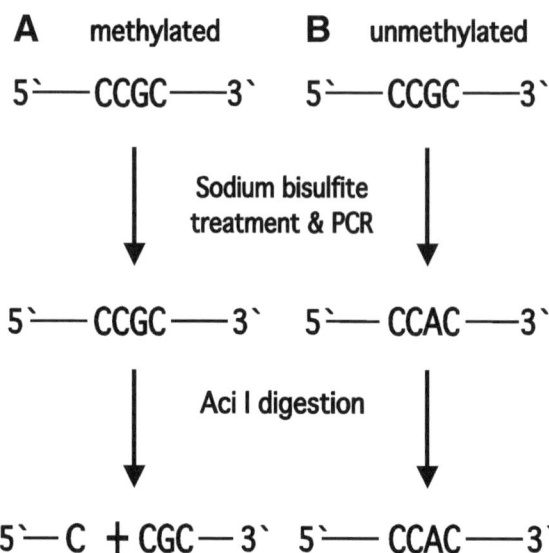

Fig. 2. Analysis of methylation states using methylation-sensitive restriction enzyme. (**A**) DNA sequence with methylated CpG dinucleotides is resistant to sodium bisulfite treatment and is then digested by the *Aci*I restriction enzyme. (**B**) Unmethylated CpG dinucleotides are converted by bisulfite treatment and the resulting PCR product is undigested by a restriction enzyme specific for the sequence.

```
                CpG1 CpG2                                                                    CpG3
5`GGTACCTCTGATTTCAGACTCCGATCGCAGGGTCCCTGCAGACTGGGGACAGGAGAGTCAGGCCTCGTCT
5`AATACCTCTAATTTCAAACTCCAATCACAAAATCCCTACAAACTAAAAACAAAAAAATCAAACCTCATCT
                CpG4          CpG5            CpG6                CpG7  CpG8
TGGGTTGAGGCATGGAACTCCGCTTGCCTCTCGAGATGTCCCCGGGGAGAGAGGCCGCTTGTCGAGCTGT
TAAATTAAAACATAAAACTCCACTTACCTCTCAAAATATCCCCAAAAAAAAAAAACCACTTATCAAACTAT
                CpG9 CpG10    CpG11                          CpG12
ATTTGGAACCTGGGGTTTTTTCCGAACGGTGCACGGAGAAGCTGCCCCTTCGTGTTGACTGCATTCACAG3`
ATTTAAAACCTAAAATTTTTTCCAAACAATACACAAAAAAACTACCCCTTCATATTAACTACATTCACAA3`
```

Fig. 3. Nucleotide sequences for a 211-bp bovine satellite DNA fragment (upper strands) and its bisulfite-converted version (lower strands). Underlined primer sequences are used for PCR amplification. These are 5'-aatacctctaatttcaaact-3' and 5'-tttgtgaatgtagttaata-3'.

3.1.2. Isolation of Genomic DNA From Preimplantation Embryos

It is important to determine how many embryos are necessary for a bisulfite conversion PCR experiment (*see* **Note 3**).

1. Place 10 blastocysts into 5 µL of ELB in a 500-µL tube and store at –20°C until use.
2. Add 50 µL of 2X extraction buffer to the tube containing 10 blastocysts.
3. Add 5 µL of proteinase K (20 mg/mL) solution and 45 µL of TE buffer to the tube.

4. Overlay mineral oil on the lysed sample.
5. Incubate the lysed sample in a water bath at 55°C overnight.
6. Remove the mineral oil from the lysed sample.
7. Add an equal volume of phenol/chloroform (1/1) to the lysed sample and gently mix by slowly shaking for 20 min.
8. Separate the two phases by centrifugation at 16,000g in a microcentrifuge for 5 min.
9. Transfer the viscous aqueous phase to a new tube.
10. Add an equal volume of chloroform to the same tube.
11. Gently mix the two phases by slowly shaking for 20 min.
12. Separate the two phases by centrifugation at 16,000g in a microcentrifuge for 20 min.
13. Transfer the upper aqueous phase to a new tube.
14. Add tRNA or glycogen (10–20 µg), 10 M ammonium acetate sufficient to give a final concentration of 3 M, and 3 vol of 100% ethanol to the tube.
15. Keep at –70°C for 30 min.
16. Precipitate genomic DNA by centrifugation at 16,000g a microcentrifuge at 4°C for 30 min.
17. Air-dry the DNA pellet and then dissolve the genomic DNA in 10 µL of distilled water.

3.1.3. Bisulfite Treatment

Bisulfite treatment is described in **Subheadings 3.1.3.1.** to **3.1.3.4.** This procedure includes (1) restriction enzyme digestion, (2) alkaline denaturation, (3) deamination, and (4) desulfonation.

3.1.3.1. RESTRICTION ENZYME DIGESTION

Genomic DNA isolated from pre-implantation embryos containing pSP70 vector (*see* **Note 4**) is digested with restriction enzyme *Bam*H I at 37°C overnight. Prepare the reaction mixture (20 µL) as follows:

Genomic DNA from the embryos	10 µL
*Bam*HI (restriction enzyme)	1 µL (10 units)
Enzyme buffer (10×)	2 µL
pSP70 vector (100–200 ng/ µL)	1 µL
DW (distilled water)	6 µL

3.1.3.2. ALKALINE DENATURATION

Bisulfite deaminates cytosine residues exclusively in single-stranded DNA. To denature the genomic DNA, add 2.2 µL of 3 M NaOH to the enzyme-digested DNA sample (20 µL) and then incubate at 37°C for 30 min.

3.1.3.3. DEAMINATION

1. Dissolve 4.2 g of sodium bisulfite in 10 mL of distilled water and adjust to pH 5.0.
2. Add 232 µL of the bisulfite solution and 13.5 µL of 10 mM hydroquinone to the denatured DNA sample.

3. Overlay the sample with 100 μL of mineral oil and incubate the reaction for 15 cycles of 94°C for 5 min followed by 55°C for 55 min on the DNA Thermal Cycler.

3.1.3.4. DESULFONATION

1. Purify the bisulfite-treated genomic DNA using the PCR product purification kit (Quiagen) according to the manufacturer's protocol and elute with 40 μL of hot distilled water (90°C).
2. Add 4.4 μL of 3 M NaOH solution to the DNA sample and incubate at 37°C for 30 min.
3. Add 12 μL of 10 M ammonium acetate, 150 μL of 100% ethanol, and 1 mg of carrier tRNA to the DNA sample and place at –70°C for 30 min.
4. Centrifuge the sample to recover the precipitate at 16,000g a micro-centrifuge at 4°C for 30 min.
5. Wash the DNA pellet with 500 μL of 70% ethanol, air-dry and dissolve in 20 μL of distilled water.

3.1.4. PCR Amplification, Cloning, and Sequencing

PCR amplification is performed in 20 μL of the reaction mixture which consists of 1X supplied incubation buffer (10 mM Tris-HCl, pH 9.0; 50 mM KCl; and 1.5 mM $MgCl_2$), 200 μM of each of four deoxynucleotides, 10 pmol of each primer, and 1.0 unit of Taq DNA polymerase. The bisulfite-treated genomic DNA is amplified using a PCR premix.

1. Amplify satellite I sequences from bisulfite-treated embryonic DNA using 40 cycles of 94°C for 1 min, 46°C for 1 min, and 72°C for 20 s using the specific primers on a DNA Thermal Cycler.
2. Clone the PCR products into the pGEM T-easy vector (Promega) according to manufacturer's instruction and sequence using ABI PRISM-377.
3. Sequence PCR products using commercial sequencing service.

3.1.5. Identification of Methylation State

To calculate the conversion rate, align the sequences of the resulting clones amplified from bisulfite-treated genomic DNA and count the converted cytosines (see **Fig. 4**).

3.2. Bisulfite Conversion PCR Digestion

The bisulfite conversion PCR digestion method includes (1) designing primers, (2) isolation of embryonic genomic DNA, (3) bisulfite treatment, (4) PCR amplification and restriction enzyme digestion, and (5) calculating band intensity. The procedures for designing primers, isolation of embryonic genomic DNA, and bisulfite treatment are the same as described in **Subheadings 3.1.2.** to **3.1.3.**

A

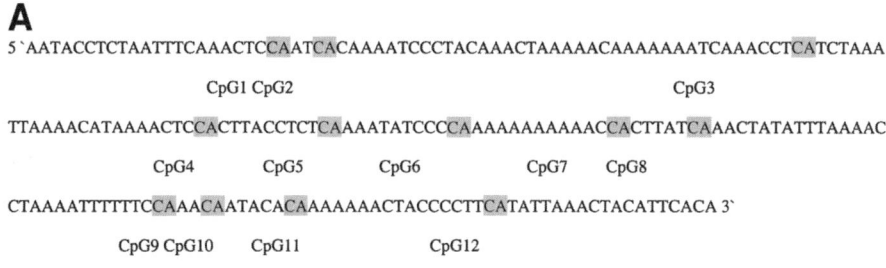

B

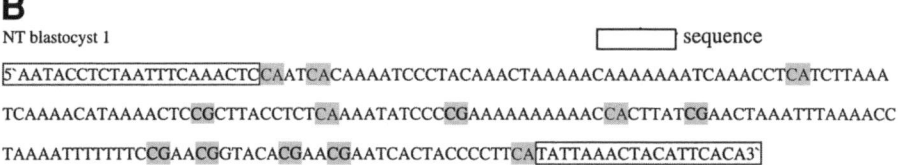

C

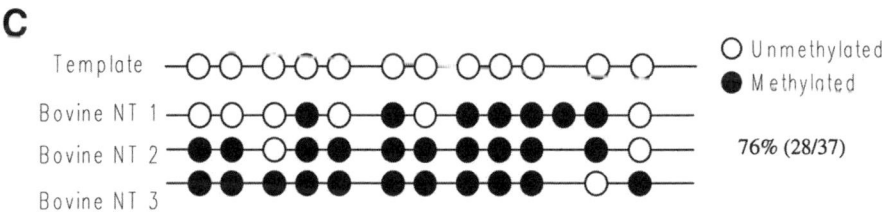

Fig. 4. Determining methylation states by the bisulfite conversion PCR sequencing method. (**A**) Converted sequences of CpG dinucleotides in the bovine satellite DNA fragment. (**B**) Sequencing data of PCR products amplified from bisulfite-treated genomic DNA of single cloned (NT) blastocysts. (**C**) Array of methylation states counted from sequencing data. "Template" indicates that the bovine satellite DNA fragment contains 12 CpG dinucleotides (open circle). A methylated CpG dinucleotide is indicated by a closed circle.

3.2.1. PCR Amplification and Restriction Enzyme Digestion

Primary PCR is described in **Suhbeading 3.1.4.**

1. Label the PCR product by reamplifing one tenth of the primary PCR product for 25 cycles under the same conditions (*see* **Subheading 3.1.4.**) in the presence of 2 μCi α-[^{32}P]dCTP in a 50-μL reaction.
2. Purify the radiolabeled PCR products using the wizard DNA purification kit.
3. Digest approx 100 ng of purified PCR product with the methylation-sensitive restriction enzyme (e.g., *Aci*I, *Taq*I, or *Acc*II) overnight at the temperature recommended by the manufacturer.
4. Resolve the fragments by 5% polyacrylamide gel electrophoresis, dry the gel, and expose to autoradiographic film at −70°C for a week.

3.2.2. Calculating Band Intensity

Band intensity or methylation percentage is calculated using an image analyzer Tina 20. An example of bisulfite conversion PCR digestion is shown in **Fig. 5**. The satellite sequence of blastocysts derived via in vitro fertilization was considerably undermethylated (9%), whereas cloned embryos were shown to be highly methylated at this sequence (65%), which was similar to the nuclear donor cells (72%) *(1)*.

4. Notes

1. It is very important to design an optimal primer set for the bisulfite conversion PCR experiments. If an inaccurate primer set is used, target sequences are not amplified by PCR. CpG dinucleotide sequences within the primer should be avoided because 5-MeC is tolerant to bisulfite treatment. If CpG dinucleotides are unavoidable within the primer, we recommend using the IUR code (R for 5' primer or Y for 3' primer) instead of a G nucleotide. The optimal size of DNA fragment amplified by bisulfite conversion PCR is less than 300 bp because longer DNA sequences are prone to cleavage by sodium bisulfite itself. If a large quantity of PCR product from genomic DNA of pre-implantation embryos is needed, nested or semi-nested PCR is recommended.
2. Before analyzing the methylation state of early stage embryos, the designed primer set should be tested using genomic DNAs purified from tissue or cultured cells which provide a large amount of genomic DNA. Approximately 250 ng of genomic DNA is usually used as a template to test for bisulfite treatment and PCR amplification.
3. The number of embryos needed for PCR amplification may depend on the embryonic developmental stages and the target sequences. Methylation can be detected in genomic DNA derived from as few as approx 100 cells (one blastocyst). Moreover, methylation profiles can be identified from genomic DNAs of single bovine blastocysts when primers specific for the repetitive sequence Bovine Satellite I are used *(1)*. It is easier to amplify repetitive sequences as opposed to single-copy gene sequences.

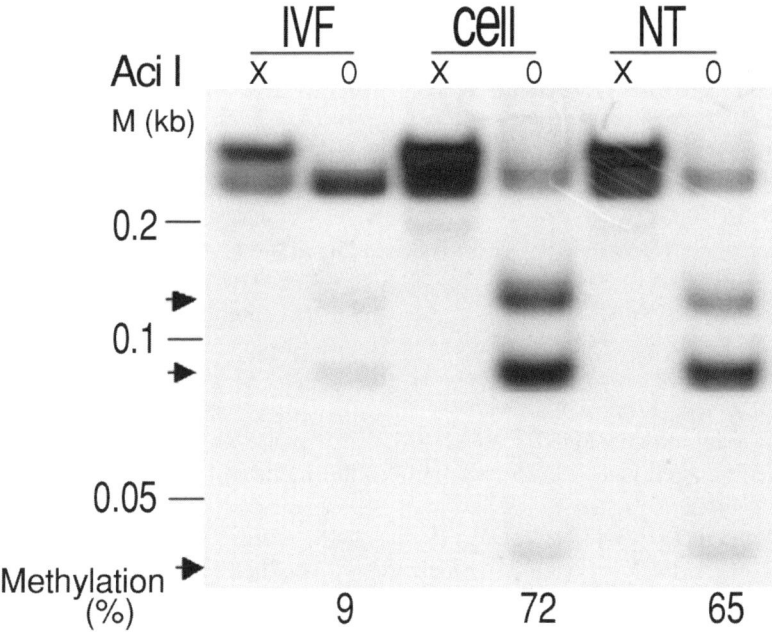

Fig. 5. An example of bisulfite conversion PCR digestion. To determine the methylation ratio, radiolabeled PCR products were amplified from bisulfite-treated genomic DNAs and digested with the *Aci*I restriction enzyme, which recognizes only unconverted 5'-CCGC-3' sequence. Complete digestion of two target sequences (CpG-4 and CpG-7) yields 35-bp, 86-bp, and 90-bp fragments. X is the PCR product without enzyme digestion and O is enzyme-digested PCR product. Arrows indicate enzyme-digested DNA fragments. Band intensity or methylation percentage is calculated by the proportion of summed band intensity of digested fragments to that of the whole fragments. DNAs were prepared from IVF-derived embryos (IVF), donor fetal fibroblasts (Cell), and cloned embryos (NT), respectively.

4. Very small amounts of genomic DNA isolated from pre-implantation embryos (~10 blastocysts) can lead to limitation of restriction enzyme digestion. Addition of pSP70 plasmid vector, which lacks the *Bam*HI recognition sequence, to the reaction will enhance the efficiency of enzyme digestion and reduce star activity *(7)*.

Acknowledgments

We thank Dr. K. Yu for critical reading. This work was supported by a grant from The Stem Cell Research Center of the 21st Century Frontier Research Program (SC2090) and a grant of the National Research Laboratory Program (NLM0050111) funded by the Ministry of Science and Technology, Republic of Korea.

References

1. Kang, Y. K., Koo, D. B., Park, J. S., Choi, Y. H., Chung, A. S., Lee, K. K., et al (2001) Aberrant methylation of donor genome in cloned bovine embryos. *Nat. Genet.* **28,** 173–177.
2. Millar, D. S., Warnecke, P. M., Melki, J. R., and Clark, S. J. (2002) Methylation sequencing from limiting DNA: embryonic, fixed, and microdissected cells. *Methods* **27,** 108–113.
3. Santos, F., Hendrich, B., Reik, W., and Dean, W. (2002) Dynamic reprogramming of DNA methylation in the early mouse embryo. *Dev. Biol.* **241,** 172–182.
4. Kang, Y. K., Koo, D. B., Park, J. S., Choi, Y. H., Kim, H. N., Chang, W. K., et al. (2001) Typical demethylation events in cloned pig embryos. Clues on species-specific differences in epigenetic reprogramming of a cloned donor genome. *J. Biol. Chem.* **276,** 39,980–39,984.
5. Oswald, J., Engemann, S., Lane, N., Mayer, W., Olek, A., Fundele, R., et al. (2000) Active demethylation of the paternal genome in the mouse zygote. *Curr. Biol.* **10,** 475–478.
6. Frommer, M., McDonald, L. E., Millar, D. S., Collis, C. M., Watt, F., Grigg, G. W., et al. (1992) A genomic sequencing protocol that yields a positive display of 5-methylcytosine residues in individual DNA strands. *Proc. Natl. Acad. Sci. USA* **89,** 1827–1831.
7. Raizis, A. M., Schmitt, F., and Jost, J. (1995) A bisulfite method of 5-methylcytosine mapping that minimizes template degradation. *Anal. Biochem.* **226,** 161–166.

19

Chromatin Immunoprecipitation Assay
for Mammalian Tissues

Fiona B. Turner, Wang L. Cheung, and Peter Cheung

Summary

In this postgenome era, understanding how a cell regulates access to information encoded in the deoxyribonucleic acid (DNA) is essential. In eukaryotic cells, DNA is bound to histone proteins to form chromatin fibers. Numerous studies have now shown that post-translational histone modifications play an important role in regulating the access of DNA-dependent proteins to the DNA template. Determining the status of histone modifications in a genomic region has proven to yield information on the chromatin structure and the regulation of a specific gene in vivo. Chromatin immunoprecipitation (ChIP) allows researchers to determine the status of both histone modifications and the nuclear effector proteins located at gene of interest. ChIP, if applied globally, can also reveal how chromatin structures are dynamically changed when cells respond to certain stimuli. In this chapter, we describe this powerful technique in detail.

Key Words: Chromatin immunoprecipitation; chromatin IP; ChIP; in vivo formaldehyde crosslinking; histone modification; protein–DNA interaction; transcription; semiquantitative PCR.

1. Introduction

Genomic DNA, the master template for replication and transcription, is at any given time physically associated with a wide variety of nuclear proteins. First, DNA is tightly wrapped around histone proteins to form chromatin, the structural format that functions to package and condense the genome through higher-order folding *(1)*. In addition, numerous DNA binding factors bind to specific regions of the genome to execute a variety of nuclear functions. Because of the intimate interactions between histones and DNA, it is now clear that chromatin structure has important regulatory roles by restricting or facilitating access of the effector DNA binding proteins to the genomic template

From: *Methods in Molecular Biology, vol. 325: Nuclear Reprogramming: Methods and Protocols*
Edited by: S. Pells © Humana Press Inc., Totowa, NJ

(2,3). Moreover, posttranslational modifications that occur on histones not only alter the contacts between histones and DNA, but recent evidence indicates that some of these modifications also function to recruit and mediate binding of nuclear proteins to specific regions of the genome *(4–6)*. Therefore, histones and histone modifications are now recognized to play a fundamental role in determining the functional organization of the genome of eukaryotic cells *(7)*.

One of the recent technical advances that greatly facilitated the functional analysis of protein–DNA interactions is the chromatin immunoprecipitation (ChIP) assay. Initially developed to probe the genomic loci that are associated with specific post-translational histone modifications *(8)*, this technique now also is used routinely to assay the binding of nuclear factors to specific DNA regions. Two facets of this technique allow it to be a powerful tool for investigating protein–DNA interactions in vivo. First, the addition of formaldehyde to intact cells results in stable crosslinks between DNA and proteins interacting in the native chromatin environment. This crosslinking process is very rapid and therefore allows researchers to obtain an almost instantaneous snapshot of proteins bound to the DNA at the time of formaldehyde addition. Second, the specificities of the antibodies used allow one to selectively enrich for chromatin directly associated with the histone modifications or protein of interest. By extracting the DNA from the enriched chromatin, one can then quantitatively analyze the extent as well as the kinetics of the protein–DNA interactions. The analysis of the immunoprecipitated chromatin can be performed one gene at a time by the use of primers to a specific gene in question (as described in this chapter), or in a genome-wide scale through the use of microarray profiling *(9–11)*. This technique has been applied successfully to several diverse cell types, including yeast *(12,13)*, *Tetrahymena* (*14*), *Arabidopsis (15)*, *Drosophila (11)*, and mammalian species *(16,17)*, and its usefulness will continue as increasingly more sophisticated data gathering and analysis technologies are developed.

2. Materials

1. Sonicator with microtip.
2. Oligonucleotide primers and PCR reagents.
3. Formaldehyde, 37% stock.
4. End-over-end rotator.
5. Agarose gel equipment.
6. Sodium dodecyl sulfate (SDS)-polyacrylamide gel electrophoresis gel equipment.
7. Type B dounce homogenizer (for tissue).
8. Protein A-agarose beads.
9. Protease inhibitors: phenylmethylsulfonyl fluoride (PMSF), 100 mM in EtOH; aprotinin, 1 mg/mL in H_2O; pepstatin, 1 mg/mL in EtOH/acetic acid (95/5); leupeptin, 1 mg/mL in H_2O.

10. Phosphatase inhibitor (optional): microcystin LR, 5 mM in H$_2$O.
11. Antihistone modification-specific antibodies.
12. Waterbath set at 65°C.
13. 3 M NaOAc, pH 5.2.
14. EtOH.
15. Glycogen (molecular biology grade).
16. Phenol/chloroform (1/1), pH 6.8.
17. RNase A, 10 mg/mL.
18. Proteinase K, 20 mg/mL.

2.1. Solutions

1. Wash buffer I*†: 0.25% Triton X-100, 10 mM ethylenediaminetetraacetic acid (EDTA), pH 8.0, 0.5 mM ethylenebis(oxyethylenenitrilo)tetraacetic acid (EGTA), pH 7.5, 10 mM HEPES, pH 7.5, 1 mM PMSF, 1 µg/mL aprotinin, 1 µg/mL pepstatin, 1 µg/mL leupeptin, 1 µM microcystin.
2. Wash buffer II*†: 200 µM NaCl, 1 mM EDTA, pH 8.0, 0.5 mM EGTA, pH 7.5, 10 mM HEPES, pH 7.5, 1 mM PMSF, 1 µg/mL aprotinin, 1 µg/mL pepstatin, 1 µg/mL leupeptin, 1 µM microcystin.
3. Lysis buffer*†: 150 mM NaCl, 25 mM Tris-HCl, pH 7.5, 5 mM EDTA, pH 8.0, 1% Triton X-100, 0.1% SDS, 0.5% sodium deoxycholate, 1 mM PMSF, 1 µg/mL aprotinin, 1 µg/mL pepstatin, 1 µg/mL leupeptin, 1 µM microcystin.
4. RIPA buffer*†: 150 mM NaCl, 50 mM Tris-HCl, pH 8.0, 0.1% SDS, 0.5% sodium deoxycholate, 1% Nonidet-P40 (NP-40), 1 mM PMSF, 1 µg/mL aprotinin, 1 µg/mL pepstatin, 1 µg/mL leupeptin, 1 µM microcystin.
5. High-salt wash*†: 500 mM NaCl, 50 mM Tris-HCl, pH 8.0, 0.1% SDS, 1% NP-40, 1 mM PMSF, 1 µg/mL aprotinin, 1 µg/mL pepstatin, 1 µg/mL leupeptin, 1 µM microcystin.
6. LiCl wash*†: 250 mM LiCl, 50 mM Tris-HCl, pH 8.0, 0.5% sodium deoxycholate, 1% NP-40, 1 mM PMSF, 1 µg/mL aprotinin, 1 µg/mL pepstatin, 1 µg/mL leupeptin, 1 µM microcystin.
7. TE buffer†: 10 mM Tris-HCl, pH 7.5, 1 mM EDTA, pH 8.0.
8. Elution buffer*: 2% SDS, 10 mM dithiothreitol, 0.1 M NaHCO$_3$.
9. 5X Proteinase K buffer: 50 mM Tris-HCl, pH 7.5, 25 mM EDTA, pH 8.0, 1.25% SDS.

Note: *Make solutions fresh each time; †add protease inhibitors directly before use, store on ice.

3. Methods
3.1. Formaldehyde Crosslinking of Cells From Culture or Tissue

In this section, we will discuss the formaldehyde cross-linking step for both cells in culture and harvested tissues *(12,18)*.

3.1.1. Cell Culture

For tissue culture cells, determine the number of cells to harvest based on the goal of using approx 1 to 3×10^6 cell-equivalents of chromatin for each immunoprecipitation (IP). In theory, the more chromatin used per IP, the lower the background signal in the final PCR step. Controls include "no antibody" and "mock" samples (*see* **Note 1**).

1. Adherent cells grown on culture plates are washed once with prewarmed 1X PBS.
2. Cells are trypsinized and resuspended in growth medium. Use 10 mL of growth medium per 150-mm dish. Take a small aliquot for counting cells and pellet the rest by centrifugation (4°C, 450g, 8 min; *see* **Note 2**).
3. Prepare 1% formaldehyde growth medium mix by diluting 270 μL of 37% formaldehyde (HCHO) stock per 10 mL of growth medium.
4. Resuspend the cells in 1% HCHO growth medium mix. Aim to keep the density of cells constant in successive experiments (for example, 5×10^5 cells per milliliter of medium) to ensure equivalent exposure to HCHO.
5. Rotate the cell suspension for 8 min at 25°C with an end-over-end rotator.
6. For suspension cultures, count an aliquot of cells and add tissue culture medium to the flask to reach the desired density (for example, 5×10^5 cells/mL). Add a stir bar to the flask and place on magnetic stir plate. Add formaldehyde (HCHO – 37% stock) to 1% final concentration. Stir for 8 min at 25°C.

3.1.2. Tissue

1. Harvest tissue from mouse/rat. Using forceps, rinse the tissue briefly in PBS, and weigh.
2. Estimate the amount of tissue needed based on using approx 30 mg of tissue for each immunoprecipitation. Place the tissue on a clean glass plate and use a razor blade to dice the tissue into small pieces.
3. Transfer the tissue into a conical centrifuge tube and add 1 mL of tissue culture medium for every 15 mg of tissue (any standard medium with or without fetal bovine serum).
4. Add formaldehyde to 1% final concentration. Rotate tube on an end-over-end rotator at 25°C for 15 min.

3.2. Preparation of Soluble Chromatin

We detail the steps to obtain soluble chromatin from formaldehyde-fixed cells. This soluble lysate will be used for the subsequent immunoprecipitation.

3.2.1. Cell Culture

1. Pellet formaldehyde-fixed cells in a centrifuge (4°C, 450g, 8 min; re-pellet cells using same conditions between washes).
2. Aspirate medium and wash with ice-cold PBS (10 mL per sample).
3. Wash with ice-cold wash buffer I (10 mL per sample). Please see materials section for the content of all the buffers used in this section.

4. Wash with ice-cold wash buffer II (10 mL per sample).
5. Resuspend the cell pellet in lysis buffer (~1 × 10^7 cells/500 µL).
6. Use a 1/16-in microtip to sonicate the cell lysate on ice with 24 × 15-s bursts and a 30-s rest period between bursts (this sonication step should be optimized for different sonicators; *see* **Note 3**).
7. Centrifuge the sonicated lysate at 4°C, ≥16,000g, for 10 min to pellet cellular debris. Collect the supernatant (containing the soluble chromatin fraction), and divide samples into aliquots of approx 1 to 3 × 10^6 cells for each IP. This is an excellent place to stop and store the aliquots at –80°C by snap-freezing in liquid nitrogen.

3.2.2. Tissue

1. Centrifuge the sample at 450g for 15 min at 4°C. Aspirate the supernatant and wash with 10 mL of PBS containing protease inhibitors. Invert tissue to mix; do not pipet up and down because of tissue loss on the inside of pipets.
2. Pour the suspension of tissue in PBS into an ice-cold type B dounce homogenizer (loose-fitting ball-on-stick model) and homogenize the tissue on ice with 20 to 25 strokes. Pour the homogenate into a clean 15-mL tube. Wash the dounce with 5 mL of PBS containing protease inhibitors and add to the sample (*see* **Note 4**).
3. Centrifuge the sample at 450g for 10 min at 4°C and wash the pellet in 10 mL of ice-cold PBS.

Continue with the protocol as described for cell culture samples in **Subheading 3.2.1.**

3.3. Immunoprecipitation of Soluble Chromatin

1. Dilute the soluble chromatin sample to 200 µL with lysis buffer containing fresh protease inhibitors.
2. Prepare protein A beads (*see* **Note 5**): transfer an aliquot of beads in storage solution (20% EtOH) to a microfuge tube. Gently pulse down the beads (always spin down beads at low speed, 1500g or 4000 rpm in microfuge) and remove the storage solution from the packed beads. Wash three times with 500 µL of lysis buffer. After the last wash, estimate the volume of packed beads, and add an equal volume of lysis buffer to make a 50% slurry (*see* **Note 6**).
3. Preclear the IP sample by adding 20 µL of 50% protein A slurry to the 200 µL of chromatin sample, rotate (end-over-end) at 4°C for 1 h, then spin (1500g) for 5 min and transfer the supernatant to a new tube (discard the protein A beads).
4. Add antibody to precleared chromatin (*see* **Note 7**) and rotate (end-over-end) at 4°C overnight. Include a "no-antibody" tube containing chromatin, but no antibody.
5. The next day, prepare and wash more protein A beads (resuspend as 50% slurry in lysis buffer) to pull down the antibody/protein/DNA complex (*see* **Note 6**).
6. Add 30 µL of protein A slurry to each IP sample, and rotate at 4°C for 1 h.
7. Pellet protein A beads at 1500g (4000 rpm), 4°C, for 5 min. Collect the supernatant from the no antibody tube for "total chromatin." Otherwise, save the beads and discard the supernatant.

8. Wash the protein A beads by resuspending the pellet in 500 µL of RIPA buffer, rotate (end-over-end) at 4°C for 10 min, and spin to pellet as in **step 7**.

9. Wash the protein A pellet (as in **step 8**) in 500 µL of high-salt buffer.

10. Wash the pellet (as in **step 8**) in 500 µL of LiCl buffer.

11. Wash the pellet (as in **step 8**) twice in 500 µL of TE buffer.

12. Elute the antibody from the protein A by adding 200 µL of elution buffer to the pellet. Rotate (end-over-end) for 15 min at room temp, centrifuge at 1500*g*, and collect the supernatant. Repeat with another 200 µL of elution buffer and pool both elution fractions.

13. Add 20 µL of 4 *M* NaCl to the 400 µL of eluate and incubate at 65°C for 4 to 6 h to reverse the formaldehyde crosslinking.

14. Add 1 mL 95% EtOH and precipitate overnight at –20°C.

15. Pellet the precipitate by centrifugation ($\geq$16,000*g* for 10 min at 4°C; one should see the pellet—mostly salt), wash once in 70% EtOH, and dry the pellet in a speedvac or on the bench. Resuspend the pellet in 180 µL of TE (the pellet should go into solution easily), add 1 µL of 10 mg/mL RNase A, and incubate at 37°C for 30 min.

16. Add 20 µL of 5X proteinase K digestion buffer, 1 µL of 20 mg/mL proteinase K, and incubate at 42°C for 1 h.

17. Add an equal volume of phenol/chloroform (1/1), pH 6.8. Vortex and spin at $\geq$16,000*g* for 5 min at 25°C. Recover the aqueous phase and precipitate the DNA with 1/10th vol of 3 *M* NaOAc, pH 5.2; 2X volume of ice-cold 95% ethanol; and 1 µL of glycogen (20 µg) as carrier.

18. Pellet the precipitated DNA by centrifugation, resuspend in 20 µL of dH$_2$O, and use 1 µL of DNA per PCR.

3.4. Analysis of the Immunoprecipitated DNA

The most common way of analyzing the immunoprecipitated DNA is by PCR. Primers designed to a gene of interest are applied to the immunoprecipitated DNA and to the total input DNA (*see* **Note 1** and this subheading). The following technique describes a semiquantitative PCR method. The key to this semiquantitative PCR is to make sure the amplification reactions do not reach saturation so that the product from the immunoprecipitated DNA may be compared to the product from total input DNA at the same cycle number. Alternatively, one may use real-time PCR to achieve quantitative results.

1. Before using the immunoprecipitated DNA samples in the PCRs, optimize the PCR conditions for each primer set using the "total input DNA" obtained from the no antibody sample as template.

2. For each primer set, determine the number of PCR cycles that amplifies a product within the linear range. When the products are viewed on an agarose gel, choose the cycle number at which the product has visibly increased with successive cycles, yet has not reached levels of saturation (*see* **Fig. 1**).

Primer set: 1   2   3

Cycle number: 26 28 30 32 33 36 26 28 30 32 33 36 26 28 30 32 33 36

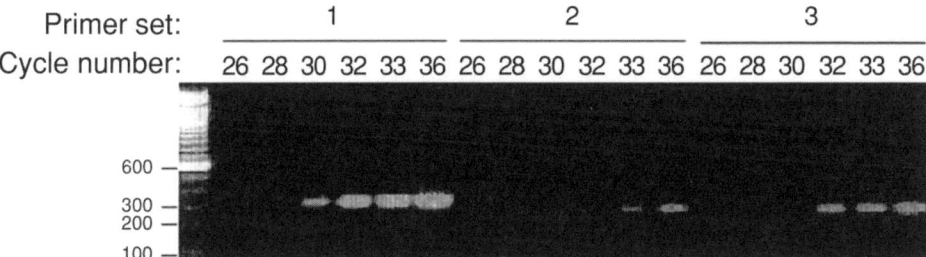

Fig. 1. Determination of cycle number within the linear range for primer pairs used in PCR-based ChIP analysis. Primers designed to three separate genomic regions were applied to 20 ng of sonicated total input DNA. The reactions were stopped at cycle numbers 26, 28, 30, 32, 33, and 36 and run on an ethidium-stained agarose gel. For each primer pair, the cycle number was chosen that produced a product that was visibly increasing with successive cycles and had not reached levels of saturation (i.e., Set 1: cycle 30, Set 2: cycle 36, Set 3: cycle 32).

3. Ideally, one should have a positive and a negative control for each antibody. For example, when using an antibody to a transcription factor that regulates constitutively active genes, one may amplify actin or histone H4 sequence from the immunoprecipitated material to serve as a positive control. Conversely, primers to the β-globin gene may serve as a negative control as β-globin has a limited pattern of expression (*see* **Note 8**). One must remember that because of the sensitivity of PCR, one may amplify a product in the negative control at a high cycle number. Therefore, it is important to choose a cycle number that produces a product in the linear range with the total DNA input, yet does not produce a significant product in the negative control.

4. Set up PCRs:

Example:

Tube No.	DNA	Primers
1	Total input	+ control
2	Total input	– control
3	Total input	Gene of interest
4	IP-ed DNA	+ control
5	IP-ed DNA	– control
6	IP-ed DNA	Gene of interest
7	No antibody	+ control
8	No antibody	– control
9	No antibody	Gene of interest

IP-ed, immunoprecipitated.

For a 50-µL reaction:

5 µL	10X PCR buffer
4 µL	25 mM MgCl$_2$
0.25 µL	20 mM dNTPs
1 µL each	20 µM primers
0.5 µL	Taq polymerase
38.25 µL	dH$_2$O
50 µL total volume	

5. Run PCR:
   a. First cycle: 4 min at 94°C, 4 min at 60°C, 1 min at 72°C.
   b. Next n cycles*: 40 s at 94°C, 40 s at 60°C, 1 min at 72°C
   c. Last cycle: 10 min at 72°C.

   *Each primer set may require a different cycle number

6. After PCR, add 150 µL of dH$_2$O to 50 µL of PCR. Extract once by phenol–chloroform and precipitate the PCR-amplified DNA by adding 1/10 vol of 3 M NaOAc with glycogen as carrier and 2 vol of ice-cold 95% ethanol.
7. Pellet the precipitated DNA, resuspend in 6 µL of 1X DNA gel loading buffer and run on a 15% polyacrylamide gel in 0.5X TBE (one should prerun the gel for 30 min at 200 V before loading samples; see **Note 9**).
8. Quantify the relative signal intensities of the PCR bands from the different samples by using a digital camera and image analysis software. For each primer set, draw a box around the product from the total input DNA sample and record the intensity value. Then, use a box of the same dimensions to determine the intensity of the immunoprecipitated DNA product. In this manner, the ratio of immunoprecipitated DNA to total input DNA is determined. Any products in the "no antibody" negative control lanes also are quantified and provide the background level of amplification for each primer set.

4. Notes

1. In the "no antibody" control, the addition of antibody to sonicated chromatin is omitted. Protein A-agarose beads are added to the "no antibody" chromatin and pelleted. The eluant from the beads serves as a negative control in PCR (see also **Note 5** and **Subheading 3.4.**). The "no antibody" supernatant is collected for "total chromatin" which is used as the total input DNA in PCR. The "mock" sample does not contain chromatin or antibody, and serves as a control for buffer contamination.
2. Fixing cells in a tube (rather than on tissue culture dishes) minimizes the loss of cells caused by the crosslinking of cells to culture dishes. Also, harvesting cells first, instead of after HCHO fixation, is easier and more efficient.
3. The purpose of this sonication step is to shear the genomic DNA to small pieces. Ideally one wants the chromatin to be approx 500 to 1000 bp in length. Aliquot a small portion of the sonicated chromatin to determine the resulting fragment

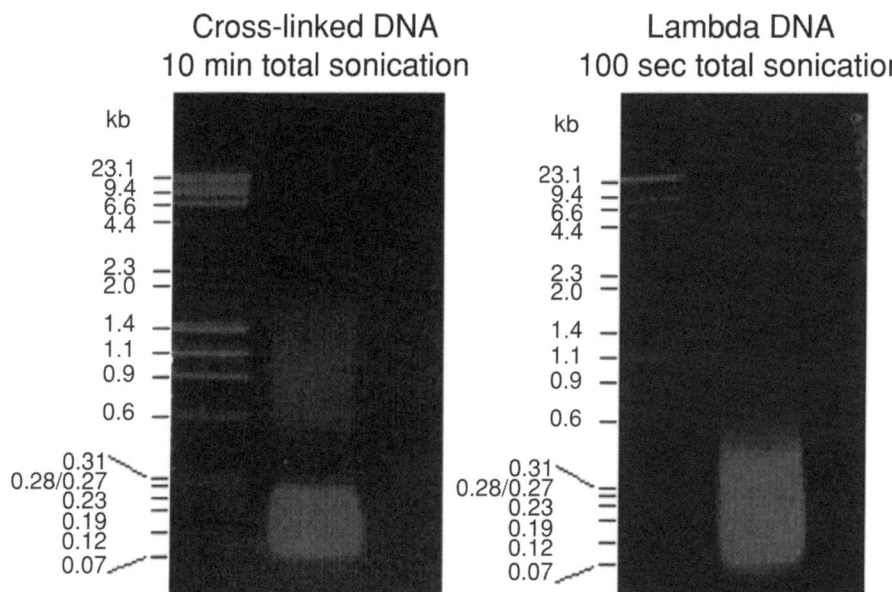

Fig. 2. Analysis of sonicated DNA fragment sizes on an ethidium-stained agarose gel. After sonication, an aliquot of DNA should be set aside to determine if the resulting DNA fragment size is approx 1 kb or less. The DNA aliquot is heat-denatured to reverse crosslinks, treated with RNase A and proteinase K, and run on an ethidium-stained agarose gel. Because formaldehyde crosslinking makes DNA resistant to sonication, the amount of sonication required to achieve fragments of ≤1 kb depends on the extent of crosslinking. After 10 min of total sonication time, crosslinked mouse genomic DNA is sheared to approx 100- to 300-bp fragment size, with a minor proportion of higher molecular weight DNA of approx 1.5 kb. Noncrosslinked lambda DNA is sheared to approx 500 bp or less with only 100 s of total sonication.

sizes. This sample must be heat-denatured to reverse the crosslinks and treated with RNase A and proteinase K before analysis on an agarose gel (*see* **Fig. 2**). *See* **Subheading 3.3.**, **steps 13–18** for more details. As an alternative to the settings suggested in the protocol, program the sonicator to sonicate 2 s on, 1 s off for 10 to 15 min of total time. In practice, sonicate as much as you can without aerosolizing or overheating the samples. Keep sonicator tip sufficiently submerged to avoid aerosolization of the sample but avoid touching the bottom of the tube. (This is tricky if one is holding the tube. *See* suggestion below). For a microtip, 400 μL is probably the smallest volume one may use while avoiding aerosolization. If aerosolization occurs, stop the sonication, spin down the sample, and attempt sonication again. If aerosolization persists, other remedies include decreasing the power setting and increasing sample volume. Microtips require lower power settings than larger sonicator tips. Check with the manufacturer to determine the maximum safe setting for your tip. As a rule, use the highest power setting possible for the tip that allows repeated pulses without aerosolizing the sample. A hint for keeping the sample still and cool during sonication: position the tube so that the sonicator tip is submerged in the liquid. Use a stand and clamp to hold the tube in place during sonication. Place an insulated cup underneath the tube and fill the cup with ice to prevent overheating of the sample. One may also perform the sonication step in a cold room (4°C).

4. If clumps of undispersed tissue remain in the homogenate, place one sheet of sterile gauze over the mouth of a 50-mL conical tube and depress the center of the gauze into the tube. Pour the homogenate through the gauze into the tube. Wash the dounce with 5 mL of PBS containing protease inhibitors and pour through the gauze to add to the sample.

5. One may also use protein G for most IgG antibodies.

6. If one finds background levels of PCR products in the negative controls, one may use blocked Protein A beads in the preclearing (**Subheading 3.3., step 3**) and antibody pull-down (**Subheading 3.3., step 6**). For each sample, aliquot 20 to 30 µL of slurry into clean tubes, and add 200 µL of lysis buffer. Add 5 µg of sonicated λ DNA (Invitrogen) and 5 µg of BSA (NEB) to each tube and rotate 1 h to overnight at 4°C. Before use, pulse down the beads, remove the blocking solution and wash the beads three times in lysis buffer. Add 10 to 15 µL of lysis buffer to packed beads to make a 50% slurry.

7. Generally, use 1 to 4 µg of antibody for each IP, but the amount of antibody varies with the amount of lysate and is different for each antibody. For example, when using 3×10^6 cell-equivalents of chromatin per IP, 2 µg of anti-acetyl Histone H3 (Lys9) or anti-dimethyl Histone H3 (Lys9) from Upstate Biotech was adequate.

8. H4 primers *(19,20)* (mouse): (5'-GACACCGCATGCAAAGAATAGCTG-3' and 5'-CTTTCCCAAGGCCTTTACCACC-3'), 230 bp product; Globin primers *(20)* (mouse): (5'-CAGCATGTGCTGAGGACTTGG-3' and 5'-ACTGCCTTCAGAGAATCGCCC-3'), 196-bp product.

9. It is also possible to quantify the PCR products on an agarose gel. However, the polyacrylamide gel method is more sensitive and better suited for low DNA yields from the PCR. In addition, polyacrylamide gels allow more precise quantification with a digital camera.

References

1. Luger, K., Mader, A. W., Richmond, R. K., Sargent, D. F., and Richmond, T. J. (1997) Crystal structure of the nucleosome core particle at 2.8 A resolution. *Nature* **389,** 251–260.

2. van Leeuwen, F. and Gottschling, D. E. (2002) Genome-wide histone modifications: gaining specificity by preventing promiscuity. *Curr. Opin. Cell Biol.* **14,** 756–762.

3. Santos-Rosa, H., Schneider, R., Bernstein, B. E., Karabetsou, N., Morillon, A., Weise, C., et al. (2003) Methylation of histone H3 K4 mediates association of the Isw1p ATPase with chromatin. *Mol. Cell* **12**, 1325–1332.

4. Jacobson, R. H., Ladurner, A. G., King, D. S., and Tjian, R. (2000) Structure and function of a human TAFII250 double bromodomain module. *Science* **288**, 1422–1425.

5. Jacobs, S. A., Taverna, S. D., Zhang, Y., Briggs, S. D., Li, J., Eissenberg, J. C., et al. (2001) Specificity of the HP1 chromodomain for the mthylated N-terminus of histone H3. *EMBO J.* **20**, 5332–5241.

6. Fischle, W., Wang, Y., Jacobs, S. A, Kim, Y., Allis, C. D, and Khorasanizadeh, S. (2003) Molecular basis for the discrimination of repressive methyl-lysine marks in histone H3 by Polycomb and HP1 chromodomains. *Genes Dev.* **17**, 1870–1881.

7. Fischle, W., Wang, Y., and Allis, C. D (2003) Binary switches and modification cassettes in histone biology and beyond. *Nature* **425**, 475–479.

8. Orlando, V., Strutt, H., and Paro, R. (1997) Analysis of chromatin structure by in vivo formaldehyde cross-linking. *Methods* **11**, 205–214.

9. Buck, M. J., and Lieb, J. D (2004) ChIP-chip: considerations for the design, analysis, and application of genome-wide chromatin immunoprecipitation experiments. *Genomics* **83**, 349–360.

10. Euskirchen, G., Royce, T. E., Bertone, P., Martone, R., Rinn, J. L., Nelson, F. K., et al. (2004) CREB binds to multiple loci on human chromosome 22. *Mol. Cell Biol.* **24**, 3804–3814.

11. Schübeler, D., MacAlpine, D. M., Scalzo, D., Wirbelauer, C., Kooperberg, C., van Leeuwen, F., et al. (2004) The histone modification pattern of active genes revealed through genome-wide chromatin analysis of a higher eukaryote. *Genes Dev.* **18**, 1263–1271.

12. Kuo, M. H. and Allis, C. D (1999) In vivo cross-linking and immunoprecipitation for studying dynamic protein:DNA associations in a chromatin environment. *Methods* **19**, 425–433.

13. Pidoux, A., Mellone, B., and Allshire, R. (2004) Analysis of chromatin in fission yeast. *Methods* **33**, 252–259.

14. Taverna, S. D., Coyne, R. S., and Allis, C. D. (2002) Methylation of histone H3 at lysine 9 targets programmed DNA elimination in *Tetrahymena*. *Cell* **110**, 701–711.

15. He, Y., Michaels, S. D., and Amasino, R. M. (2003) Regulation of flowering time by histone acetylation in Arabidopsis. *Science* **302**, 1751–1754.

16. Boggs, B. A., Cheung, P., Heard, E., Spector, D. L, Chinault, A. C., and Allis, C. D. (2002) Differentially methylated forms of histone H3 show unique association patterns with inactive human X chromosomes. *Nat. Genet.* **30**, 73–76.

17. Janicki, S. M., Tsukamoto, T., Salghetti, S. E., Tansey, W. P., Sachidanandam, R., Prasanth, K. V., et al. (2004) From silencing to gene expression: real-time analysis in single cells. *Cell* **116**, 683–698.

18. Weinmann, A. S., Bartley, S. M., Zhang, T., Zhang, M. Q., and Farnham, P. J. (2001) Use of chromatin immunoprecipitation to clone novel E2F target promoters. *Mol. Cell Biol.* **21,** 6820–6832.
19. Cheung, P., Tanner, K. F., Cheung, W. L., Sassone-Corsi, P., Denu, J. M., and Allis, C. D. (2000) Synergistic coupling of histone H3 phosphorylation and acetylation in response to epidermal growth factor. *Mol. Cell* **5,** 905–915.
20. Thomson, S., Clayton, A. L., and Mahadevan, L. C. (2001) Independent dynamic regulation of histone phosphorylation and acetylation during immediate-early gene induction. *Mol. Cell* **8,** 1231–1241.

20

Histone Modifications and Transcription Factor Binding on Chromatin

ChIP–PCR Assays

Jaejoon Won and Tae Kook Kim

Summary

Chromatin, the eukaryotic template of genetic information, is subject to a diverse array of posttranslational modifications that largely impinge on the N-termini of histones, such as acetylation, methylation, phosphorylation, and ubiquitination. Distinct histone modifications generate synergistic or antagonistic interaction affinities for nonhistone proteins, which in turn dictate dynamic transitions between transcriptionally active or silent states of chromatin. Besides transcription, numerous biological processes, including DNA replication, DNA repair, and recombination, are regulated by chromatin-associated factors. The chromatin immunoprecipitation (ChIP) technique provides us with an exquisite tool to investigate the interplay between the structural or regulatory proteins and DNA and its role in regulating diverse cellular processes in vivo by formaldehyde crosslinking of proteins to proteins and proteins to DNA, followed by immunoprecipitation of the fixed material and detection of the associated DNA. Here we illustrate the overall experimental procedure by taking ChIP analysis of the human telomerase reverse transcriptase promoter as an example.

Key Words: Chromatin; immunoprecipitation; formaldehyde; crosslinking; sonication; PCR; hTERT; histone; transcription factor.

1. Introduction

Eukaryotic DNA is packaged into nucleosomes, which consist of approx 146 base pairs of DNA wrapped around an octamer core of pairs of histones H2A, H2B, H3, and H4. These nucleosomes pack to create higher-order structures to form chromatin. The N-terminal tails of histones are subject to post-translational modifications by acetylation, methylation, phosphorylation, and ubiquitination *(1–5)*. Histone deacetylases (HDACs) and histone

From: *Methods in Molecular Biology, vol. 325: Nuclear Reprogramming: Methods and Protocols*
Edited by: S. Pells © Humana Press Inc., Totowa, NJ

acetyltransferases (HATs) determine the pattern of histone acetylation, which together with other dynamic posttranslational modifications represent a "code" that can be recognized by nonhistone proteins forming complexes involved in transcriptional regulation. In addition, numerous chromatin-associated factors are involved in other essential cellular processes, such as DNA replication, DNA repair, and recombination.

Understanding how all these factors are organized in the chromatin and how their functions are regulated and coordinated in vivo is a challenging task. Chromatin immunoprecipitation (ChIP) is a powerful approach that allows one to investigate the interaction of diverse factors with a specific chromatin region in living cells, thereby providing a snapshot of the factors bound to genes in different functional states *(6–8)*. In brief, ChIP involves treating cells briefly with formaldehyde to crosslink proteins to proteins and proteins to DNA. An antibody against a protein suspected to bind a chromatin region is then used to immunoprecipitate chromatin fragments. Analysis of the immunoprecipitate with primers flanking the DNA region of interest by using polymerase chain reaction (PCR) reveals whether the protein is associated with this DNA region in vivo.

The catalytic subunit of telomerase, known as human telomerase reverse transcriptase (hTERT), is highly expressed in stem cells, germ cell lines, and most human tumors *(9–11)*. The expression of hTERT induces telomerase activity, which prevents the mortality checkpoints evoked by programmed telomere shortening that occurs during each round of cell division. In contrast, hTERT expression is repressed in most mortal human somatic cells *(12,13)*. Analyses of the hTERT promoter with antiacetylated histone antibodies using ChIP have shown that histone acetylation and deacetylation play a critical role in the activation and repression of *hTERT*, respectively *(14)*. In addition, analysis of the hTERT promoter using ChIP has been exploited successfully in revealing the in vivo promoter occupancy by diverse transcription factors directly involved in its activation and repression *(14–17)*. Here, we illustrate the overall experimental procedure using the ChIP assay for analyzing in vivo occupancy of the hTERT promoter by the transcription factor Sp1, which plays an important role in the activation and repression of *hTERT* in tumor and normal cells, respectively. This protocol can be adopted for the detection of the association of other proteins (including modified histones) with other chromatin region.

2. Materials

1. 293 Cells (ATCC; Manassas, VA).
2. Dulbecco's modified Eagle's medium (DMEM) supplemented with 10% fetal bovine serum, 120 µg/mL penicillin G, and 200 µg/mL streptomycin.

3. 37% Formaldehyde. *Caution:* carcinogen and highly toxic by inhalation and contact with the skin. Handle with great care and only in a chemical fume hood.
4. Phosphate-buffered saline (PBS).
5. Cell scraper.
6. Protease inhibitor cocktail (Roche Molecular Biochemicals; Mannheim, Germany).
7. Lysis buffer: 1% sodium dodecyl sulfate (SDS), 10 mM ethyelenediaminetetraacetic acid (EDTA), and 50 mM Tris-HCl, pH 8.1.
8. Sonicator.
9. 5 M NaCl.
10. Phenol/chloroform/isoamyl alcohol (25/24/1).
11. Chloroform.
12. 7.5 M Ammonium acetate.
13. Ethanol.
14. 20 mg/mL Glycogen. Stored at –20°C.
15. DNA gel electrophoresis equipment.
16. Dilution buffer: 0.01% SDS, 1.1% Triton X-100, 1.2 mM EDTA, 16.7 mM Tris-HCl, pH 8.1, and 167 mM NaCl at 4°C.
17. Salmon sperm DNA/protein A agarose (Upstate Biotechnology; Lake Placid, NY) at 4°C.
18. Rotating wheel/platform.
19. Antibodies for chromatin immunoprecipitation: anti-Sp1 antibody (Santa Cruz Biotechnology; Santa Cruz, CA), and normal rabbit IgG (Santa Cruz Biotechnology).
20. Low-salt wash buffer: 0.1% SDS, 1% Triton X-100, 2 mM EDTA, 20 mM Tris-HCl, pH 8.1, and 150 mM NaCl at 4°C.
21. High-salt wash buffer: 0.1% SDS, 1% Triton X-100, 2 mM EDTA, 20 mM Tris-HCl, pH 8.1, and 500 mM NaCl at 4°C.
22. LiCl wash buffer: 0.25 M LiCl, 1% NP-40, 1% deoxycholate, 1 mM EDTA, and 10 mM Tris-HCl, pH 8.1, at 4°C.
23. TE buffer: 10 mM Tris-HCl, 1 mM EDTA, pH 8.0.
24. 1X SDS gel-loading buffer: 50 mM Tris-HCl, pH 6.8, 100 mM dithiothreitol, 2% SDS, 0.1% bromophenol blue, and 10% glycerol. Store 1X SDS gel-loading buffer lacking dithiothreitol at room temperature, and add dithiothreitol from a 1 M stock just before use.
25. Equipments for SDS-PAGE and Western blot.
26. Elution buffer: 1% SDS, 0.1 M NaHCO$_3$. Make fresh as required.
27. 0.5 M EDTA, pH 8.0.
28. 1 M Tris-HCl, pH 6.5.
29. 10 mg/mL proteinase K.
30. Oligonucleotide primers.
31. 10 mM dNTP mix.
32. EF-Taq DNA polymerase (SolGent; Daejeon, Korea), 10X EF-Taq reaction buffer, and 5X Band Doctor solution.
33. PCR machine.

3. Methods

The methods described below outline (1) optimization of DNA shearing, (2) immunoprecipitation of DNA–protein complexes, (3) Western blot to detect immunoprecipitated protein, and (4) PCR to amplify DNA associated with the immunoprecipitated protein.

3.1. Optimization of DNA Shearing

Establish optimal conditions required for shearing crosslinked DNA to approx 500 bp in length by following steps in **Subheadings 3.1.1.** to **3.1.3.** These steps include crosslinking of proteins to DNA, DNA shearing by sonication, and phenol–chloroform extraction. If DNA shearing conditions have been optimized, proceed to **Subheading 3.2.** after completing **Subheadings 3.1.1.** through **3.1.2.**

3.1.1. Crosslinking of Proteins to DNA

1. Grow approx 1×10^6 cells on a 10-cm dish.
2. Crosslink proteins to DNA by replacing culture medium (DMEM supplemented with 10% fetal bovine serum, 120 µg/mL penicillin G, and 200 µg/mL streptomycin) with that containing 1% formaldehyde and incubate at 37°C for 15 min (*see* **Notes 1–4**).

3.1.2. DNA Shearing by Sonication

1. Aspirate medium and wash cells twice using ice-cold PBS.
2. Scrape the cells into a microfuge tube, and pellet the cells for 4 min at 2000*g* at 4°C. Warm the lysis buffer to room temperature to dissolve precipitated SDS and add protease inhibitor cocktail (*see* **Note 5**).
3. Resuspend the cell pellet in 200 µL of lysis buffer and incubate for 10 min on ice (*see* **Note 6**).
4. Sonicate the lysate to fragment DNA into segments approx 500 bp in length, being sure to keep samples ice-cold (*see* **Notes 7** and **8**). Vary the power setting or the number of 20-s pulses of sonication. In our experience, DNA is sheared to the appropriate length with three sets of 20-s pulses using a Sonics & Materials, 500 Watt Ultrasonic Processor (VCX500) equipped with a 3 mm tip and set to 30% of maximum power.

3.1.3. Phenol–Chloroform Extraction

1. Add 8 µL of 5 *M* NaCl and reverse crosslinks at 65°C for 4 h.
2. Add an equal volume of phenol/chloroform/isoamyl alcohol (25/24/1) and mix vigorously by vortexing for 1 min.
3. Centrifuge the mixture at 15,000*g* for 10 min at room temperature.
4. Carefully transfer the upper, aqueous phase to a fresh microfuge tube. Discard the interface and organic phase.

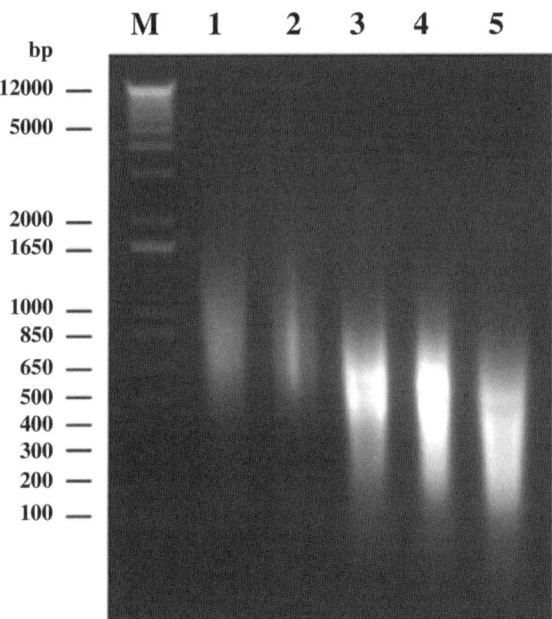

Fig. 1. Optimization of DNA shearing. Lysates from 1 × 10⁶ 293 cells were sonicated with 1, 2, 3, 4, and 5 sets of 20-s pulses (lanes 1, 2, 3, 4, and 5) using a Sonics &
Materials 500 Watt Ultrasonic Processor (VCX500) equipped with a 3-mm tip and set
to 30% of maximum power.

5. Add an equal volume of chloroform and mix vigorously by shaking for 1 min.
6. Centrifuge the mixture at 15,000*g* for 5 min at room temperature.
7. Carefully transfer the upper, aqueous phase to a fresh microfuge tube. Discard
   the interface and organic phase.
8. Add 0.5 volume of 7.5 *M* ammonium acetate and 2.5 µL of 20 mg/mL glycogen
   (*see* **Note 9**). Mix the solution well. Add 2 vol of ethanol and mix by inverting
   four or five times.
9. Store the mixture at –20°C overnight to allow the precipitate of DNA to form.
10. Pellet the DNA by centrifugation at 20,000*g* for 40 min at 4°C.
11. Carefully remove the supernatant. Take care not to disturb the DNA pellet.
12. Carefully fill the tube with 1 mL of 70% ethanol and centrifuge at 20,000*g* for 5
    min at 4°C.
13. Carefully remove the supernatant. Take care not to disturb the DNA pellet.
14. Leave the tube open at room temperature until the last traces of fluid have
    evaporated.
15. Dissolve the DNA pellet in 100 µL of H₂O.
16. Run a sample in a 1.4% agarose gel to visualize shearing efficiency (**Fig. 1**).

3.2. Immunoprecipitation of DNA–Protein Complexes

Described below are the steps for the immunoprecipitation of the protein of interest crosslinked to DNA.

1. Centrifuge the sonicated samples (**Subheading 3.1.2.**, **step 4**) at 16,000*g* for 10 min at 4°C.
2. Aliquot 100 µL of sonicated sample to fresh microfuge tubes and dilute 10-fold in dilution buffer supplemented with protease inhibitor cocktail (*see* **Notes 5** and **10**).
3. To reduce nonspecific background, preclear the 1 mL diluted sonicated sample with 40 µL of salmon sperm DNA/protein A agarose slurry for 30 min at 4°C with rotation.
4. Pellet the agarose by centrifugation at 1000*g* for 1 min at 4°C and combine the supernatant fraction in a conical tube. Dispense 1 mL of the combined supernatant to each fresh microfuge tube (*see* **Note 11**).
5. Add 5 µg of immunoprecipitating antibody (*see* **Note 12**) and incubate overnight at 4°C with rotation.
6. Add 60 µL of salmon sperm DNA/protein A agarose slurry and incubate for 2 h at 4°C with rotation.
7. Pellet the agarose by centrifugation at 1000*g* for 1 min at 4°C. Carefully remove the supernatant that contains unbound, nonspecific DNA–protein complexes. Wash the precipitates for 5 min on a rotating platform with 1 mL of each of the buffers listed in the order below: (1) Wash once with low-salt wash buffer; (2) wash once with high-salt wash buffer; (3) wash once with LiCl wash buffer; and (4) wash twice with TE buffer.

3.3. Western Blot for the Detection of Immunoprecipitated Protein

After washing the beads (*see* **Subheading 3.2.**, **step 7**), immunoprecipitated proteins can be analyzed by using Western blot (*see* **Fig. 2**). Add 25 µL of 1X SDS gel-loading buffer per sample and boil for 5 min. Load 20 µL per lane of SDS-PAGE and perform Western blot by standard methods *(18)*.

3.4. PCR to Amplify DNA That Is Associated With the Immunoprecipitated Protein

The steps described in **Subheadings 3.4.1.** and **3.4.2.** outline the procedure for elution of DNA from the immunoprecipitated protein and PCR to amplify DNA associated with the immunoprecipitated protein

3.4.1. Elution of DNA Associated With the Immunoprecipitated Protein

1. Prepare fresh elution buffer.
2. Elute the protein–DNA complex from the antibody by adding 250 µL of freshly prepared elution buffer to the pelleted protein A agarose/antibody/target protein/ DNA complex from **Subheading 3.2.**, **step 7**. Mix by brief vortexing and incu-

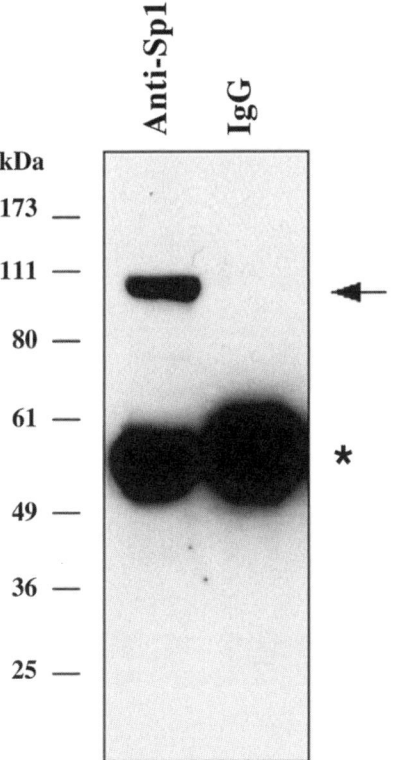

Fig. 2. Detection of the immunoprecipitated protein by Western blot. Sheared chromatin from the lysate of 293 cells was immunoprecipitated with anti-Sp1 antibody or normal rabbit IgG. The immunoprecipitates were separated by 10% SDS-polyacrylamide gel electrophoresis and transferred to nitrocellulose membrane. The membrane was blocked with 5% nonfat milk and probed with anti-Sp1 antibody. The membrane was then incubated with horseradish peroxidase-conjugated anti-rabbit IgG and visualized using an enhanced chemiluminescence detection system. The position of Sp1 is marked with an arrow. Asterisk indicates the immunoprecipitating antibodies.

bate at room temperature for 15 min with rotation. Spin down the agarose and carefully transfer the supernatant fraction (eluate) to a fresh microfuge tube. Repeat the elution and combine the eluates.

3. Add 20 μL of 5 *M* NaCl to the 500 μL of combined eluate and reverse protein–DNA crosslinks by heating at 65°C for 4 h. At this step the sample can be stored at –20°C and the protocol continued the next day.

4. Add 10 μL of 0.5 *M* EDTA; 20 μL of 1 *M* Tris-HCl, pH 6.5; and 2 μL of 10 mg/mL proteinase K to the combined eluate and incubate at 45°C for 1 h.

5. Recover the DNA by phenol/chloroform extraction as described in **Subheading 3.1.3**, **steps 2–15**.

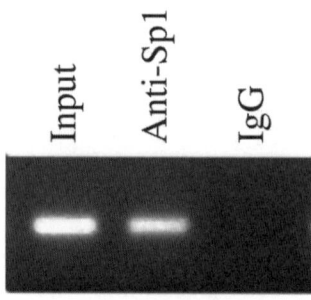

Fig. 3. Detection of DNA associated with the immunoprecipitated protein. An approx160-bp fragment in the hTERT proximal promoter was amplified using the primers 5'-TGCCCCTTCACCTTCCAG-3' and 5'-CAGCGCTGCCTGAAACTC-3'. PCR was carried out as follows: 1 cycle at 94°C for 3 min; 34 cycles at 94°C for 30 s, 53°C for 30 s, 72°C for 20 s; and 1 cycle at 72°C for 1 min. The amplified DNA was separated on a 2% agarose gel and visualized with ethidium bromide. 0.1% of the total amount of chromatin used in the immunoprecipitations (input) and normal rabbit IgG served as positive and negative controls, respectively.

3.4.2. PCR to Amplify DNA Associated With the Immunoprecipitated Protein

1. In a thin-walled 0.2-mL PCR tube, add and mix as follows (*see* **Note 13**).

10X EF-Taq buffer	5 µL
5X Band Doctor	10 µL
10 m*M* dNTP	2 µL
10 µ*M* Forward primer	3 µL
10 µ*M* Reverse primer	3 µL
Template	26 µL
EF–Taq DNA polymerase (2.5 U/µL)	1 µL

2. Amplify DNA as follows (*see* **Note 14**): 1 cycle at 94°C for 3 min; 34 cycles at 94°C for 30 s, 53°C for 30 s, 72°C for 20 s; and 1 cycle at 72°C for 1 min.
3. Run the PCR product on an agarose gel (*see* **Fig. 3** and **Note 15**).

4. Notes

1. Replace the media immediately after mixing 37% formaldehyde with fresh media to make a concentration of 1%.
2. Culture dishes containing formaldehyde should be placed in a sealed bag if returned to a humidified 37°C incubator to prevent exposing other cells in the incubator to formaldehyde.
3. The crosslinking efficiency is different for different proteins. In particular, when one immunoprecipitates the protein associated with DNA through another DNA binding protein, a longer crosslinking time is required. Titration of incubation time is suggested.

4. In cases in which excess crosslinking should be prevented, after the incubation with formaldehyde, add glycine to a final concentration of 0.125 M (from 2.5 M stock) and incubate for 5 min to quench the formaldehyde.

5. Add protease inhibitor cocktail just before use. Instead of commercial protease inhibitor cocktail, 1 mM phenylmethylsulfonyl fluoride, 1 µg/mL aprotinin, and 1 µg/mL pepstatin A can be used. Likewise, these protease inhibitors should be added to the lysis buffer just before use.

6. The 200 µL of lysis buffer is per 1 × 10⁶ cells; if more cells are used, the resuspended cell pellet should be divided into 200-µL aliquots so that each 200-µL aliquot contains approx 1 × 10⁶ cells.

7. Ear protection must be worn during sonication.

8. Be sure to keep the sample on ice at all times because the sonication generates heat that will denature the DNA.

9. The addition of an inert carrier, such as 50 µg/mL of glycogen or 20 µg/mL of yeast tRNA, improves the recovery of small quantities of DNA and helps visualize the DNA pellet.

10. A portion of the supernatant (1%, 10 µL) should be kept to quantify the amount of DNA present in different samples by PCR at **Subheading 3.4.** This sample is considered to be the input material and needs to have the protein–DNA crosslinks reversed and extracted with phenol/chloroform (*see* **Subheading 3.1.3.**).

11. By combining the supernatants and redistributing 1-mL aliquots of the combined supernatant, possible variations in crosslinking, degree of shearing, and the clearance of nonspecific background can be minimized.

12. The proper amount of antibody can vary, for example, according to the affinity of the immunoprecipitating antibody. The quality of the immunoprecipitating antibody is very important in ChIP analysis.

13. Many genomic regions have a high GC content. For amplification of GC-rich DNA sequences, the use of glycerol (5%), dimethyl sulfoxide (DMSO; 5%), or formamide (5%), which promote separation of DNA at lower temperatures, yields good results. Band Doctor is the supplement provided by SolGent (Daejeon, Korea) to improve the PCR amplification of GC-rich regions. PCR without this reagent yields poor amplification of the proximal hTERT promoter, since this region is very GC-rich.

14. This PCR protocol is for the amplification of the proximal hTERT promoter. PCR conditions must be determined empirically in each case. The temperature used for the annealing step depends on the melting temperature of the oligonucleotide primers. Polymerization time needs to be adapted to suit the length of the amplified DNA region.

15. In some cases, the amount of amplified product may be too small to be detected by conventional staining with ethidium bromide. In this case, stain the DNA in the gel with SYBR Gold or perform PCR in the presence of 40 µM each of dATP, dGTP, dTTP, 4 µM of dCTP, and 1 µCi of [α-³²P]dCTP instead of unlabeled 400 µM dNTP, run the PCR product on an 8% polyacrylamide gel, and visualize by autoradiography.

Acknowledgments

This work was supported by the Brain Korea 21 Project, the Molecular and Cellular BioDiscovery Research Program, the Chemical Genomics R&D Program, the Korea Health 21 R&D Project, the Korea Research Foundation Program, R&D Program for Fusion Strategy of Advanced Technologies, and Growth Engine Technology Program.

References

1. Agalioti, T., Chen, G., and Thanos, D. (2002) Deciphering the transcriptional histone acetylation code for a human gene. *Cell* **111**, 381–392.
2. Zhang, Y. and Reinberg, D. (2001) Transcription regulation by histone methylation: interplay between different covalent modifications of the core histone tails. *Genes Dev.* **15**, 2343–2360.
3. Richards, E. J. and Elgin, S. C. (2002) Epigenetic codes for heterochromatin formation and silencing: rounding up the usual suspects. *Cell* **108**, 489–500.
4. Jenuwein, T. and Allis, C. D. (2001) Translating the histone code. *Science* **293**, 1074–1080.
5. Spotswood, H. T. and Turner, B. M. (2002) An increasingly complex code. *J. Clin. Invest.* **110**, 577–582.
6. Braunstein, M., Sobel, R. E., Allis, C. D., Turner, B. M., and Broach, J. R. (1996) Efficient transcriptional silencing in *Saccharomyces cerevisiae* requires a heterochromatin histone acetylation pattern. *Mol. Cell Biol.* **16**, 4349–4356.
7. Luo, R. X., Postigo, A. A., and Dean, D. C. (1998) Rb interacts with histone deacetylase to repress transcription. *Cell* **92**, 463–473.
8. Parekh, B. S. and Maniatis, T. (1999) Virus infection leads to localized hyperacetylation of histones H3 and H4 at the IFN-β promoter. *Mol. Cell* **3**, 125–129.
9. Hahn, W. C. and Meyerson, M. (2001) Telomerase activation, cellular immortalization and cancer. *Ann. Med.* **33**, 123–129.
10. Shay, J. W. and Bacchetti, S. (1997) A survey of telomerase activity in human cancer. *Eur. J. Cancer* **33**, 787–791.
11. Mason, P. J. (2003) Stem cells, telomerase and dyskeratosis congenita. *Bioessays* **25**, 126–133.
12. Meyerson, M., Counter, C. M., Eaton, E. N., Ellisen, L. W., Steiner, P., Caddle, S. D., et al. (1997) hEST2, the putative human telomerase catalytic subunit gene, is up-regulated in tumor cells and during immortalization. *Cell* **90**, 785–795
13. Wright, W. E. and Shay, J. W. (2002) Historical claims and current interpretations of replicative aging. *Nat. Biotechnol.* **20**, 682–688.
14. Xu, D., Popov, N., Hou, M., Wang, Q., Bjorkholm, M., Gruber, A., et al. (2001) Switch from Myc/Max to Mad1/Max binding and decrease in histone acetylation at the telomerase reverse transcriptase promoter during differentiation of HL60 cells. *Proc. Natl. Acad. Sci. USA* **98**, 3826–3831.

15. Won, J., Yim, J., and Kim, T. K. (2002) Sp1 and Sp3 recruit histone deacetylase to repress transcription of human telomerase reverse transcriptase (hTERT) promoter in normal human somatic cells. *J. Biol. Chem.* **277,** 38230–38238.
16. Won, J., Yim, J., and Kim, T. K. (2002) Opposing regulatory roles of E2F in human telomerase reverse transcriptase (hTERT) gene expression in human tumor and normal somatic cells. *FASEB J.* **16,** 1943–1945.
17. Lin, S. Y. and Elledge, S. J. (2003) Multiple tumor suppressor pathways negatively regulate telomerase. *Cell* **113,** 881–889.
18. Sambrook, J. and Russell, D. W. (2001) *Molecular Cloning: A Laboratory Manual,* Third ed. Cold Spring Harbor Laboratory Press, Cold Spring Harbor, New York.

21

In Vivo Genomic Footprinting Using LM–PCR Methods

Hiromi Tagoh, Peter N. Cockerill, and Constanze Bonifer

Summary

Epigenetic regulatory proteins such as transcription factors, chromatin components, and chromatin modification activities alter gene activity during development. The means by which alterations in these factors influence gene expression is poorly understood, but information of this kind is essential if we want to reprogram the epigenotype of specific cell types in a directed fashion. To facilitate chromatin structure–function analysis, we have developed a relatively simple procedure that uses magnetic beads to perform ligation-mediated polymerase chain reaction in solid phase. In this chapter, we describe detailed procedures for the examination of chromatin fine-structure and nucleosome positioning as well as changes in transcription factor binding-site occupancy during cellular differentiation.

Key Words: Chromatin; in vivo footprinting; transcription factors; nucleosome positioning; DNaseI; micrococcal nuclease; ligation-mediated PCR.

1. Introduction

Eukaryotic cells contain on average more than one meter of DNA in nuclei of only approx 20 μm in diameter. To make this possible, DNA is packaged into chromatin. The fundamental unit of chromatin is the nucleosome, which contains approx 200 base pairs (bp) of DNA. Nucleosomes consist of an octamer core composed of two molecules of each of the histone proteins H2A, H2B, H3, and H4. A total of 146 bp of DNA is wrapped around the nucleosome core, and the remaining linker DNA flanking the nucleosome often exists associated with histone H1 protein. Gene expression is regulated by the assembly and disassembly of transcription factor complexes at specific regulatory regions of individual genes. As a result of chromatin compaction, most gene loci are inaccessible to the binding of transcription factors. Hence, the transcriptional activation of specific gene loci is largely dependent on the modification and remodelling of chromatin structure. It is not, therefore, possible to understand

From: *Methods in Molecular Biology, vol. 325: Nuclear Reprogramming: Methods and Protocols*
Edited by: S. Pells © Humana Press Inc., Totowa, NJ

how the expression of specific genes is regulated in living cells without studying transcription factor binding and the underlying chromatin structure in vivo.

In vivo genomic footprinting using chemical or enzymatic agents whose action is modified by chromosomal proteins is an important method of investigating in vivo features of chromatin structure at nucleotide resolution. In vivo genomic footprinting consists of two steps. The first step is to create DNA lesions by chemical or enzymatic DNA cleavage, and the second step is to make these lesions visible. By comparing the position and frequency of DNA modifications conducted in living cells to naked DNA modified in vitro, it is possible to draw conclusions about in vivo chromatin conformation, including the positions of transcription factors interacting with DNA, translational and rotational nucleosome positioning, and changes in chromatin compaction. DNA lesions are visualized by a polymerase chain reaction (PCR)-based amplification technique called ligation-mediated PCR (LM-PCR).

Despite being a powerful and highly informative technique, in vivo genomic footprinting has not been used extensively, probably because it is perceived as technically challenging. However, we have recently modified and simplified the LM-PCR method with improved sensitivity by using magnetic beads to capture primer extension products *(1,2)*. This method also allows us to use much less starting material. Furthermore, numerous alternative DNA modifying agents can be used for chromatin structure studies. We will only describe here the use of dimethyl sulfate (DMS), deoxyribonuclease I (DNase I), and micrococcal nuclease (MNase), although other agents, such as restriction enzymes, $KMnO_4$, or ultraviolet irradiation also can be used with LM-PCR. The general outline of such experiments is depicted in **Fig. 1**. In vivo footprinting assays will not yield unambiguous information about the identity of transcription factors binding to a specific recognition sequence, nor will they yield information about histone modifications. However, combined with chromatin immunoprecipitation assays, they yield exquisitely detailed information about the chromatin organization of active and inactive genes.

Fig. 1. *(opposite page)* Overview of the different in vivo footprinting methods. DNA is purified from DMS-treated cells or nuclease-treated nuclei. DMS-treated DNA has to be cloven by hot-piperidine treatment. DNA with single-strand breaks or double-strand breaks is subjected to LM-PCR amplification. DNA with single-strand breaks is subjected to the primer extension using biotinylated gene-specific primer (first primer). This primer extension will create blunt-ended double-stranded DNA which is ready to be ligated to the synthesised linker (LP25–21). The ligated products are captured on streptavidin-coated magnetic beads and the rest of the genomic DNA is discarded. DNA after restriction enzyme digestion (blunt end) is directly subjected to the linker ligation. As MNase digestion creates a 5'-OH end, DNA has to be phosphorylated

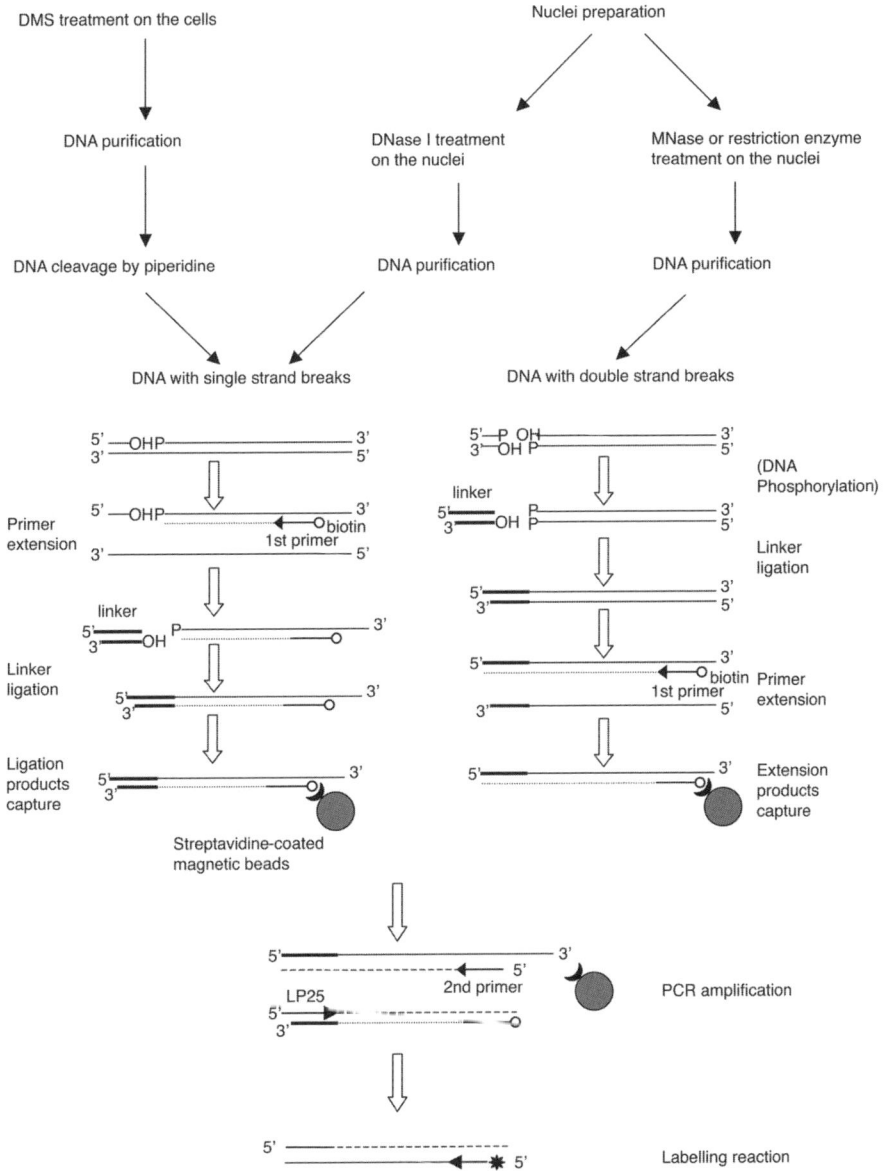

Fig. 1. *(continued)* prior to the ligation. Although it is possible to subject these ligated products directly to PCR amplification, we incorporate a primer extension step with a biotinylated primer to enable enrichment of specific gene sequences on magnetic beads. This bound DNA is amplified using LP25 primer and gene-specific primer (second primer). These amplified products are visualised by linear amplification using end-labeled gene-specific primer (third primer).

1.1. LM-PCR

LM-PCR was developed to identify in vivo DNA–protein interactions *(3)* as well as cytosine methylation patterns *(4)*. LM-PCR methods allow the visualization of single-strand as well as double-strand breaks in DNA. Single-strand breaks are generated by reagents such as DNaseI, or by piperidine treatment of DMS-modified DNA, whereas reagents such as MNase or restriction enzymes generate double-strand breaks. The different LM-PCR steps are outlined in **Fig. 1**. The first step for the detection of single-strand breaks is a primer extension reaction using a sequence-specific primer that creates a double-stranded end ready for ligation to a nonphosphorylated synthetic blunt end linker. Blunt-ended double-stranded DNA breaks can be ligated directly without this step. However, in the case of MNase-treated DNA, genomic DNA needs to be phosphorylated first because this enzyme creates 5'-OH ends (*see* **Subheading 3.5.1.2.**). Linker-ligated DNA is then amplified by PCR using both linker-specific and target sequence-specific nested primers. Amplified PCR products are then visualized by linear amplification using a sequence-specific labeled primer. In this reaction there is, however, still some PCR amplification of the labeled products as a result of the residual linker primer still present from the previous step. The combination of nested primers and PCR amplification makes this method highly sensitive as well as highly specific. Moreover, extension product capture using biotinylated primers and streptavidine-coated magnetic beads enriches specific primer extended products and helps to increase the specificity and sensitivity of PCR amplification *(1)*.

1.2. DMS In Vivo Footprinting

DMS in vivo genomic footprinting has been the method used most commonly to study in vivo DNA–protein interactions. DMS induces the formation of N-7-methylguanine (approx 70%) and N-3-methyladenine (approx 30%) in DNA. The frequency with which this occurs is modified by DNA–protein interactions, as the same residues participate in sequence-specific DNA recognition and are not available for modification once a transcription factor is stably bound. In addition, in the vicinity of a transcription factor complex guanines often become DMS-hyper-reactive, as compared with naked DNA. As a result of this, hypo- or hyper-reactive sites are defined as footprints *(5)*. Nucleosome–DNA interactions are transparent for DMS-footprinting as they mostly involve the phosphate backbone of DNA.

The big advantage of DMS for in vivo footprinting is that DMS is a very small hydrophobic compound that can readily enter intact cells and modify nuclear DNA. DNA is then purified from these cells and single-strand breaks with a 5'-phosphoryl group are introduced at the site of N-methylguanine modification by using hot piperidine treatment (*see* **Subheading 3.2.**). In this chapter, we describe DMS modification in combination with hot piperidine treatment, which introduces cleavage without significant non-specific digestion. It may, however, be possible to substitute other alkalis for piperidine at this stage.

1.3. MNase Footprinting

MNase (S7 Nuclease) is a 16.8-kDa endonuclease from *Staphylococcus aureus* that catalyzes cleavage of both DNA and ribonucleic acid (RNA) to yield fragments with terminal 5'-OH ends. Its activity is strictly dependent on Ca^{2+}. A unique characteristic of MNase that makes it useful for chromatin structure analysis is its ability to introduce double-strand DNA breaks preferentially into nucleosomal linker or nucleosome-free regions *(6)*. Combined with LM-PCR methods, this allows fine mapping of the positions of nucleosomes at specific locations *(7)*. The detection of MNase-dependent DNA cleavage by LM-PCR is conducted by treating isolated DNA with kinase, ligating the double-stranded linker directly to the MNase-digested materials followed by LM-PCR. One of the difficulties of enzyme modification is that enzymes cannot penetrate intact cells making it necessary to either permeabilize cells or isolate intact nuclei, which creates a time lag between cell membrane lysis and nuclease reaction. We solved this problem by crosslinking cells with formaldehyde *(8)* prior to nuclei isolation to minimise the disruption of DNA–protein interactions and preserve chromatin structure. Nuclei isolation is based on a previously published protocol *(9)*.

1.4. DNase I Footprinting

DNase I is a relatively large (31-kDa) endonuclease that recognizes the minor groove of DNA and cleaves phosphodiester bonds, preferentially adjacent to a pyrimidine nucleotide yielding 5'-phosphate-terminated polynucleotides with a free hydroxyl group on the 3' side. In the presence of Mg^{2+}, DNase I preferentially cleaves one strand of DNA. The accessibility of DNase I to chromosomal DNA is influenced by protein binding and nucleosome contacts. In general, the DNase I-generated cleavage pattern of chromosomal DNA as compared with naked DNA provides information about a) protein binding in the form of protection from DNase I digestion or hyper-reactivity at the edge of the protein binding sites, (2) general accessibility (condensed vs noncondensed chromatin), and (3) DNA–nucleosome contacts indicated by a 10-bp periodicity of hyper-reactive sites. In addition to studies of protein occupancy, DNase I footprinting can be used to analyze rotational and translational nucleosome positioning together with MNase footprinting. A protocol that uses permeabilized cells for DNase I treatment has been published *(10)*. This protocol has the advantage of reducing any stress and time-lag resultinf from nuclei isolation. However, nuclease reactivity is inhibited by some cytoplasmic components, such as actin. With actin-rich cells such as muscle cells, fibroblasts, or macrophages, it is very difficult to titrate the appropriate concentration of enzyme so that DNA is neither overdigested nor underdigested. To be able to apply DNaseI in vivo footprinting to a wide range of cells and to digest the same samples with different nucleases, we isolate nuclei from formaldehyde-crosslinked cells before digestion as described for MNase digestion experiments. DNaseI digestion causes chromatin to unfold, thus progressively increasing enzyme accessibility and this artefact is avoided by using crosslinked chromatin.

2. Materials

2.1. Chemical DNA Sequencing Reaction

1. DMS (*see* **Note 1**).
2. Piperidine (stored at 4°C; *see* **Note 1**).
3. 2X DMS buffer: 100 mM sodium cacodylate, pH 8.0, 2 mM ethylene diamine tetra-acetic acid (EDTA).
4. DMS stop buffer: 1.5 M sodium acetate, pH 7.0, 1 M 2-mercaptoethanol (freshly prepared).
5. 3 M Sodium acetate, pH 5.2.
6. Iso-butanol.
7. 100% Ethanol (kept in freezer).
8. 70% Ethanol.
9. 0.1X TE: 1 mM Tris-HCl, 0.1 mM EDTA, pH 7.4.

2.2. In Vivo DMS Treatment

1. Ice-cold phosphate-buffered saline (PBS).
2. DMS.
3. Lysis buffer 1: 20 mM Tris-HCl, pH 8.0, 20 mM NaCl, 20 mM EDTA, 1% sodium dodecyl sulphate (SDS).
4. Lysis buffer 2: 150m M NaCl, 10m M EDTA, pH 8.
5. Proteinase K: make up with water at 50 mg/mL (this concentration can vary). Store at –20°C.
6. DNase-free RNase A, 10 mg/mL: DNase free RNase A can be purchased or prepared by heating to 100°C for 15 min in 10 mM Tris-HCl, pH 7.5, 15 mM NaCl. Store at –20°C.
7. Phenol: saturated with 50 mM Tris-HCl, pH 8.0. Stored at 4°C.
8. Chloroform/isoamyl alcohol (24/1; CIAA).
9. Lambda DNA (0.4–1 µg/µL; Invitrogen). Stored at –20°C.
10. 20 mg/mL Glycogen (Roche Applied Science). Stored at –20°C.
11. 5 M NaCl.
12. Isopropanol.
13. 70% Ethanol.
14. 0.1X TE.
15. Piperidine (stored at 4°C).
16. 3 M Sodium acetate, pH 5.2.

2.3. Nuclear Preparation and Nuclease Treatment

1. Ice-cold PBS.
2. 38% formaldehyde solution (Sigma). Stored at room temperature.
3. Quenching solution: 2 *M* glycine in PBS, freshly prepared.
4. Nuclei preparation buffer: 300 m*M* sucrose, 10 m*M* Tris-HCl, pH 7.5, 15 m*M* NaCl, 60 m*M* KCl, 5 m*M* MgCl$_2$, 0.1 m*M* ethylenediaminetetraacetic acid, 0.15 m*M* spermine, 0.5 m*M* spermidine, 0.1% Nonidet-P40, 0.5 m*M* phenylmethyl sulfonyl fluoride (PMSF); freshly prepared or stored at –20°C, except spermine, spermidine, and PMSF (*see* **Note 2**).
5. Nuclease digestion buffer: 300 m*M* sucrose, 10 m*M* Tris-HCl, pH 7.5, 15 m*M* NaCl, 60 m*M* KCl, 5 m*M* MgCl$_2$, 0.1 m*M* ethylenediaminetetraacetic acid, 0.15 m*M* spermine, and 0.5 m*M* spermidine.
6. Enzyme dilution buffer: Nuclease digestion buffer plus 1 m*M* CaCl$_2$, 0.1 mg/mL bovine serum albumin (BSA).
7. Lysis buffer: 50 m*M* Tris-HCl, pH 8.0, 20 m*M* EDTA, 1% SDS, 500 µg/mL proteinase K.
8. DNase-free RNase A, 10 mg/mL (*see* **Subheading 2.2.6.**).
9. Phenol, saturated with 50 m*M* Tris-HCl, pH 8.0.
10. Chloroform/isoamyl alcohol (24/1).
11. 5 *M* NaCl.
12. Isopropanol.
13. 70% Ethanol.

2.4. LM-PCR (see *Note 3*)

1. 2 U/µL Vent$_R$®(Exo-) DNA polymerase (New England Biolabs) stored at –20°C.
2. 10X ThermoPol buffer: 100 m*M* KCl, 200 m*M* Tris-HCl, pH 8.8 at 25°C, 100 m*M* (NH$_4$)$_2$SO$_4$, 20 m*M* MgSO$_4$, 1% Triton X-100, stored at –20°C.
3. Pfu turbo DNA polymerase (Stratagene) stored at –20°C.
4. 10X Cloned Pfu buffer: 200 m*M* Tris-HCl, pH 8.8, 20 m*M* MgSO$_4$, 100 m*M* KCl, 100 m*M* (NH$_4$)$_2$SO$_4$, 1% Triton® X-100, 1 mg/mL nuclease-free bovine serum albumin (BSA), stored at –20°C.
5. 3 U/µL T4 DNA ligase (Promega), stored at –20°C.
6. T4 DNA Ligase 10X reaction buffer: 300 m*M* Tris-HCl, pH 7.8 at 25°C, 100 m*M* MgCl$_2$, 100 m*M* dithiothreitol (DTT), and 10 m*M* ATP. Store at –20°C in small aliquots.
7. 10 U/µL T4 polynucleotide kinase (T4-PNK; New England Biolabs). Store at –20°C.
8. 10X PNK buffer: 700 m*M* Tris-HCl, pH 7.6, 100 m*M* MgCl$_2$, 50 m*M* DTT. Store at –20°C.
9. 10 mg/mL Dynabeads (Dynal M-270 or M-280), stored at 4°C; do not freeze.
10. 2X Binding and washing buffer (B&W buffer): 10 m*M* Tris-HCl, pH 7.5, 1 m*M* EDTA, 2.0 *M* NaCl.
11. 5 U/µL Taq DNA Polymerase in Storage Buffer B (Promega), stored at –20°C.
12. 10X Taq DNA Polymerase reaction buffer with MgCl$_2$: 500 m*M* KCl, 100 m*M* Tris-HCl, pH 9.0 at 25°C, 1% Triton® X-100, 15 m*M* MgCl$_2$ stored at –20°C.

13. LP25 oligonucleotide (GCG GTG ACC CGG GAG ATC TGA ATT C; 200 μ*M* in 1X TE).
14. LP21 oligonucleotide (GAA TTC AGA TCT CCC GGG TCA); 200 μ*M* in 1X TE.
15. Biotinylated first primer (*see* **Subheading 3.6.**); 200 μ*M* in 1X TE stored at –20°C in small aliquots.
16. Second primer: 200 μ*M* in 1X TE, stored at –20°C in small aliquots.
17. Third primer: 200 μ*M* in 1X TE, stored at –20°C in small aliquots.
18. 50% polyethylene glycol (PEG) 6000, in water.
19. 100 m*M* ATP stored at –20°C in small aliquots.
20. 1 *M* Tris-HCl, pH 7.5.
21. 1 *M* MgCl$_2$, stored at –20°C.
22. 1 *M* DTT, stored at –20°C in small aliquots.
23. 10 mg/mL BSA, stored at –20°C.
24. 5 *M* Betaine (in water), stored at –20°C.
25. Dimethyl sulfoxide (DMSO), stored at room temperature.
26. dNTP mix (25 m*M* each stored at –20°C in small aliquots).
27. 370 MBq/mL Redivue™ [γ-^{32}P] ATP (~110 TBq/mmol; Amersham Biosciences).
28. MicroSpin™ G-25 column (Amersham Biosciences). (Spin column containing Sephadex G25 DNA Grade F equilibrated in water with 0.05% Katon™.)
29. Formamide–EDTA–XC–BPB gel loading buffer (10 m*M* EDTA, 0.1% bromophenol blue, 0.1% xylen cyanol FF in formamide).
30. 8 *M* Urea–6% acrylamide(acrylamide/bis, 19/1) in 1X TBE.
31. 1X TBE (89 m*M* Tris-borate, 2 m*M* EDTA). For the stock solution, i.e., 5X, autoclaving is recommended to avoid creating precipitates by storage.
32. 25% Ammonium persulfate (in water) stored at –20°C. Once thawed, do not refreeze.
33. TEMED (*N,N,N',N'*-tetramethylethylenediamine; Sigma). Stored at 4°C.
34. Gel fixation buffer: 10% methanol, 10% acetic acid in water.

3. Methods

3.1. G Reaction With DMS (Naked DNA Control)

1. Ethanol precipitate an appropriate amount (50–100 μg) of genomic DNA purified by standard SDS–proteinase K–phenol extraction methods. DNA purification must be conducted with care to minimize the generation of nicks as these will show up as background in any LM-PCR.
2. Dissolve the DNA into 100 μL of water; allow plenty of time for it to dissolve because genomic DNA is very viscous.
3. Gently mix the DNA with 100 μL of 2X DMS buffer.
4. Add 1 μL of DMS or 10 μL of 10% DMS freshly diluted in 100% ethanol (0.5% final; *see* **Note 4**; **Fig. 2**).
5. Incubate for 3 min at room temperature.
6. Stop the reaction by adding 30 μL of DMS stop and 750 μL of prechilled 100% ethanol.
7. Precipitate the DNA for 30 min in a –75°C freezer.

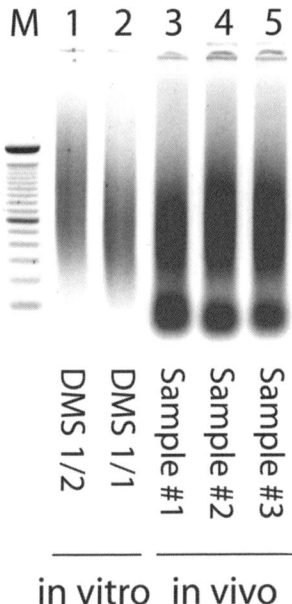

M 1 2 3 4 5

DMS 1/2
DMS 1/1
Sample #1
Sample #2
Sample #3

in vitro in vivo

Fig. 2. Optimal size distribution of DNA after DMS treatment and piperidine cleavage. After the DMS and piperidine treatment, the DNA should show a smear on the gel
that has the peak at approx 600 bp. In the experiment depicted here, in vitro (lanes 1
and 2) and in vivo (lanes 3–5) DMS-treated DNA was cloven with hot piperidine.
After the removal of piperidine, DNA was run on a 1.5% agarose gel stained with
ethidium bromide. DMS for in vitro treatment was used at 0.25% (lane 1) and 0.5%
(lane 2). In vivo DMS treatment was performed in 0.2% DMS in PBS (lanes 3–5). The
marker (M) is a 100-bp ladder.

3.2. Cleavage of DNA at the Sites of Chemical Modification

3.2.1. Piperidine Treatment

1. Pellet the DNA by centrifugation for 5 min at 15,000g in a micro-centrifuge.
2. Rinse the DNA with 75% ethanol.
3. Dissolve the DNA in 99 µL of 0.1X TE with gentle pipetting or tapping the tube.
   Do not vortex (*see* **Note 5**).
4. Add 1 µL of piperidine (10 *M*) and mix well but gently (final concentration should
   be 0.1 *M*).
5. Incubate for 10 min at 90°C.
6. Transfer the tube to ice water.

3.2.2. Removal of Piperidine by Lyophilization

1. Freeze the piperidine-cloven DNA solution in dry ice or liquid nitrogen.
2. Lyophilize the samples in a Speed Vac (*see* **Note 6**).

3. Resuspend the sample in 100 μL of water and repeat the lyophilization two more times.
4. Dissolve the DNA in an appropriate vol of 0.1X TE.

3.2.3. Removal of Piperidine Using Iso-Butanol Extraction

If no Speed Vac is available, follow this alternative way of removing piperidine:

1. Add 300 μL of water.
2. Extract the samples twice using butanol by shaking with 2 vol (800 μL) of iso-butanol (or 1.5 vol of *n*-butanol) and spin in a microcentrifuge. Because butanol also acts to concentrate DNA samples, take care that the samples are not extracted to the point where no aqueous phase remains (if this occurs, then add 100 μL of water to the tube, shake well, and respin). Two extractions remove greater than 99% of the piperidine.
3. Remove the upper butanol phase and extract the aqueous phase once with chloroform to remove excess butanol. (Chloroform extraction can be substituted by an ether extraction, which also removes butanol.)
4. Transfer the DNA solution to a fresh tube and add 1/10 vol of 3 *M* sodium acetate and 2 vol of ethanol. Precipitate the DNA at –20°C for more than 30 min. DNA molecules at this point are small and take longer to precipitate. The addition of glycogen (1 μL) also helps to maximize the efficiency of the precipitation.
5. Dissolve the DNA in an appropriate volume of 0.1X TE (ideally, 1 μg/μL; it should not be too dilute because only a small volume of DNA [2–4 μL, *see* **Subheading 3.5.2.2.**] can be applied to LM-PCR).

3.3. In Vivo DMS Treatment

3.3.1. DMS Treatment

3.3.1.1. Cells Grown as Monolayer

1. Discard the culture medium and wash the cells twice with PBS (*see* **Note 7**).
2. After the washing with PBS, add 0.2% DMS/PBS (*see* **Note 8**) to cover the cell layer (e.g., in the case of using a 10-cm dish, 5 mL is enough to cover the cells. From a 10-cm dish, you can get 3 to 10 × 10^6 cells depending on the size of the cells. From 10^6 cells, you can normally recover approx 10 μg of DNA).
3. Incubate for 5 min at room temperature.
4. Discard the DMS–PBS (into a solution of 3 *M* sodium acetate to neutralize the DMS) and wash the cells three times with ice-cold PBS (*see* **Note 9**).
5. After the final wash, carefully discard the PBS and add lysis buffer 2 with proteinase K (600 μg/mL) to the cells (*see* **Note 10**).
6. Add an equal volume of lysis buffer 1 and gently swirl the dish.

3.3.1.2. Suspension Cells

1. Harvest the cells and wash them twice with PBS (*see* **Note 7**).
2. Suspend the cells in 0.2% DMS/PBS (*see* **Note 8**) by gentle pipetting. The volume of DMS/PBS solution can vary depending on the number of cells and size of

the cell pellet. However, it is the concentration and the incubation time that is important, not the volume. For example, we add 500 to 1000 μL of 0.2% DMS–PBS to 1 to 5 × 10⁷ cells in a 15-mL conical tube. A number of relatively small cells (1 × 10⁶) can be resuspended in 100 μL or, alternatively, the cell pellet is suspended in 50 μL of PBS and an equal volume of 0.4% DMS–PBS is added. When fewer than 1 × 10⁶ cells are handled, a 2-mL conical tube is useful as one can see the cell pellet easily.

3. Incubate for 5 min at room temperature.
4. Dilute the DMS by adding 10 to 20 times volume of ice-cold PBS and pellet the cells. Neutralize the DMS in the discarded supernatant by adding NaOH to approx 1 *M*.
5. Wash the cells twice with ice-cold PBS.
6. Remove the PBS from the cell pellet after the final wash and suspend the cells in Lysis buffer 2 with proteinase K (600 μg/mL; *see* **Note 10**).
7. Add an equal volume of Lysis buffer 1 and invert the tube gently.

3.3.2. Isolating DNA From DMS-Treated Cells

1. Incubate the cell lysate overnight at room temperature.
2. Add RNase A (DNase free) at a final concentration 100 μg/mL and incubate for 0.5 to 1 h at 37°C.
3. Phenol–chloroform–isoamyl alcohol: extract the samples twice (*see* **Note 11**). Minimize the generation of nicks by shearing as these will show up as background in any LM-PCR reaction. For example, do not pipet DNA through the narrow end of a pipet tip, do not vortex at any stage, and do not freeze the DNA.
4. Chloroform–isoamyl alcohol: extract once (*see* **Note 12**).
5. Add 1/10 volume of 5 *M* NaCl and an equal volume of isopropanol. Use 20 μg (1 μL) of glycogen as precipitation carrier if necessary.
6. Perform precipitation at 4°C or –20°C.
7. Precipitate the DNA by centrifugation and rinse the pellet with 70% ethanol.
8. Briefly dry the pellet and dissolve it into 100 μL of 0.1X TE.

3.3.3. Piperidine Cleavage of DMS-Modified DNA

1. Essentially as described in **Subheading 3.2.**
2. The conditions used here are for up to 100 μg of DNA. If more DNA needs to be treated this should be done in several reactions. Note, however, that LM-PCR is highly sensitive and that 1 μg of DNA is sufficient for one reaction.

3.3.4. Confirmation of the Quality of DMS-Treated DNA

1. The size distribution of DMS-treated DNA fragments is examined by electrophoresis on nondenaturing 1.5% agarose gels in any suitable buffer, such as TBE.
2. The gel should show a smear of DNA between 200 bp and 1.5 kb with a peak at approx 600 bp. An example of such an analysis is shown in **Fig. 2**. DNA treated with DMS in vitro (lanes 1 and 2) should show the same size distribution as DNA treated with DMS in vivo (lanes 3–5).

cell number   2x10⁵/250µl   6x10⁵/250µl   2x10⁶/250µl

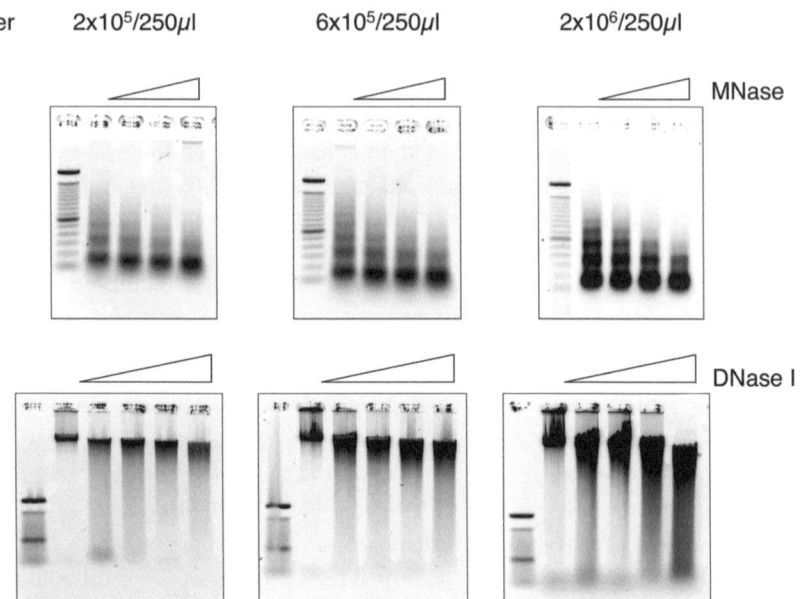

Fig. 3. The efficiency of nuclease digestion on isolated formaldehyde-crosslinked nuclei. With the same concentration of nuclease, the number of nuclei does not affect the degree of nuclease digestion. Nuclei from crosslinked cells were isolated according to the protocol as described above. The isolated nuclei were resuspended in nuclease digestion buffer at a concentration of 0.8×10^6/mL , 2.4×10^6/mL , and 8×10^6/mL. A total of 250 µL of nuclei suspension was mixed with 5 µL of 100 mM CaCl$_2$, and 25 µL of 10X enzyme solution and incubated for 15 min on ice for DNase I treatment and at 37°C for MNase treatment. The reaction was stopped by adding 250 µL of lysis buffer. After the reversal of crosslinking, the DNA was purified and run on an agarose gel (1.5% for MNase-treated DNA [upper panel], 0.8% for DNase I-treated DNA [lower panel]). The amount of nuclease used here was as follows, MNase: 50, 100, 150, 200 U/mL , DNase I: 20, 30, 40, 50 U/mL. The first lane for each gel analysis of DNase I-digested samples contains untreated DNA from isolated nuclei.

3.4. Nuclear Preparation and Nuclease Treatment

3.4.1. Nuclei Preparation (see **Note 13**)

1. Add 1/38 volume of formaldehyde solution to the cells in culture medium (1% at final concentration) and incubate them for 5 min at room temperature with gentle shaking or rocking.
2. Add 1/16 volume of the quenching solution (0.125 M glycine at final concentration).
3. Harvest the cells and wash them with ice-cold PBS (*see* **Note 14**).
4. Estimate the total number of cells and calculate the necessary volume of reagents (*see* **Note 15**).

5. Resuspend the cells in 5 mL of nuclei preparation buffer using a plastic pipet and then add nuclei preparation buffer up to 30 mL. The best way to do this is to first partially suspend the cell pellet in the small volume of remaining PBS by gentle tapping of the tube before adding the buffer.
6. Incubate for 3 min on ice water.
7. Collect the nuclei by centrifugation for 5 min at 500g at 4°C.
8. Draw off most but not all of the supernatant using a 10- or 25-mL pipet. Take great care because the nuclei pellet is very loose. Do not aspirate the buffer by suction too quickly or try to remove all of the buffer at this stage, or the pellet may be disrupted and the nuclei lost.
9. Centrifuge (for approx 1 min) to collect all the remaining buffer at the bottom of the tube.
10. Remove the remaining buffer using a P-1000 micro-pipet.
11. *Optional:* when cells with a high level of actin such as fibroblasts or macrophages are handled, additional washing with 30 mL of nuclease digestion buffer can help to reduce the amount of actin interfering with DNaseI activity.
12. Resuspend the nuclei in nuclease digestion buffer (~10^7 cells/500 µL; *see* **Note 16**).

3.4.2. Nuclease Digestion

Using 500 µL of nuclei suspension for each concentration of nuclease (*see* **Note 17** and **Fig. 3**).

1. Prepare a dilution series of 10X nuclease stock (DNaseI or MNase) in enzyme dilution buffer containing different amounts of enzyme. We normally use 50 to 250 U/mL of MNase and 10 to 100 U/mL of DNase I. Typical 10X concentrations would be as follows: MNase : 0, 500, 1000, 1500, 2000, 2500 U/mL; DNase I: 0, 100, 200, 300, 400, 500, 600, 700, 800, 1000 U/mL.
2. Dispense 50 µL of 10X nuclease and 10 µL of 100 mM CaCl$_2$ into precooled tubes. (Conical tubes, 2-mL or 15-mL, can be used; *see* **Note 18**.)
3. Put 500 µL of nuclei suspension into individual tubes with nuclease and CaCl$_2$ and mix them by gentle pipetting.
4. For DNase I treatment tubes are kept on ice water. Start the incubation time when the nuclei are mixed with enzyme/CaCl$_2$ and stop the reaction at the end of 15-min incubation by adding 500 µL of lysis buffer.
5. For MNase treatment, transfer the tubes to a 37°C water bath, and start timing. The reaction is finished as for DNase I treatment.

3.4.3. Preparation of the DNA

1. Incubate the nuclei in lysis buffer overnight at 65°C. This also reverses the crosslinking.
2. Add DNase-free RNaseA at 100 µg/mL (final concentration) and incubate for 0.5 to 1 h at 37°C.
3. Extract twice with phenol–chloroform–isoamyl alcohol and once with chloroform–isoamyl alcohol (*see* **Note 12**).

4. Precipitate the DNA by adding 1/10 volume of 5 *M* NaCl and an equal volume of isopropanol and incubating for 30 min on ice water.
5. Centrifuge for 20 min at 15,000*g* at 4°C.
6. Rinse the pellet with 70% ethanol.
7. Briefly dry the pellet and dissolve in an appropriate volume of 0.1X TE.

3.5. LM-PCR

3.5.1. Using LM-PCR to Detect Double-Stranded DNA Breaks

3.5.1.1. PREPARATION OF LP25-21 LINKER

1. Prepare the mixture of LP21 and LP25 oligonucleotides in 0.1X TE.

			Final concentration
200 μ*M*	LP21	10 μL	20 μ*M*
200 μ*M*	LP25	10 μL	20 μ*M*
Water		80 μL	
		100 μL	

2. Perform the annealing by incubating the oligo mixture at 95°C for 15 min and cool down gradually down to 4°C. This step can be done on a thermal cycler as per the following program:

> 95°C, 3 min
> 65°C, 15 min (slope 0.01°C/s)
> 55°C, 15 min (slope 0.01°C/s)
> 45°C, 15 min (slope 0.01°C/s)
> 37°C, 15 min (slope 0.01°C/s)
> 25°C, 15 min (slope 0.01°C/s)
> 4°C, 1 min

3.5.1.2 DNA Phosphorylation

This step is not necessary for restriction enzyme-digested DNA with blunt ends.

1. Prepare 30 μL of reaction mixture on ice.

				Final concentration
10 X	PNK buffer	3.0 μL	Tris-HCl (pH 7.6)	70 m*M*
			MgCl$_2$	10 m*M*
			DTT	5 m*M*
10 m*M*	ATP	3.0 μL		1 m*M*
10 U/μL	T4-PNK	2.0 μL		20 units
Water		22.0 – *x* μL		
MNase-digested DNA (10 μg)		*x* μL		
		30.0 μL		

2. Incubate for 1 h at 37°C and stop the reaction and inactivate the enzyme by incubation for 10 min at 75°C.
3. Adjust the volume to 50 μL.

3.5.1.3. Ligation

1. Prepare the ligation buffer.

		Final concentration
Water or 50% PEG6000	338.04 μL	
1 *M* Tris-HCl, pH 7.5	35.20 μL	82.8 m*M*
1 *M* MgCl$_2$	9.76 μL	23 m*M*
1 *M* DTT	24.00 μL	56 m*M*
100 m*M* ATP	12.00 μL	2.8 m*M*
10 mg/mL BSA	6.00 μL	0.14 mg/mL
	425.00 μL	

Freshly prepared or stored at –20°C in small aliquots to avoid repeated freeze and thaw cycles (*see* **Note 19**).
2. Prepare the reaction mixture.

			Final concentration
Ligation buffer	10.0 μL	Tris-HCl (pH 7.5)	30 m*M*
		MgCl$_2$	8.5 m*M*
		DTT	20 m*M*
		ATP	1.0 m*M*
		BSA	0.05 m*M*
LP25-21 linker	5.0 μL		100 pmol
3 U/μL T4 Ligase	2.0 μL		6 Weiss units
	17.0 μL		
Phosphorylated DNA (2 μg)	10.0 μL		
	27.0 μL		

We normally make our own ligation buffer. However, it is possible to replace it with the reaction buffer supplied with T4 DNA ligase. Even in that case, however, the buffer has to be kept at –20°C in small aliquots. When using a 10X ligase buffer, use this following alternative recipe:

			Final concentration
Water	5.5 μL		
Ligation buffer	2.5 μL	Tris-HCl (pH 7.8)	30 m*M*
		MgCl$_2$	10 m*M*

		DTT	10 mM
		ATP	1 mM
LP25-21 linker	5.0 µL		100 pmol
3 U/µL T4 Ligase	2.0 µL		6 Weiss units
	15.0 µL		
Phosphorylated DNA (2 µg)	10.0 µL		
	25.0 µL		

3. Incubate overnight at 16 to 17°C.
4. Add 50 µL of 1X TE and precipitate the DNA by adding 8 µL of 3 *M* NaAcetate and 200 µL of ethanol.
5. Rinse the pellet with 70% ethanol.
6. Dissolve the pellet into 10 to 50 µL of 0.1X TE.

3.5.1.4. PRIMER EXTENSION

1. Prepare the reaction mixture on ice.

Final concentration

10X	ThermoPol buffer	3.0 µL	Tris-HCl	20 mM
			(pH 8.8)	
			KCl	10 mM
			$(NH_4)_2SO_4$	10 mM
			$MgSO_4$	2 mM
			Triton X-100	0.1%
25 mM	dNTPs	0.3 µL		250 µM
20 µM	Biotinylated first primer	0.1 µL		2 pmol
2 U/µL	Vent polymerase Exo-	1.0 µL		2 units
DMSO		1.5 µL		5%
DNA		n µL		
Water		$24.1 - n$ µL		
		30.0 µL		

See **Notes 20** and **21**.

2. Perform primer extension by PCR. Keep the reaction mixture on ice and transfer the tubes (or plates) when the Thermal cycler is set at 95°C.

PCR program:

Denaturation	95°C	5 min
Annealing	Tm – approx 3 to 5°C	10 min
Extension	72°C	10 min
		×1 cycle
Denaturation	95°C	45 s
Annealing	Tm – approx 3 to 5°C	5 min
Extension	72°C	10 min
		×11 cycles
Complete extension	72°C	10 min

3.5.1.5. Capture of Primer Extension Products

1. Aliquot Dynabeads M-270 or M-280 (15 µL/reaction).
2. Wash twice with 2X B&W buffer.
3. Suspend the beads in 2X B&W buffer (30 µL/sample). At least 1 *M* sodium salt is necessary to observe maximum binding to the beads. Therefore, an equal volume of beads suspension in 2X B&W buffer has to be added to each sample.
4. Add 30 µL of beads to the primer extension products.
5. Incubate them for at least 4 h at room temperature with rotation.
6. Place the tubes in a magnetic apparatus to bind the beads to the walls of the tubes and discard the supernatant.
7. Wash the beads once with 100 µL of 2X B&W buffer.
8. Wash the beads twice with 100 µL of 1X TE.
9. Suspend the beads into 10 µL of 0.1X TE.
10. Release the DNA from the beads by heating the beads suspension at 95 to 100°C for 15 min

3.5.2. Use of LM-PCR to Detect Single-Stranded DNA Breaks

3.5.2.1 PREPARATION OF LP25-21 LINKER

Preparation of the LP25-21 linker has been discussed previously in **Subheading 3.5.1.1.**

3.5.2.2. Primer Extension

1. Prepare 5 µL of reaction mixture on ice, where "*n*" is the volume of DNA used.

				Final concentration
10X	ThermoPol buffer	0.5 µL	Tris-HCl (pH 8.8)	20 mM
			KCl	10 mM
			$(NH_4)_2SO_4$	10 mM
			$MgSO_4$	2 mM
			Triton X-100	0.1%
25 mM	dNTPs	0.05 µL		250 µM
20 µM	biotinylated 1st primer	0.05 µL		1 pmol
2 U/µl.	Vent polymerase Exo-	1.0 µL		2 units
DMSO	0.25 µL			5%
DNA	*n* µL			
Water		3.15 − *n* µL		
		5.0 µL		

See **Notes 20** and **21**.

In the aforementioned protocol, a maximum of 3.15 µL of DNA can be used in one reaction. We confirmed that increasing the total volume of the reaction mixture to up to 6.0 µL does not affect the performance. In that case, the pipetting protocol will be as follows:

				Final concentration
10X	ThermoPol buffer	0.6 µL	Tris-HCl (pH 8.8)	20 mM
			KCl	10 mM
			$(NH_4)_2SO_4$	10 mM
			$MgSO_4$	2 mM
			Triton X-100	0.1%
25mM	dNTPs	0.06 µL		250 µM
20 µM	biotinylated 1st primer	0.05 µL		1 pmol
2 U/µL	Vent polymerase Exo-	1.0 µL		2 U
DMSO		0.3 µL		5%
DNA		n µL		
Water		4.05 – n µL		
		6.0 µL		

2. Perform primer extension by PCR. Keep the reaction mixture on ice and transfer the tubes (or plates) when the Thermal cycler is set at 95°C.

PCR program:

Denaturation	95°C	15 min
Annealing	Tm – approx to 5°C	20 min
Extension	72°C	20 min
		1 cycle

3.5.2.3. Ligation

1. Prepare the ligation buffer.
2. Prepare the reaction mixture

			Final concentration
Ligation buffer	4.25 µL	Tris-HCl (pH 7.5)	29 mM
		$MgCl_2$	8.0 mM
		DTT	20 mM
		ATP	1.0 mM
		BSA	0.05 mM
LP25-21 linker	2.00 µL		40 pmol
3 U/µL T4 Ligase	0.96 µL		2.88 units
	7.21 µL		
Primer extension products	5.00 µL		
	12.21 µL		

Note that Vent polymerase works at pH 8.8 and ligase works at pH 7.5. The 6.25 mL of ligase mixture has a high concentration of Tris (60 mM, pH 7.5) to lower the pH.

3. Incubate overnight at 16°C (*see* **Note 22**).

3.5.2.4. CAPTURE OF PRIMER EXTENSION PRODUCTS (*SEE* SUBHEADING **3.5.1.5.**, STEPS **1** AND **2**)

1. Suspend the beads in 2X B&W buffer (12.5 μL/sample).
2. Add 12.5 μL of beads to linker-ligated products.
3. Incubate for at least 4 h at room temperature with rotation.
4. Place the tubes in a magnetic apparatus to bind the beads to the walls of the tubes and discard the supernatant.
5. Wash the beads once with 100 μL of 2X B&W buffer.
6. Wash the beads twice with 100 μL of 1X TE.
7. Suspend the beads in 10 μL of 0.1X TE.
8. Release the DNA from the beads by heating the beads suspension at 95 to 100°C for 15 min

3.5.2.5. PCR Amplification With a Second Nested Primer and LP25

1. Prepare the reaction mixture on ice.

				Final concentration
Water		18.5 μL		
10X	Cloned Pfu buffer	5.0 μL	Tris-HCl (pH 8.8)	20 mM
			KCl	10 mM
			$(NH_4)_2SO_4$	10 mM
			$MgSO_4$	2 mM
			Triton® X-100	0.1%
			BSA	0.1 mg/mL
5 M	Betaine	14.0 μL		1.4 M
25 mM	dNTPs	0.5 μL		250 μM
20 μM	LP25 primer	0.5 μL		10 pmol
20 μM	2nd primer	0.5 μL		10 pmol
2.5 U/μL	Pfu turbo			
	DNA polymerase	1.0 μL		2.5 units
Linker ligated primer				
Extension products		10.0 μL		
		50.0 μL		

See **Note 23**.

2. Perform PCR amplification. Keep the mixture on ice and transfer the tubes when the thermal cycler is set at 95°C.

> PCR program:
> | Denaturation | 95°C | 5 min |
> |---|---|---|
> | | | 1 cycle |
> | Denaturation | 95°C | 45 s |
> | Annealing | Tm – approx 3 to 5°C | 3 min |
> | Extension | 72°C | 5 min |
> | | | ×22 cycles |
> | Complete extension | 72°C | 10 min |

Optional: Check the PCR products by performing PCR with nested primers The efficiency of LM-PCR can be examined before the PCR products are labelled. When the LM-PCR amplification is efficient and enough material was used, it is possible to see 5 μL of PCR products on an agarose gel. However it is also possible to subject only 0.5 to 1 μL of PCR products to a PCR amplification using the third (labeling) primer and a primer recognising the opposite strand. If no products are observed after 15 cycles of amplification, it is quite likely that the LM-PCR failed. It is important to have a positive control such as genomic DNA to make sure this PCR amplification is working. If the signal intensities of individual samples vary, it suggests that either individual samples were not handled in the same way or that the amount of starting material varied. This information can also be used to adjust the loading amount on the sequencing gel. Alternatively, the LP25 linker primer can be used in combination with the nested primer. This can be useful in the case of reactions using MNase-treated or restriction enzyme-treated DNA, where the size of amplified products can be small and it may be difficult to design the primer on the opposite strand.

1. Prepare the reaction mixture on ice.

				Final concentration
PCR products		0.5 to 1.0 μL		
Water		16.4 to 16.9 μL		
10X	Taq buffer	2.0 μL	Tris-HCl (pH 9.0)	10 m*M*
			KCl	50 m*M*
			MgCl$_2$	1.5 m*M*
			Triton® X-100	0.1%
25 m*M*	dNTPs	0.2 μL		250 μ*M*
20 μ*M*	Third primer	0.1 μL		2 pmol
20 μ*M*	Primer for the opposite strand or LP25 primer	0.1 μL		2 pmol
5 U/μL	Taq DNA polymerase	0.2 μL		1 U
		20.0 μL		

2. Perform PCR amplification. Keep the mixture on ice and transfer the tubes when the thermal cycler is set at 95°C.

PCR program:

Denaturation	95°C	5 min
		1 cycle
Denaturation	95°C	40 s

Annealing	Tm – approx 3 to 5°C	1 min
(take lower annealing temper-		
ature of two primers)		
Extension	72°C	1 min
		×15 cycles
Complete extension	72°C	10 min

3. Check the yield of PCR products by agarose gel (1.5–2%) electrophoresis (the band intensity can be quantified by using the PhosphorImager system and used to adjust the loading amount for polyacrylamide DNA-sequencing gel electrophoresis).

3.5.2.6. LABELING OF THIRD PRIMER WITH ^{32}P

1. Prepare the reaction mixture.

				Final concentration
Water		1.0 μL		
10 x	PNK buffer	1.0 μL	Tris-HCl (pH 7.6)	70 mM
			MgCl$_2$	10 mM
			DTT	5 mM
10 U/μL	T4 Kinase	1.0 μL		10 units
20 μM	3rd primer	2.0 μL		40 pmol
0.37 MBq/μL	[γ -^{32}P] ATP	5.0 μL		1.85 MBq
		10.0 μL		

2. Incubate for 60 min at 37°C or overnight at room temperature.
3. Prepare a MicroSpin G-25 column by resuspending the resin, snapping off the bottom closure, placing the column in a 1.5-mL microcentrifuge tube, and spinning in a microcentrifuge at 735g for 1 min. Discard the buffer in the 1.5-mL collection tube, and repeat the centrifugation again, discarding the buffer in the collection tube.
4. Add 40 μL of water to the radiolabeled primer and apply onto the centre of the column bed.
5. Remove the unincorporated [γ-^{32}P] ATP by centrifugation of the column at 735g for 2 min (unincorporated radioactivity will remain on the column). Discard the column into radioactive waste.

3.5.2.7. PERFORMING DIRECT LABELING USING RADIOACTIVELY LABELED THIRD PRIMER

1. Prepare the reaction mixture.

				Final concentration
10X	Cloned Pfu buffer	0.4 μL	Tris-HCl (pH 8.8)	20 mM
			KCl	10 mM
			(NH$_4$)$_2$SO$_4$	10 mM

			MgSO$_4$	2 mM
			Triton® X-100	0.1%
			BSA	0.1 mg/mL
5 M	Betaine	1.0 µL		1.25 M
2.5 mM	dNTPs	0.4 µL		250 µM
2.5 U/µL	Pfu turbo			
	DNA polymerase	0.2 µL		0.5 units
Radiolabeled third primer		2.0 µL		1.6 pmol
		4.0 µL		

PCR products: 5.0 to 10.0 mL.

This normally works fine with 5 µL of PCR products and no significant differences were found between the experiments using 5.0-µL and 10-µL PCR products.

2. Perform labeling by PCR. Keep the mixture on ice and transfer the tubes when the thermal cycler is set at 95°C.

PCR program:

Denaturation	95°C	5 min
		1 cycle
Denaturation	95°C	45 s
Annealing	Tm – 3 to 5°C	3 min
Extension	72°C	5 min
		×6 cycles
Complete extension	72°C	10 min

See **Note 24**.

3.5.2.8. POLYACRYLAMIDE DNA-SEQUENCING GEL ELECTROPHORESIS

1. Prepare the sequencing gel with 0.4-mm spacers and comb (6% polyacrylamide, 8 M urea, acrylamide/bis, 19/1; *see* **Note 25**).
2. Assemble the gel apparatus and start the prerun with 1X TBE.
3. Add formamide loading buffer to the labeled PCR products (1/1) and denature the DNA by incubation for 5 min at 95 to 100°C and quickly chill on ice.
4. Perform a prerun for more than 60 min and confirm that the gel temperature is greater than 50°C.
5. After the prerun, each well will be filled with urea that has leaked from the gel. This has to be cleaned by flushing using a syringe with a needle. Once the wells are cleaned, load the denatured samples (using flat-ended gel loading tips is the easy way to do this). The electrophoresis is performed at 80 to 90 W (for a 40-cm × 40-cm × 0.4-mm gel). If a long and a short run are performed on one plate, the second series of samples should be loaded when the BPB dye front has reached approx 10 cm from the bottom of the gel. BPB has approximately the same mobility as the primers, so stop the electrophoresis when the BPB has just run out from the gel.

6. Separate the glass plates and fix the gel with fixation buffer for approx 10 min. Blot the gel onto 3MM Whatman paper, cover the gel with Saran Wrap, and dry the gel on a gel dryer at 80°C (it can vary, but it normally takes 30 to 60 min).
7. When the gel is dried, expose overnight to X-ray film or a PhospoImager screen and develop the next day.

3.6. Designing the Primers and Optimization of the Annealing Temperature

3.6.1. Primer Design

We use the Oligo™ 5.1 computer program to design the primers. The following conditions are important: the lengths of the primers are normally between 21 bp and 25 bp. Generally, longer primers perform better. The most critical primer is the first primer. Primers must be nested. We do not find it important whether or not the first and second primers overlap. However, the second and third primers ideally *should* overlap by at least 15 bases to avoid interference between primers. We do not recommend using a smaller overlap. Ideally, the Td of the primer should increase in the order 1 < 2 < 3. Ideal Tds (nearest neighbor method) of the primers are 52 ± 6 for primer 1, 70 ± 3 for primer 2, and 77 ± 5 for primer 3. Because the amount of the first primer is ten times less than the second primer, the residual first primer should not interfere with the PCR amplification with the second primer. In practice, it may be difficult to design the first primer with a high specificity and with a lower melting temperature than the second primer, but this usually is not a problem. However, it is important for the second primer to have a lower Td than the overlapping third primer, or there will be interference between these two primers in the labeling reaction. Interference can be minimized by keeping the overlap to approx 15 bp and the difference in the Td between the second and third primers to 7 ± 4°C. The primers are also designed such that their 3' ends do not have a high probability of being engaged in stable secondary structure with themselves (hairpin and primer dimer), with the linker primer, or with nonspecific sequences (i.e., the %G of the 3'-end is as low as possible).

Even when the primers are designed as described, some primer sets still do not work efficiently. It is possible to improve the performance by modifying the annealing temperature. There are two ways of doing this, as follows.

3.6.2. Primer Optimization 1

Determine the best annealing temperature by performing normal PCR using the test primer and one on the opposite strand (with Td 5 degrees higher than the primer to be tested). Perform PCR in increasing steps of 2 to 3°C across the Tm for 50 mM NaCl. The highest temperature at which efficient PCR still occurs is defined as Tm and use Tm – 2°C as the annealing temperature for the LM-PCR.

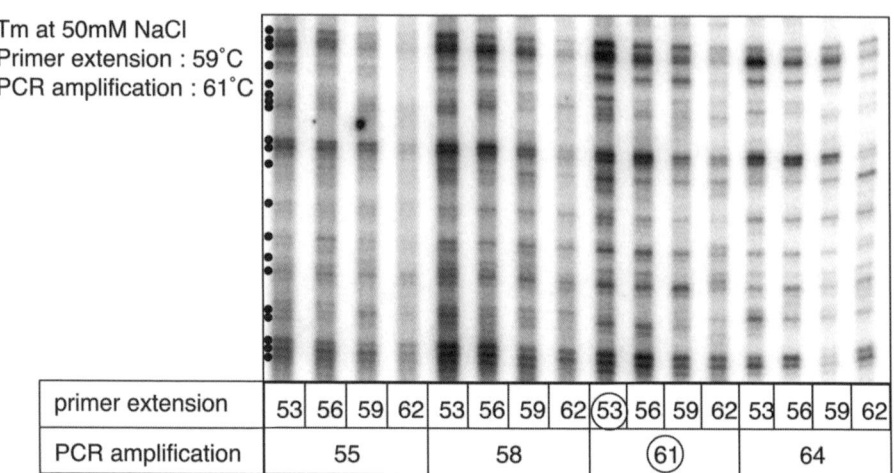

Fig. 4. An example for annealing temperature optimization. The annealing temperature for the primer extension step was tested between 53 and 62°C at 3°C intervals, and for PCR between 55 and 64°C at 3°C intervals. From the signal intensity and accuracy of the specific amplification, 53°C for primer extension and 61°C for PCR amplification were chosen for the future experiments (circled temperature). Note that only bands marked with black dots represent guanines; all other signals represent either methylated adenines or background bands caused by polymerase-stops. The theoretically determined Tm at 50 m*M* NaCl for each primer is shown on the upper left corner.

3.6.3. Primer Optimization 2

It is also possible to optimize the annealing temperatures by performing LM-PCR at a range of temperatures. Using a G reaction sample as a template, four to six tubes of LM-PCR with the newly designed primers are set up for each extension temperature to be tested. The annealing temperatures for primer extension are arranged to span the theoretical Tm (at 50 m*M* NaCl) at intervals of 2 to 3°C (9 to 10°C separation between highest and lowest temperatures.) The LM-PCR procedure is performed as usual up to the PCR amplification. The extension products are then used in 50-µL PCRs with the second primer at four to six different temperatures. The test temperatures are selected as in the primer extension step. An example of such an optimization is shown in **Fig. 4**, where the optimum temperatures are circled. From this analysis, it can be seen that the optimal temperature for the PCR primer was the same as the theoretical Td, whereas in contrast, the optimal temperature for the extension primer is 6°C lower than the theoretical Td. This highlights the need to empirically determine the ideal annealing temperatures used for each primer, which in practice can vary considerably higher or lower than the theoretical Td.

3.7. Representative Example of In Vivo Genomic Footprinting at the c-fms Promoter

Figure 5 represents an example of in vivo genomic footprinting at the mouse c-fms promoter. c-fms is expressed in macrophages and their progenitors cells but not in other cell types. We compared patterns generated in fibroblasts that do not express c-fms, and in macrophages that do express c-fms. **Figure 5A** shows a result of a DMS in vivo footprinting experiment. The DMS reactivity of the c-fms locus in the non-expressing cell (fibroblasts) is the same as observed for naked DNA, suggesting that there is no significant binding of factors. In contrast, the DMS reactivity in c-fms-expressing cells (macrophages) shows a different pattern. Hyperreactivity (–130 bp) and hyporeactivity (–127 to 129 bp) of specific bases in the region of a PU.1 binding site is the most prominent footprint seen in this region. Significant hyperreactivity (–173 bp) at a second PU.1 binding site was also observed. In addition, a weak but reproducible hyperreactivity was found at –103 bp located at a CAGGAA motif (X). **Figure 5B** shows a DNase I footprinting experiment. In the RAW264 macrophage cell line, protection from DNase I digestion compared with naked DNA and in 3T3 fibroblasts was found between –103 bp and –130 bp. In addition, hyperreactivity to DNase I was found at around –88 bp. These data suggest that factors are bound between –100 bp and –130 bp. This region overlaps a PU.1 binding site and the CAGGAA motif (X). Because DNase I is a large protein, and is more prone to steric hindrance from bound factors than small molecules such as DMS, DNaseI footprints typically span much broader regions than DMS footprints. **Figure 5C** represents the results of MNase nucleosome mapping. Strong digestion at –70 bp was found both in vivo for digestions of nuclei, and in vitro for the naked DNA control. For the digestions of nuclei, we also saw strong bands at –86 bp and –133 bp, which is indicative of positioned nucleosomes. The strong reactivity at –133 was observed only for 3T3 cells. In contrast, the reactivity at –86 bp was much stronger in RAW264 than 3T3 cells. These data suggest that the nucleosome border is situated at –86 bp in RAW264 cells and mainly at –133 bp in 3T3 cells.

4. Notes

1. Hydrazine, DMS, and piperidine are highly toxic. Perform all manipulations in fume hoods. All tips used with DMS should be rinsed in 1 *M* NaOH, and all waste solutions containing DMS should be treated with NaOH before disposal to inactivate the DMS.

2. The half-life of PMSF in aqueous buffers at 20 to 37°C is only approx 15 to 30 min; therefore, it is important to add PMSF (100 m*M* in ethanol or isopropanol) to the buffer just before use.

3. All enzymes and reaction buffers are stored at –20°C. ATP, DTT, and dNTPs are very sensitive to repeated freeze–thaw cycles. Therefore, they must be stored

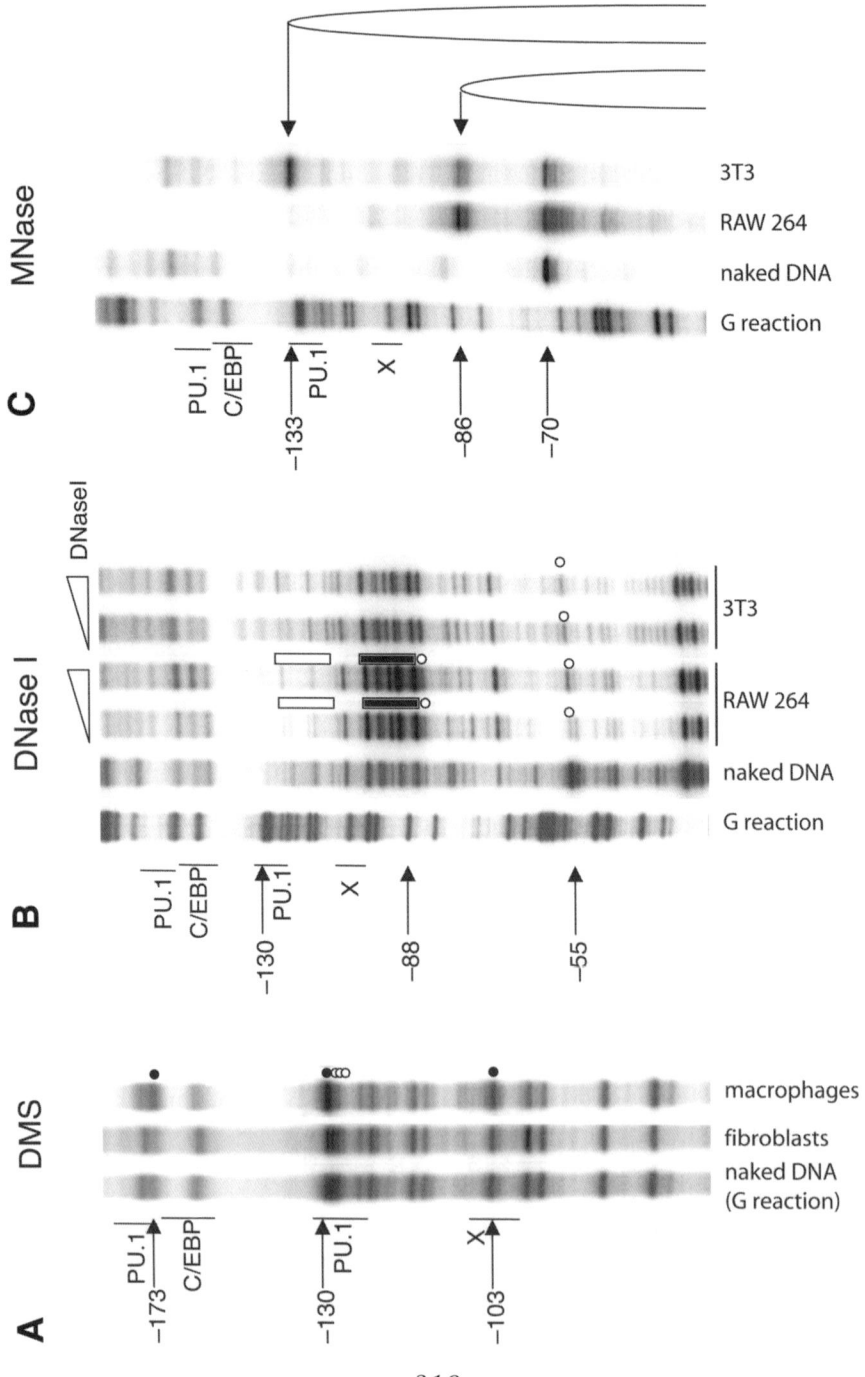

310

frozen in small aliquots. Oligonucleotides are especially unstable in dilute solution. To minimize any degradation, they must be reconstituted with 1X TE at a concentration of 200 μ*M* or more and stored at –20 or –80°C in small aliquots. We also recommend that you check the length and purity of the oligonucleotide by running on a 20% polyacrylamide gel after radioactive labelling using T4-polynucleotide kinase.

4. It is sometimes difficult to obtain in vitro DMS-treated DNA with the digestion at a proper degree. It sometimes helps if a few batches of G reactions treated with different concentrations of DMS (e.g., 0.125, 0.25, 0.5%) are prepared and are mixed afterwards (*see* **Fig. 2**, lanes 1 and 2).

5. DNA with DMS treatment shows much less viscosity than untreated genomic DNA.

6. It normally takes approx 4 to 6 h depending on the efficiency of the suction of the pump connected to the speed vac. This method is not recommended if the pump is not strong enough and it takes too long to dry the DNA. In this case, DNA is exposed to concentrated piperidine at high temperature, which causes nonspecific piperidine cleavage and high background.

Fig. 5. *(opposite page)* Representative example of in vivo genomic footprinting at the c-fms promoter. The number on the left indicates the nucleotide position relative to the ATG codon. Transcription factor-binding sites are indicated as a line. Black circles indicate hypermethylated G residues; open circles are hypomethylated G residues compared with DMS-treated naked DNA (G). The nested primers situated downstream of the c-fms promoter were used to amplify the c-fms promoter region. (**A**) DMS footprinting with primary fibroblasts (c-fms nonexpressing cells) and macrophages (c-fms-expressing cells). DNA isolated from the intact cells after treatment with DMS was subjected to LM-PCR. Naked DNA treated with DMS (G reaction) was used as a control. DMS hyper-reactive (black circles) and hyporeactive guanines (white circles) were seen at PU.1 binding sites at around –130 bp and at –173 bp. In addition, weak but reproducible hyperreactivity was observed at –103 bp (labeled as X). (**B**) DNase I footprinting with 3T3 (mouse fibroblast cell line, c-fms non-expressing) and RAW264 (mouse macrophage cell line, c-fms expressing). DNA isolated from the nuclei after DNase I digestion was subjected to LM-PCR. DNase I-digested naked DNA was used as a control. G reaction was used to show the sequence of this region. Protection from DNase I digestion (white bars) was observed at the PU.1 binding site located at –130 bp as compared with naked DNA. Hyperreactivity (black bars) was found around –88 bp. Hyporeactivity (white circles) at –55 bp may indicate protection from digestion by contact with a nucleosome. (**C**) MNase nucleosome mapping with 3T3 cells and RAW264 cells. DNA isolated from the nuclei after MNase digestion was subjected to LM-PCR. MNase-digested naked DNA was used as a control. G reaction was used to show the sequence of this region. The specific MNase cleavage seen at –86 bp was stronger in RAW264 cells than in 3T3 cells. In contrast, the cleavage at –133 bp was observed only in 3T3 cells but not in RAW264 cells. This may indicate that the nucleosome at the c-fms promoter in this cell line is positioned differently.

7. Washing with PBS before the DMS treatment can be omitted when time is tight, such as in time-course experiments. However, because DMS can react with proteins as well, all serum in the medium has to be removed.

8. This solution has to be prepared just before use. In general, DMS should be dissolved in ethanol and diluted with buffers because it is not a water-soluble agent, but in the case of low concentrations, such as 0.2%, vigorous mixing is sufficient to dissolve it.

9. The PBS has to be added gently and carefully because the cells are easily detached from the dish after DMS treatment.

10. In the case of using a small number of cells, a carrier such as 5 μg of λ DNA or transfer ribonucleic acid and/or 20 μg of glycogen should be added (2).

11. When harvesting DNA supernatants from a small number of cells, re-extract the organic phase with an extra 50 μL of lysis buffer #2, recentrifuge, and pool the aqueous phases.

12. DNA extraction can be done by using commercial DNA purification kits.

13. It is very important to keep the nuclei and every solution cold during the procedure and avoid bubbles/foams to prevent any physical damage to the nuclei.

14. For adherent cells, wash the dishes with ice-cold PBS and scrape them into ice-cold PBS with 0.5 mM PMSF in the cold room.

15. The protocol below is ideal for 5 to 10×10^7 cells. When you use fewer cells you can decrease the volume of buffers.

16. Depending on the number of cells used, this volume can vary. When you plan the experiments, you have to keep in mind that DNA from 10^6 cells is approx 5 to 10 μg and it is necessary to use several dilutions of DNaseI as the digestion by DNaseI is quite dynamic.

17. Although most previously published protocols using DNaseI or MNase treatment suggest that the ratio between the amount of enzyme and nuclei is important to control the nuclease reaction, we have observed that with our crosslinked material that it appears as if the concentration of the nuclease in the reaction solution is more important than the total amount of enzyme. An example of such a titration analysis is depicted in **Fig. 3**.

18. Although Ca^{2+} is not absolutely necessary for DNase I reactions, maximum enzyme activity is obtained when both Mg^{2+} and Ca^{2+} are present (10 times stronger than in the presence of Mg^{2+} only).

19. PEG has been reported to enhance the ligation efficiency (11). However, the quality of PEG is extremely variable. Purchase a good quality PEG and check it in a ligation reaction before to use. In most cases, no significant differences were found in reactions either with or without PEG.

20. The required amount of DNA is totally dependent on the performance of the primer sets and sequence of the target region. It has to be tested for each primer set.

21. DMSO is the most commonly used agent that improves yield and specificity (12) by disrupting base paring and reducing the melting temperature. This is not absolutely necessary in this reaction. However, in cases of having to amplify a difficult sequence for the primer extension, such as a GC-rich sequence, adding DMSO may help.

22. Most fragments ligate within 2 h, but some DNA ends ligate more efficiently than others. Leaving them overnight at 16°C will therefore help you to get more uniform band patterns. Blunt-end ligations work best at 16°C but can also work in the range of 4 to 22°C.

23. Betaine monohydrate (*N,N,N*-trimethylglycine) is a zwitterionic osmoprotectant found to alter DNA stability such that G/C-rich regions melt at temperatures more similar to A/T-rich regions and eliminate pausing by DNA polymerases *(13)*. It is not absolutely necessary to have this component in the PCR reaction. However we observed reproducible and significant improvement in the efficiency and accuracy of the amplification by adding betaine. Within the loci we have investigated, no significant improvement by adding DMSO in addition to betaine was found. However the use of the combination of chemicals such as DMSO, betaine, formamide, glycine, or tetramethylammonium chloride may be helpful to optimize PCR conditions.

24. The number of cycles for the labelling can be changed depending on the amplification efficiencies and amount of starting materials.

25. An example of the procedure to prepare a polyacrylamide DNA-sequencing gel is as follows: (1) Prepare a 40% acrylamide solution by dissolving 380 g of acrylamide, 20 g of *N,N′*-methylenebisacrylamide in 1 L of water and filtering the solution through a nitrocellulose 0.45-μm filter. Note that acrylamide powder is toxic and there are number of commercially available premade solutions that are easier and safer to use. (2) Prepare a 6% acrylamide gel solution (75 mL of 40% acrylamide solution, 100 mL of 5X TBE, 240 g urea/500 mL) and filtrate through a nitrocellulose 0.45-μm filter. This final solution can also be purchased (e.g., SequaGel system from Flowgen), in which case, make up the solution according to the manufacturer's instructions. (3) Assemble the glass plates. Lay the larger plate flat on the bench, put spacers along the sides and place the smaller plate on top of it. (4) Add 50 μL of 25% ammonium persulfate and 50 μL of TEMED to 50 mL of 6% acrylamide gel solution. (5) Pour the gel solution between two glass plates while tapping the top glass to avoid formation of bubbles. (6) Insert the comb and clamp the comb and glass plates using binder clips.

Acknowledgments

Research in C. Bonifer's laboratory is supported by grants from the Wellcome Trust, the Biochemical Biotechnological Research Council, the Leukaemia Research Fund, Yorkshire Cancer Research (YCR), the Medical Research Council and the Association for International Cancer Research (AICR). Hiromi Tagoh holds a Kay Kendall Fund Fellowship in Leukaemia Research. Research in Peter Cockerill's laboratory is supported by grants from the YCR and the AICR. We wish to thank Prof. Art Riggs for scientific discussions and generous help regarding in vivo footprinting using LM-PCR methods and Dr. Joanna Kontaraki for establishing this system in our laboratory.

References

1. Kontaraki, J., Chen, H. H., Riggs, A., and Bonifer, C. (2000) Chromatin fine structure profiles for a developmentally regulated gene: reorganization of the lysozyme locus before trans-activator binding and gene expression. *Genes Dev.* **14,** 2106–2122.

2. Tagoh, H., Himes, R., Clarke, D., Leenen, P. J., Riggs, A. D., Hume, D., et al. (2002) Transcription factor complex formation and chromatin fine structure alterations at the murine c-fms (CSF-1 receptor) locus during maturation of myeloid precursor cells. *Genes Dev.* **16,** 1721–1737.

3. Mueller, P. R. and Wold, B. (1989) In vivo footprinting of a muscle specific enhancer by ligation mediated PCR. *Science* **246,** 780–786.

4. Pfeifer, G. P., Steigerwald, S. D., Mueller, P. R., Wold, B., and Riggs, A. D. (1989) Genomic sequencing and methylation analysis by ligation mediated PCR. *Science* **246,** 810–813.

5. Pfeifer, G. P., Tanguay, R. L., Steigerwald, S. D., and Riggs, A. D. (1990) In vivo footprint and methylation analysis by PCR-aided genomic sequencing: comparison of active and inactive X chromosomal DNA at the CpG island and promoter of human PGK-1. *Genes Dev.* **4,** 1277–1287.

6. Telford, D. J. and Stewart, B. W. (1989) Micrococcal nuclease: its specificity and use for chromatin analysis. *Int. J. Biochem.* **21,** 127–137.

7. McPherson, C. E., Shim, E. Y., Friedman, D. S., and Zaret, K. (1993) An active tissue-specific enhancer and transcription factors exist in a precisely positioned nucleosomal array. *Cell* **75,** 387–398

8. Fragoso, G. and Hager, G,L. (1997) Analysis of in vivo nucleosome positions by determination of nucleosome-linker boundaries in crosslinked chromatin. *Methods* **11,** 246–252.

9. Cockerill, P. N. (2000) Identification of DNaseI hypersensitive sites within nuclei. *Methods Mol. Biol.* **130,** 29–46.

10. Pfeifer, G. P. and Riggs, A. D. (1991) Chromatin differences between active and inactive X chromosomes revealed by genomic footprinting of permeabilized cells using DNase I and ligation-mediated PCR. *Genes Dev.* **5,** 1102–1113.

11. Pheiffer, B. H. and Zimmerman, S. B. (1983) Polymer-stimulated ligation: enhanced blunt- or cohesive-end ligation of DNA or deoxyribooligonucleotides by T4 DNA ligase in polymer solutions. *Nucleic Acids Res.* **11,** 7853–7871.

12. Varadaraj, K. and Skinner, D. M. (1994) Denaturants or cosolvents improve the specificity of PCR amplification of G + C rich DNA using genetically engineered DNA polymerases. *Gene* **140,** 1–5.

13. Mytelka, D. S. and Chamberlin, M. J. (1996) Analysis of DNA polymerase pauses and suppression associated with a trinucleotide consensus. *Nucleic Acids Res.* **24,** 2774–2781.

22

Analyzing Histone Modification Using Crosslinked Chromatin Treated With Micrococcal Nuclease

Pascal Lefevre and Constanze Bonifer

Summary

Epigenetic processes involve alterations in the covalent modification of histones, which are driven by sequence-specific transcription factors recruiting the enzymatic machinery performing these reactions. Such histone modifications alter the charge or biochemical surface of the chromatin fiber and also serve as "docking" sites for transcription factor and chromatin remodelling complexes. Examining histone modification patterns is vital if we want to understand how transcription complexes establish patterns of gene expression. Histone modification can be quite localized and sometimes only involves a few nucleosomes. We therefore established a chromatin immunoprecipitation procedure that uses crosslinked chromatin allowing the isolation of small chromatin fragments while simultaneously minimizing histone movement during chromatin preparation.

Key Words: Chromatin/nucleosomes; histone modifications; chromatin immunoprecipitation assay; ChIP; formaldehyde crosslinking; real-time PCR; micrococcal nuclease.

1. Introduction

The histone tails protruding from the nucleosome surface are subject to enzyme-catalyzed modifications of selected amino acids. These modifications, including acetylation, methylation, phosphorylation, and ubiquitination, provide an important source of epigenetic information (1). It has been suggested that specific tail modifications or combinations thereof constitute a code that defines actual or potential transcription states (2). More insight into the complexity of histone tail modification patterns has only been possible after the development of new techniques such as the chromatin immunoprecipitation assay (ChIP). ChIP is a powerful approach that identifies the interaction of proteins with specific chromosomal sites in living cells, thereby providing a snapshot of the native chromatin structure and factors bound to genes in differ-

From: *Methods in Molecular Biology, vol. 325: Nuclear Reprogramming: Methods and Protocols*
Edited by: S. Pells © Humana Press Inc., Totowa, NJ

A Formaldehyde crosslinking and nuclei   **C** Immunoprecipitation
   preparation

B MNase treatment of chromatin   Supernatant   Precipitate

   D Reversal of crosslinks, extraction of DNA

Lysate

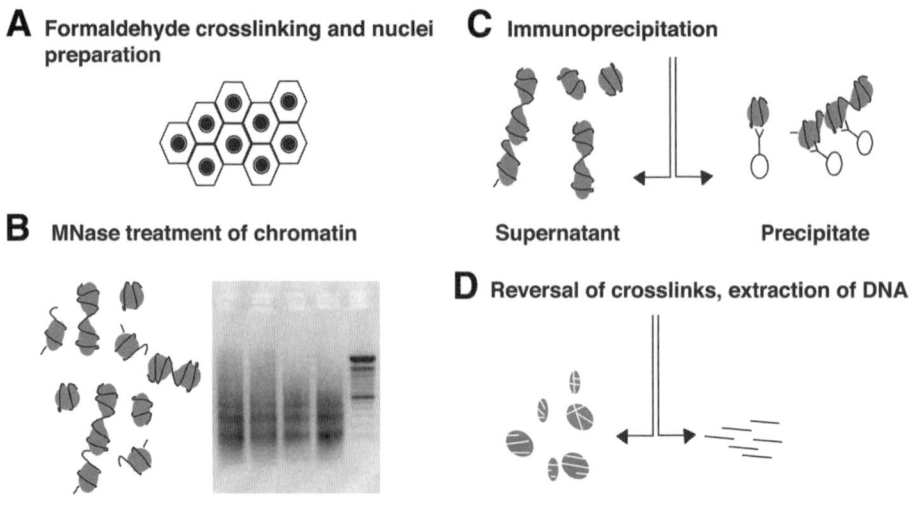

Fig. 1. Schematic outline of the ChIP procedure. (**A**) Cells are crosslinked with formaldehyde. (**B**) Chromatin released from the cells is then fragmented by micrococcal nuclease treatment. A 2% agarose gel shows four different chromatin preparations from the chicken macrophage cell line HD11 prepared several weeks apart (compare the left four lanes with 100-bp ladder, rightmost lane). (**C**) Chromatin containing a particular histone mark is recovered from the cell lysate by addition of specific antibodies and adsorption to protein A agarose beads. (**D**) DNA is isolated from samples after reversal of crosslinks by heat treatment. Purified DNA is then used as template for PCRs.

ent functional states (*3*). The ChIP technique has been applied to a large variety of biological systems from yeast to mammalian tissue culture cells and embryos. A schematic outline of the method is shown in **Fig. 1**. ChIP involves treating cells or tissue for an empirically predetermined time with formaldehyde (HCHO, **Fig. 1A**). HCHO is a very reactive dipolar compound in which the carbon atom acts as a center for nucleophilic attack. Amino groups of amino acids and of DNA react readily with HCHO, leading to the formation of a Schiff's base. This intermediate can further react with a second amino group and condense to give the final adduct. When the purpose of the investigation is to look at histone tail modifications, formaldehyde crosslinking essentially is used to prevent redistribution of chromatin components during chromatin preparation, as histone octamer mobility in low salt has been demonstrated.

Upon crosslinking, the cells are lysed, and chromatin released from the cells must be turned into smaller fragments. The smaller the fragments, the higher the resolution. Sonication is a rapid and simple way to shear chromatin fragments. However, this technique becomes difficult when manipulating small numbers of cells and often generates chromatin preparations of variable quality. For this reason, we have developed a ChIP protocol using micrococcal nuclease (MNase) treatment to fragment chromatin (**Fig. 1B**). MNase is an enzyme that catalyses the endonucleolytic cleavage of DNA into 3'-phosphomononucleotides and 3'-phospho-oligonucleotides. It can cause hydrolysis of double- or single-stranded DNA or RNA. When mixed with chromatin, micrococcal nuclease cleaves unprotected DNA between nucleosomes. MNase generates fragments mainly between 140 and 600 bp, containing mono-, di-, and trinucleosome fractions. In contrast to sonication, MNase-treated chromatin preparations that are run on an agarose gel show highly homogeneous patterns, thus limiting the variability of experimental results (**Fig. 1B**).

After MNase treatment, the chromatin immunoprecipitation experiment is performed similarly to conventional ChIP. An antibody against a specific histone modification is then used to immunoprecipitate chromatin fragments (**Fig. 1C**). Finally, protein–DNA crosslinking in the immunoprecipitated material is reversed, and the DNA fragments are purified (**Fig. 1D**). Analysis of the immunoprecipitate with primers flanking the *cis*-element using polymerase chain reaction (PCR) reveals genomic regions associated in vivo with specific modified histones. In this chapter, we will describe a method based on the real-time PCR technique to analyse immunoprecipitation products.

2. Materials

2.1. In Vivo Crosslinking and Chromatin Preparation

1. Phosphate-buffered saline (PBS): 140 mM NaCl, 2.5 mM KCl, 8.1 mM Na$_2$HPO$_4$, 1.5 mM KH$_2$PO$_4$, pH 7.5.
2. 37% Formaldehyde solution (Merck).
3. Quenching solution: 2 M glycine (Sigma) in PBS.
4. Lysis buffer: 10 mM Tris-HCl, pH 7.4, 10 mM NaCl, 5 mM MgCl$_2$, 0.2% NP40.
5. Glycerol buffer: 10 mM Tris-HCl, pH 7.4, 0.1 mM ethylenediaminetetraacetic acid (EDTA), 5 mM MgAc$_2$, and 25% glycerol.
6. 2X MNase buffer: 50 mM KCl, 8 mM MgCl$_2$, 2 mM CaCl$_2$, and 100 mM Tris-HCl, pH 7.4, stored at room temperature.
7. Immunoprecipitation (IP) buffer: 25 mM Tris-HCl, pH 8.0, 2 mM EDTA, 150 mM NaCl, 1% Triton X100, 0.1% sodium dodecyl sulfate (SDS).
8. Protease inhibitor cocktail (Sigma), stored in aliquots at –20°C.
9. 50 mM Phenylmethyl sulfonyl fluoride (PMSF) in isopropanol, stored at room temperature.
10. 20 mg/mL Proteinase K, stored at –20°C.
11. 10 mg/mL DNAse-free RNase A, stored at –20°C.
12. EDTA 500 mM, pH 8.0, stored at room temperature.

2.2. Immunoprecipitation and DNA Isolation

1. Antibodies: Although monoclonal antibodies can be used, it is better to use polyclonal (mostly rabbit) antibodies, which ideally should have a concentration of approx 1 mg/mL and ideally should be affinity purified. Antibodies are stored at −20°C for long-term storage. Most of the antibodies need to be aliquoted and stored at 4°C for short-term use after being defrosted (*see* **Note 1**).
2. Affinity matrix: 50% (v/v) suspension of protein A-agarose beads in PBS (Sigma), stored at 4°C.
3. 10 mg/mL Bovine serum albumin (BSA), stored at −20°C.
4. 5 mg/mL Sonicated salmon sperm DNA, stored at −20°C.
5. Washing buffer I: 20 m*M* Tris-HCl, pH 8.0, 2 m*M* EDTA, 1% Triton X100, 0.1% SDS; 150 m*M* NaCl, stored at room temperature.
6. Washing buffer II: 20 m*M* Tris-HCl, pH 8.0, 2 m*M* EDTA, 1% Triton X100, 0.1% SDS, 500 m*M* NaCl, stored at room temperature.
7. LiCl buffer: 0.25 *M* LiCl, 0.5% NP40, 0.5% Na deoxycholate, 1 m*M* EDTA, 10 m*M* Tris-HCl, pH 8.0, stored at room temperature.
8. TE: 10 m*M* Tris-HCl, pH 8.0, 1 m*M* EDTA, stored at room temperature.
9. Elution buffer: 1% SDS, 0.1 *M* NaHCO$_3$ (freshly prepared).

3. Methods

3.1. In Vivo Crosslinking and Chromatin Preparation

Nuclei isolation should be done on ice in the cold room and great care must be taken to handle the nuclei gently. Before starting, it is important to precool the centrifuges to 4°C and to place lysis-, glycerol-, and IP buffers on ice. Protease inhibitor cocktail and PMSF must be added to the different buffers including the 2X MNase buffer (1 µL of protease inhibitor cocktail and 10 µL of 50 m*M* PMSF per milliter of solution; *see* **Note 2**). Nuclei isolation needs to be completed within the shortest time possible.

1. Each immunoprecipitation experiment uses approx 10^6 cells. For easy manipulation, cells need to be resuspended in 1 mL of medium and transferred into 1.5-mL Eppendorf tubes (*see* **Note 3**). Up to 5 × 10^6 cells/mL can be used at the crosslinking step to allow several immunoprecipitations from the same chromatin preparation (*see* **Note 4**).
2. Add 27 µL of 37% formaldehyde per milliliter of cell suspension (final concentration formaldehyde 1%) to the cells. Mix rapidly and incubate for 5 min at room temperature on a rotating wheel (*see* **Notes 5** and **6**). Avoid skin contact or breathing formaldehyde fumes.
3. Quench the crosslinking reaction by adding 65 µL of 2 *M* glycine per milliliter of cell suspension to get 0.125 *M* glycine final concentration.
4. Spin down the cells for 5 min at 1500*g*. Remove the supernatant. Resuspend the cells in 1 ml of ice cold PBS and spin down for 5 min at 1500*g*.

5. Remove the supernatant and resuspend the cells in 1 mL of lysis buffer. Incubate for 60 min at 4°C on rotating wheel (*see* **Note 7**). Spin down for 15 min at 500*g* (*see* **Note 8**).

6. Remove supernatant and drain the tube well. Add 20 µL of glycerol buffer per 10^6 cells. Resuspend the nuclei pellet carefully and count the nuclei (*see* **Notes 9** and **10**). At this stage the nuclei can be stored at –80°C and used later for MNase treatment.

7. To generate nucleosomal material, digestions are conducted by adding 1 vol of 2X MNase buffer. The volume of the nuclei preparation has to be evaluated. Add MNase at the final concentration of 500 U/mL and incubate for 10 min at 37°C. Stop the reaction by adding EDTA to a final concentration of 10 m*M* (*see* **Note 11**).

8. Add IP buffer to a final volume of 260 µL for 10^6 cells and 200 µL more per each additional 10^6 cells (example 5 × 10^6 cells = 1060 mL). Centrifuge for 2 min at 1500*g* to remove debris (*see* **Note 12**).

9. Take 50 µL as input DNA control (*see* **Notes 10, 13,** and **14**).

10. Add 1 µL of 10 mg/mL RNase A solution to the 50-µL input. Incubate for 30 min at 37°C. Add 1 µL of 20 mg/mL proteinase K and reverse the crosslinking with an overnight incubation at 65°C. For DNA purification, *see* **Subheading 3.4.3.** in Chapter 21.

3.2. Immunoprecipitation and DNA Isolation

1. Prepare 40 µL of 50% protein A agarose suspension per sample analyzed. Rinse the protein A beads twice in TE and collect them at each washing step by a 2-min centrifugation at 1500*g* (*see* **Note 15**).

2. Add 5 µL of 5 mg/mL salmon sperm DNA and 10 µL of 10 mg/mL BSA per 100 µL of washed 50% bead suspension. Keep at 4°C for the duration of the experiment (*see* **Note 16**).

3. To reduce nonspecific background, preclear the chromatin solution with 20 µL of salmon sperm DNA/BSA/Protein A 1-agarose slurry for 1 h at 4°C on a rotating wheel.

4. Pellet the protein A agarose beads for 20 s at 1500*g* and collect the supernatant.

5. Use 200 µL for each individual immunoprecipitation.

6. Add an appropriate amount of antibody (usually 1–5 µL) to the chromatin solu tion. Incubate overnight at 4°C on a rotating wheel (*see* **Note 17**).

7. Add 20 µL of salmon sperm DNA/BSA/protein A agarose slurry. Incubate for 2 h at 4°C on a rotating wheel.

8. Pellet the protein A agarose beads with the immunoprecipitate for 20 s at 1500*g*. Remove the supernatant; this can be kept for analysis. Between the different washing and elution steps protein A agarose beads will be centrifuged using the same conditions.

9. Wash the beads once with 300 µL of washing buffer I. Incubate for 5 min at room temperature on a rotating wheel. Discard the supernatant.

10. Wash the beads once with 300 µL of washing buffer II. Incubate for 5 min at room temperature on a rotating wheel (*see* **Note 18**). Discard the supernatant.

11. Wash the beads once with 300 μL of LiCl buffer. Incubate for 5 min at room temperature on a rotating wheel. Discard the supernatant.
12. Wash the beads twice with 300 μL of TE. Incubate for 5 min at room temperature on a rotating wheel. Discard the supernatant. Recentrifuge the tube and draw off the remaining liquid.
13. Elute the immune complexes by adding 50 μL of elution buffer and incubate for 15 min at room temperature with vortexing every 5 min (*see* **Note 19**).
14. Pellet the protein A agarose beads and collect the eluted fraction. Repeat **step 13** and pool the two fractions together (*see* **Note 20**).
15. Reverse the crosslink at 65°C for 4 h to overnight (*see* **Note 21**).
16. Purify the DNA using Qiaquick columns (Qiagen) as indicated by the manufacturer except that samples are first mixed with agitation for 30 min with PB buffer (*see* **Note 22**).
17. Resuspend the pellet in 50 μL of TE. Store at 4°C (*see* **Note 23**).

3.3. Real-Time Quantitative PCR Analyses

1. Real-time quantative PCR has been a major breakthrough in the analysis of ChIP products. The basic principle of real-time PCR is that the amount of PCR product synthesized is measured after each cycle with fluorescent probes so that an amplification curve for each reaction is established. This curve allows the quantification of the initial amount of material in the log phase of the reaction relative to a calibration standard curve.
2. Several types of fluorescent probes can be used. The nonspecific SYBR Green dye, which detects any double-stranded DNA, is the cheapest and easiest to use and is in general sensitive enough. However, it is important to stress that with SYBR Green, it is absolutely necessary to control that only the correct PCR product is synthesized. Therefore, primer pairs must be carefully designed and their specificity has to be tested using melting curve analysis, agarose gel electrophoresis, and occasionally even sequencing of the obtained product. With the ABI Prism 7700 sequence detection system (Applied Biosystems), annealing and polymerization steps are combined in a single 60°C step. Primer design software (Primer Express™) is provided with the machine to match the specificity of the standard PCR program. Characteristics of primer design will vary depending on the specific equipment used for the real-time PCR. In any case, it is recommended that one restrict the size of the PCR product length to approx 80 to 120 bp and to make sure that the %G at the 3' end of the primers is as low as possible (maximally 40% C or G in the last 5 bp of the 3' end) to decrease the possibility of mispriming.
3. **Figure 2** illustrates the different steps of real-time PCR analysis. Quantification is performed by measuring the number of cycles necessary to reach a threshold of fluorescence value for each sample. This value has to be in the exponential phase of the reaction (**Fig. 2A**). A dissociation curve obtained by progressively increasing the temperature from 60 to 94°C allows measurement of the double-stranded DNA denaturation kinetics. The kinetics for a specific PCR fragment are depen-

dent on nucleotide composition and size. Therefore, if the PCR generates a single clean product, only one peak will be detected (**Fig. 1B**). A standard curve obtained with known amounts of genomic DNA will help in converting data expressed as number of cycles into nanograms of DNA (**Fig. 1A,C**). Results are expressed as a percentage of the input material and compared with ChIP experiments using a nonspecific antibody as a control to determined the background level of the experiment. **Fig. 2D** represents a typical example of results obtained using this method. The presence of histone H3 acetylated at lysine 9 and histone H3 tri-methylated at lysine 4 is a mark of a transcriptionally active locus. In this example, a ChIP experiment was conducted using chromatin from a chicken macrophage cell line and immunoprecipitation was performed using both histone H3 modification-specific antibodies, respectively *(4)*. As expected, the level of histone H3 lysine 9 acetylation and lysine 4 trimethylation was high at both the promoter and the 5' coding region of the macrophage-specific chicken lysozyme gene. In contrast, the histones at neither the border region 10-kb upstream of the transcription start site nor the promoter of the liver-specific gene ApoVLDL2 are acetylated, or tri-methylated at lysine 4. For histone modification studies, proper internal negative and positive controls are important. They are necessary to test the precipitation efficiency of a given antibody and to compare absolute values obtained with your DNA region of interest with another to be known to harbor/ lack the modification to be tested.

4. Notes

1. The quality of the antibodies used is of prime importance for the outcome of the experiment. Ideally, the antibodies should possess a high affinity for the antigen, they should withstand the most stringent conditions during the immunoprecipitation, and they should only minimally crossreact with other proteins. Most of the antibodies that are available commercially are rabbit polyclonal antibodies, but some monoclonal antibodies and some from other species also have been shown to give satisfactory ChIP results.

2. PMSF is highly toxic, and great care must be taken during handling. As crystallization of PMSF in isopropanol occurs below freezing point, the solution should be kept at 4°C or at room temperature. PMSF is an unstable serine protease inhibitor and must be added to aqueous solutions just before use. Because PMSF is insoluble in water, the aqueous solution must be mixed vigorously upon addition of the stock solution or the PMSF will precipitate.

3. Chromatin is most easily prepared from cells cultured in suspension but also can be recovered from adherent cell cultures and from homogenized animal tissues. For adherent cells, we sometimes grow cells in bacteriological dishes to allow cell harvesting by pipeting. Trypsin or other techniques generating cell stress must be avoided. If gentle collecting of adherent cells is not possible or cells need to grow on tissue culture-grade dishes, we recommend crosslinking *in situ* before harvesting the cells.

4. It is possible to prepare chromatin from large numbers of cells by increasing the volume. In our hands, crosslinking and nuclei preparation can be performed with

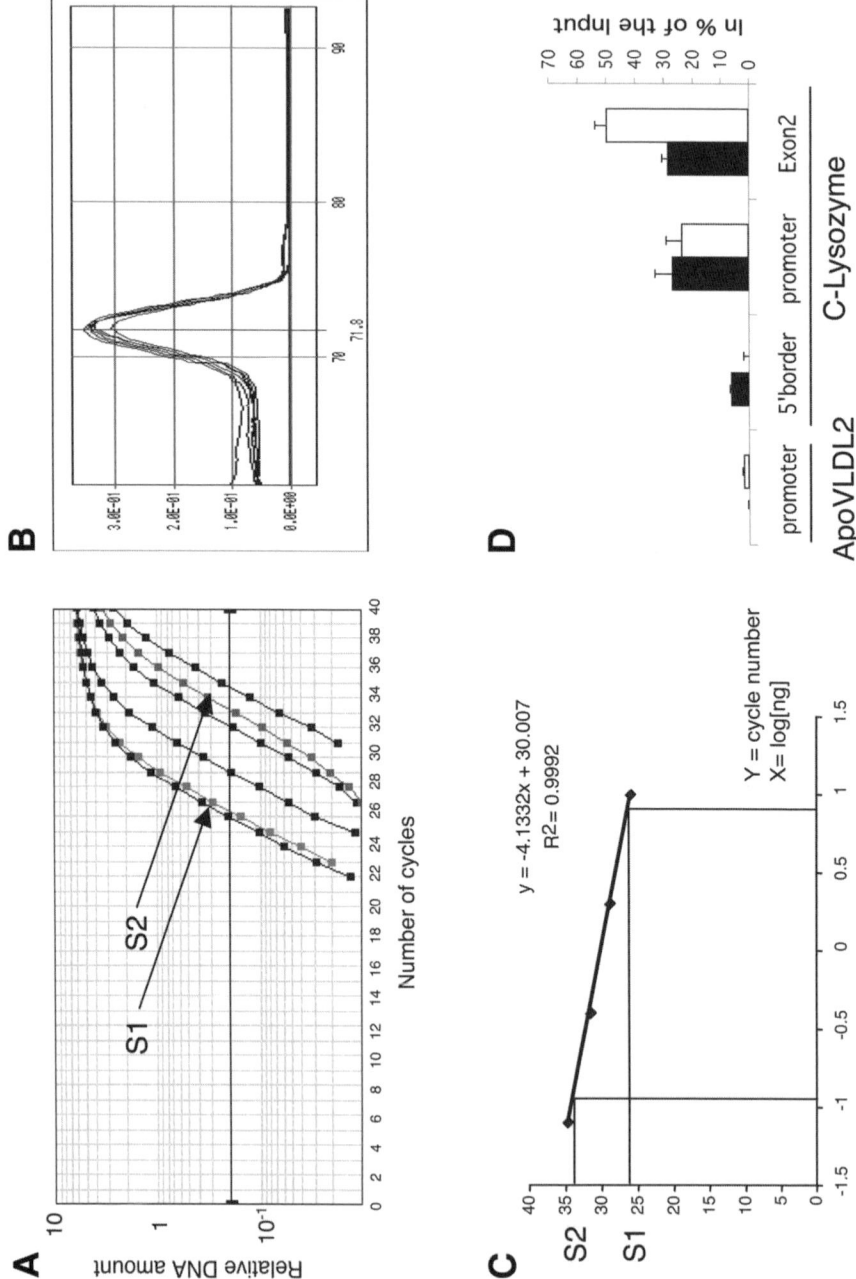

322

up to 5×10^6 cells/mL without a noticeable decrease in nuclei preparation quality compared with lower concentrations, but this has to be confirmed for each cell type.

5. One parameter that requires optimization is the length of the crosslinking step. Increasing the crosslinking period increases the number of protein–protein crosslinks. Extensive crosslinking may be useful to facilitate detection of proteins that interact indirectly with DNA through multiple protein–protein interaction layers, but it also can lead to modification of the epitope recognized by the antibody. In our hands, histone tail modifications such as phosphorylation are very sensitive to crosslinking and a short fixation time is recommended.

6. For histone tail modification studies, formaldehyde crosslinking is not absolutely necessary. However, it prevents nucleosome movement during chromatin preparation. It is also useful for looking at transient events, allowing one to "freeze" specific histone patterns at specific time points.

7. Nuclei should be prepared in buffers containing Mg^{2+}. It may be necessary to optimize the procedure for different cell types. Increasing ionic strength or including 0.1% SDS in the lysis buffer raises the stringency and decreases nonspecific precipitation .

8. Extensive clumping of the nuclei will reduce the efficiency of the MNase treatment. Clumping usually occurs as a result of centrifugation at too high speed or from vigorous homogenization or resuspension of the nuclei.

9. To count the nuclei, we recommend using methylgreen–pyronin (Sigma). The nuclei will then appear green under the microscope. With cells incubated with formaldehyde, you will obtain some permeabilized cells and nuclei trapped within fragments of cytoplasm. Methylene blue can also be used.

Fig. 2. *(opposite page)* The different steps of real-time PCR analysis. **(A)** Serial dilution of genomic DNA (10, 2, 0.4, and 0.08 ng; in black) and two samples (S1 and S2 in grey) were amplified using SYBR Green PCR reagents, the ABI Prism 7700 sequence detection system (Applied Biosystems) and chicken lysozyme promoter-specific primers. **(B)** An example of a dissociation curve indicating the presence of only one product. **(C)** Analysis of experimental results obtained in (A). The number of cycles necessary to reach a threshold fluorescence value is measured for each sample. The serial dilution is converted into a graph with y = cycle number and x = log [amount of DNA]. This graph allows calculation of the amount of specific product for each immunoprecipitation tested. **(D)** Example of histone modification determined using crosslinked chromatin treated with MNase. Chromatin preparations from the chicken macrophage cell line HD11, treated with LPS for 30 min, were immunoprecipitated with antibodies against trimethylated histone H3 Lys 4 (white bars) and acetylated H3 Lys 9 (black bars), respectively. Real-time PCR using primers specific for the chicken lysozyme exon 2, the promoter, and the region 10 kb upstream of the transcription start site not containing any *cis*-regulatory elements. The promoter of the liver-specific ApoVLDL2 gene was used as a negative control. Results are expressed using the following formula (specific signal/Input – IgG/Input)*100.

10. It is sometimes difficult to get an accurate count of nuclei. We recommend quantifying the input before performing a ChIP experiment. Using the same amount of chromatin for each IP is important to limit variability in the efficiency of the immunoprecipitation.

11. Micrococcal nuclease digestion efficiency has to be tested using several concentrations of the enzyme every time a new batch of enzyme is started. In our hands, digestion efficiency is mostly dependent on the enzyme concentration rather than on chromatin concentration and crosslinking time. MNase is less active at lower temperatures and at $CaCl_2$ concentrations less than approx 0.1 mM or if there is an excess of EDTA or if EDTA concentration exceeds $CaCl_2$ concentration.

12. The presence of micrococcal nuclease in the IP buffer does not interfere with the efficiency of the chromatin immunoprecipitation.

13. If your count of nuclei and/or input DNA control quantification gives you significant variability between your different chromatin samples, it is necessary to adjust the volume of IP buffer to work with both the same volume and concentration by adjusting the volume to that of the least concentrated sample.

14. Linker regions between phased nucleosomes, or nucleosome-free regions, will lead to an amount of input DNA amplified by PCR noticeably different from regions where nucleosomes are positioned randomly. Because results are expressed as a percentage of input DNA, this leads to variable results when the amplified product is too low. In this case it is necessary to design several primers in the region of interest to avoid having a linker region in the middle of your amplicon. To facilitate this, the same chromatin template can be used to determine nucleosome positions (*5*) by LM-PCR.

15. Protein A agarose is used with antibodies from rabbit, but protein G should be used for antibodies that bind more efficiently to this reagent (see the data sheet for commercially available antibodies).

16. It is very important to use the same amount of protein agarose bead slurry for each sample. We recommend preparing the salmon sperm DNA/BSA/protein A agarose mix in excess and resuspending the beads frequently. The base of the pipet tip needs to be widened by cutting it with a clean razor blade, as done for pipetting genomic DNA solutions.

17. It appears that the chromatin immunoprecipitation method works with a broad range of antibody concentrations, although clearly the signal strength is proportional to the amount of antibody used at lower antibody concentrations. The main issue is that sufficient material for detection is immunoprecipitated and that the abundance of bound and unbound at different chromosomal regions relative to each other is not altered in response to changes in the antibody concentration.

18. Some antibodies may give abnormally, high nonspecific signals. To optimize the washing step, adjust the detergents and NaCl concentration of washing buffer II.

19. If the phenol–chloroform extraction technique is preferred for purifying your DNA fragments, it is recommended that you elute with 2 × 250 μL of elution buffer instead of 2 × 50 μL.

20. At this step make sure no beads have been collected with the eluted fraction. This can cause variation of the background level. You can introduce an additional centrifugation step of 20 s at 1500*g* of the eluted fraction and transfer this fraction to a new Eppendorf tube.

21. At this step you can add 20 μg of proteinase K and immediately start reversing the crosslink at 65°C overnight. However, depending on the cell type, it may be necessary to incubate for 1 h at 55°C before the overnight incubation because proteinase K digests better at 55°C.

22. Any other techniques of DNA purification can be used including phenol–chloroform extraction. In this case, DNA has to be precipitated by adding 20 μg of glycogen as carrier; 1/10 volume of 3 *M* sodium acetate, pH 5.2; and 2 vol of 100% ethanol. Incubation at –20°C must be performed for at least 2 h to overnight to make sure complete precipitation will occur.

23. Storage at –20°C can cause precipitation and DNA degradation problems and is not recommended.

Acknowledgments

Research in C. Bonifer's laboratory is supported by grants from the Wellcome Trust, the Biochemical Biotechnological Research Council, the Leukaemia Research Fund, Yorkshire Cancer Research, the Medical Research Council, and the Association for International Cancer Research. The authors thank Dr. Peter Cockerill for critical comments on the manuscript.

References

1. Spotswood, H. T. and Turner, B. M. (2002) An increasingly complex code. *J. Clin. Invest.* **110,** 577–582.

2. Jenuwein, T. and Allis, C. D. (2001) Translating the histone code. *Science* **293,** 1074–1080.

3. Orlando, V., Strutt, H., and Paro, R. (1997) Analysis of chromatin structure by in vivo formaldehyde cross-linking. *Methods* **11,** 205–214.

4. Lefevre, P., Melnik, S., Wilson, N., Riggs, A. D., and Bonifer, C. (2003) Developmentally regulated recruitment of transcription factors and chromatin modification activities to chicken lysozyme cis-regulatory elements in vivo. *Mol. Cell Biol.* **23,** 4386–4400

5. Fragoso, G. and Hager, G. L. (1997) Analysis of in vivo nucleosome positions by determination of nucleosome-linker boundaries in crosslinked chromatin. *Methods* **11,** 246–252.

Index